# Origin of Mind

## a history of systems

Camelo Hundere Castillo

Allardice Press

Published 2017 by Allardice Creek Press LLC.
219 N. 2nd Avenue, Tucson, AZ 85705

*Origin of Mind: A History of Systems.*



ISBN-13: 9780692230862
ISBN-10: 0692230866

For information, contact
Allardice Creek Press LLC.
www.allardicepress.com

Cover photograph: Yukon River Delta, Science Photo Library
Author photograph by Antoine Hunt

*To Molly who made me robust*
*To Charlie who made me curious*

*We feel clearly that we are only now beginning to acquire reliable material for welding together the sum total of all that is known into a whole; but, on the other hand, it has become next to impossible for a single mind fully to command more than a small specialized portion of it. I can see no other escape from this dilemma (lest our true aim be lost forever) than that some of us should venture to embark on a synthesis of facts and theories, albeit with second-hand and incomplete knowledge of some of them - and at the risk of making fools of ourselves.*

*Erwin Schrödinger (1944, p. 1)*

# Acknowledgments

*I'd like to thank Gottfried Leibniz, Charles Darwin, Ludwig von Bertalanffy, Michael Tracy, Stephen Jay Gould, Terrence Deacon, Stuart Kauffman, Daniel Dennett, Richard Leakey, George Lakoff, Ian Stewart, Per Bak, Alain Avila and Michael "Death" Miller for their shoulders.*

# Origin of Mind

## Contents

# List of Figures and Tables

# Where is My Mind?

*"Where is my mind?*
*Where is my mind?*
*Wheeeee-ere is my mind..."*

*Pixies (1988)*

The riddle of the human mind has tantalized us for centuries. While the universe has steadily surrendered her secrets to bigger telescopes and better microscopes, the problem of the human mind has remained recalcitrant. The current approach to this problem is through the brain. Where else would the mind live? How else could it exist? Unfortunately, despite more than twenty years of intensive investment in the modern brain sciences, we still know precious little about how the brain creates subjective human experience. How and why neurons sustain the flux and reflux of hopes, fears and desires in the forms that imbue our daily experience with oh-so-human vim and vigor remains one of humanity's deepest mysteries. Perhaps we just need a better microscope, or perhaps, we need another approach.

This investigation tracks down the 'mind' in an unexpected way. It starts with the Big Bang and, step by step, it shows how each major universal transformation shares common features with previous transformations. Taken together, these common features constitute a pattern. This investigation shows how the emergence of the human mind fits into this pattern. What a mind could be, its features and capacities, becomes evident when its emergence makes sense in a universe that has been producing different systems the very same way for billions of years.

This book is the pursuit of human nature through deep time. It is a historical explanation for how the human mind evolved. It proposes an adaptation of Darwin's logic of natural selection and then applies this new version of an old logic to the natural history of the universe. Thus, the focus of this investigation will be much broader than the canonical human brain. Its larger focus will be on the evolution of *order* in general.[1] Currently, there is no theory for how the universe gained order. This investigation proposes that emergent order is always built the same way. It demonstrates how the structures at each level of historical emergence are unique, but the process by which they emerge has always been the same. Essentially, the proposal is that *emergence* is a process that is common to each of the major transitions the universe has experienced. The candidate definition of *emergence* proposed here is:

*Selection for symmetric persistence from expanding spectrums of interaction.*

The universe is an *expanding spectrum of interaction* which means that it has always been interacting with itself, so that it continually samples new interactions. These new interactions are subject to a kind of broad scale *selection* process, much like Darwinian natural selection, where only those interactions that are *symmetric*[2] end up populating our universe. Thus, only symmetric relationships are selected to *persist*. One could reformulate this hypothesis in layman's terms as, 'anything that can happen does happen, but only those things that happen regularly, persist.'[3]

Unpacking this hypothesis will be the Leitdifferenz (guiding difference) of this journey from the primordial Big Bang to the origin of mind. At each step we will see how cutting edge research from apparently unrelated fields is producing a body of evidence that supports the principle of selection for symmetric persistence from expanding spectrums of interaction.

This investigation itself is organized into chronological sections that examine how different types of systems have emerged over the history of the universe. One of the challenges of such an extensive project is establishing a balance between consistency, rigor, and readability. The first challenge is to establish a lexicon that is capable of not only unifying concepts from

---

1 See Appendix 1 for a definition of order in the context of energy, entropy, information and complexity.

2 The term *'symmetry'* will be defined in the traditional mathematical and physical way as, 'transformation with preservation of structure.' This definition is discussed in the chapter "Symmetry."

3 Hat off to Jeff Forshaw and Brian Cox (2011) for the phrase, "anything that can happen, does happen."

different fields, but of remaining readable and thus capable of navigating the vast and subtle conceptual landscapes that different academic domains have developed. Error on the side of too much rigor leads to extensive philosophical tracts that can bog down the reader, while error on the side of too much readability leads to dismissal on grounds of frivolity. The first section walks the tightrope between these two towers of error by introducing a toolbox of concepts that are rigorous enough for experts, yet useful and consistent enough to allow the reader to navigate foreign scientific landscapes. In this section ancient concepts like *transformation, thought* and *time* will be reexamined, newly adapted concepts like *symmetry* will be expanded, and completely new concepts like the *reactive face* and *propagating symmetry* will be introduced. Readers who are uninterested in philosophical foundations are invited to start with "Section 2: Universe," but they may find themselves backtracking to nail down terms.[4]

The second section reviews cosmic origins. It starts thirteen point eight billion years ago with the Big Bang and shows how *quantum, chemical* and *astronomic* structures are related through a common process of emergence that is reminiscent of Darwinian evolution. This section proposes a new working definition for emergence and it shows how the historical transitions from quantum to chemical to astronomical systems embody this emergence.

The third section reviews how Earth conditions gave rise to the emergence of *life* in the last four billion years. It shows how the major transitions in biology are related to the major transitions in cosmology. In this section group selection is revised, expanded and compared to John Maynard Smith and Eörs Szathmáry's list of *Major Transitions in Evolution.* The evolutionary histories of autocatalysis, binary fission, mitosis and meiosis are compared and the result is the establishment of a series of biological group selection events that represent the same process of emergence established in the previous section. Basically, the major biological transitions revolutionized their environments through the same process that structured the major universal transitions.

---

[4] Defining terms for such a project is difficult. The option of putting these forty or so pages at the beginning was an act of triage. The risk of losing readers' interest in an early philosophical discussion was deemed preferable to the possibility of losing their confidence in the latter process of operation and recovery.

The fourth section reviews the evolution of *humans* and *symbolic language* over the last two million years. It demonstrates how the transition from proto-humans to symbolic humans fits seamlessly into cosmological and biological history (e.g. Sections 2 and 3). In this section we examine the evolution of vocal/aural senses in bats and whales (e.g. echolocation) and show how they share design features with the evolution of language. The result is that both the emergence of Homo sapiens and symbolic language can be interpreted as fitting in to a long cosmological and biological history of emergent group selection events.

The fifth and last section focuses on the origin of the *human mind* in the last hundred thousand years as it could be anticipated by billions of years of similar universal and biological transformations. The universe has a long history of selectively maintaining novel communicative interaction that reproduces novel levels of cooperative systems. The human mind is part of this long history. Binary fission is the language of bacterial persistence. Mitosis is the language of protist persistence. Meiosis is the language of metazoan persistence. Social symbolism is the language of human persistence. In each case a new way to store and access what-has-worked-before has been added to the emergent organisms' network of maintenance and reproduction. These new reproductive languages are preserved (however and whenever they occur) because they extend the adaptive futures of their emergent organisms. Many of the current mysteries of the mind can be clarified by understanding how the human mind emerges, maintains and reproduces itself. This section demonstrates how, in many ways, the transition to mind is no different than the preceding transitions.

The journey from the Big Bang to mind will require tracking current thinking in a number of scientific disciplines. Fred Spier's *Big History and the Future of Humanity* and David Christian's *Maps of Time* have introduced us to the 'Big History' format, the fourteen billion year grand vista of what happened when. Unfortunately, 'why' is an altogether different question. Frustrated with the slow progress cosmology is making on the question of 'Why this universe and no other?' physicists like Lee Smolin and Robert Laughlin are breaking new ground by expanding the reductionist context to include evolutionary and communal emergence concepts. At another level, biologists like Stuart Kauffman are expanding the biological context by demonstrating how life could emerge from chemical diversity - in his words, 'order for free.' Physicist Per Bak even showed us exactly what the conditions for self-organizing 'order for free' look like. In different buildings, all the way across campus, conceptual tools from the cutting edge of all these sciences are being adopted by cognitive scientists. Both Terrence Deacon's seminal work on symbolic origins and Daniel Dennett's valiant attempts to naturalize the human mind synthesize concepts from chaos, complexity and evolutionary theories.

One way or another, all of these researchers are addressing the fundamental question of 'Why are we the way we are?' This investigation is based on research from all these sources, but it takes a big history view. The answer to how the universe makes a mind is founded in the broader question of how the universe makes order.

This investigation proposes a hypothesis for how the universe makes order. This hypotheses is based on a somewhat well known, but rarely applied, fundamental conceptual transition: *things are processes.* Looked at from the point of view of things, (like quarks, suns and cells), the universe looks unconnected. These things are our objects of study. Their structural differences are the foundations of our separate academic disciplines. When one looks at the different categories of our objects of study, the universe looks like a grab bag of unrelated and mysteriously emergent structures. Examined from the point of view of processes, however, the universe presents a consistent pattern of emergence. This pattern is not evident in the systems themselves, but in the way these systems came to persist - in their process of emergence.

In principle, the hypothesis of emergence proposed here is applicable to all major universal transitions, past and future (e.g. quantum, chemical, astronomical, biological, etc.). If any major new type of universal system could be shown to emerge outside of this pattern, this hypothesis would have to be discarded. Much of this investigation will be dedicated to demonstrating how the main types of systems that constitute our universe emerged from this pattern. However, once the pattern is established it will give us an evolutionary platform from which we can understand the origin of the human *mind.*

One of the most curious results of this investigation, at least for the cognitive sciences, is that the brain loses its exclusive power to explain the mind. From an evolutionary point of view the human brain is but one relatively recent link in a long chain of emergence. More important than the wet wiring of the brain (e.g. the neural correlates of consciousness or the connectome)[5] is the *process* that the mind is engaged in - its function within the unfolding of the biosphere and the unfolding of the universe. Ultimately, understanding the neural structure of the brain will give us as little information about how our minds fit into this unfolding process as the number 233 gives us about the Fibonacci sequence.[6] Alone, both the number and the brain tell us very little about the systems they help maintain and reproduce. In the case of the number 233, the structure of the number is meaningless taken out of its developmental context. It is the pattern itself - the *process* of its transformation - that makes the number 233 meaningful within the Fibonacci sequence. Likewise, it is not the structure of the human brain that determines what it means to have a mind. While a brain is necessary, it is also far from sufficient. What is sufficient is the process that the human brain has learned to maintain and reproduce. The human mind makes sense only within the transformational logic of how it is emerging into this universe. Only within the long and illustrious history of this transformational logic can we find what it means to have a mind, and to be human.

This investigation is a scale-free, state of the art enquiry into natural origins. It is a new point of view, a substantial expansion of the Darwinian paradigm. It places human evolution in perspective with the history of life-on-Earth, and within the history of the universe. The aim is to show that universal evolution has been a process of *selection for symmetric persistence from expanding spectrums of interaction* for almost fourteen billion years. This

[5] The 'neural correlates of consciousness' is a reference to the search for the neural underpinnings of conscious thoughts as suggested by Crick and Koch (1990). The 'connectome' refers to the recent effort to map the brain's neural architecture as proposed by Sporns, Tonnoni, & Kötter (2005).

[6] The Fibonacci sequence is a mathematical sequence first recorded by Leonardo Fibonacci where the addition of the last two numbers gives the next number in the sequence (e.g. 0,1,1,2,3,5,8,13, … 233, …). The Fibonacci sequence is closely related to Adolf Zeising's famous 'golden ratio' and is particularly evident in the ontological development of many organisms. See Goodwin (1994).

process has ordered the emergence of all matter/energy since the Big Bang and, at a higher resolution, human emergence and the human mind are parts of this very same universal process.

> *"From so simple a beginning endless forms most beautiful and most wonderful have been, and are being evolved."*
>
> *Charles Darwin (1859, p. 490)*

# Section 1:
# Origins

This section lays out some of the fundamental definitions and philosophical terms that will be necessary to navigate the cosmological, biological, anthropological and cognitive science sections to come.

# Transformation

*"Nothing endures but change."*

*Heraclitus (535 BC - 475 BC)*

*Now this interconnection, relationship, or this adaptation of all things to each particular one, and of each one to all the rest, brings it about that every simple substance has relations which express all the others and that it is consequently a perpetual living mirror of the universe.*

*Gottfried Wilhelm Leibniz (1962, p. 263)*

The universe has always been transforming by acting on itself. This transformation is varied, ubiquitous, simultaneous and continuous (or nearly so down to Planck levels). Both Aristotle and Newton struggled to explain the nature of transformation (i.e. motion) and both of their solutions focused on reducing the complexity of constant ubiquitous interaction to simple causes and effects. Part of the reason modern science adopted the Aristotelian and Newtonian solutions was because they reduced the messy problem of holistic transformation to calculable 'particulate' formulas of cause and effect.

Take Aristotle's logic of the 'unmoved mover' for example. To paraphrase him, 'Things move and all things that move have been set in motion by something else. Because there cannot be an infinite chain of causes, there must have been a first cause, a mover that was itself unmoved.' Like most arguments, the unmoved mover argument includes a presumption. Aristotle's presumption is that all phenomena can be adequately described as effects that have a preceding cause. Scholars consider Aristotle one of the founding fathers of the scientific method, and for good reason. Scientific thought is built on Aristotelian logic - the linear logic of cause and effect (e.g. if this, then that).

Two centuries later, Sir Isaac Newton redefined Aristotle's particulated concept of transformation into modern scientific terms. His first law of motion states that, "Every body preserves in its state of rest, or of uniform motion in a right line, unless it is compelled to change that state by forces impressed thereon" (Newton, 1848, p. 83). For Newton the natural state of material structure is frozen. It "preserves in its state of rest … unless it is compelled to change."

Newton does not mention an unmoved mover, but he does state that there is no change of motion without a changer (e.g. forces). The Newtonian universe is mathematically frozen and divided into 'the movers' and 'the moved.' No 'thing' can change without a changer. All phenomena can be reduced to lines of cause and effect, the mover and the moved, the changer and the changed.

Each of us has inherited ideas from our parents. The most benign ideas are something like 'vote Republican.' Other ideas like 'women are inferior' are odious. At some point in our lives most of us acquire the ability to reassess these beliefs and make up our own minds. This is one of the delights of becoming an adult. The same idea applies to the larger context of inherited cultural assumptions. The problem with the kind of reasoning we have inherited from Aristotle and Newton is threefold. First of all, it fails to recognize the constraints that language and thought inherently force upon understanding. This is so fundamental and important that the entire next chapter is dedicated to it. Secondly, we are coming to realize that things, in the Aristotelian and Newtonian sense of static particulate phenomena, do not exist. And last but not least, Aristotle's aversion to the idea of infinite causality is an aesthetic choice, not a logical impossibility. Everything always interacting is nothing to be afraid of at all.

As mentioned, the constraints that language's serial and digital structure impose on thought will be addressed in the next chapter. As for the idea of a 'thing' (in the sense of 'material structure' or Newton's 'body') Ludwig von Bertalanffy (1968) concisely summarizes the modern consensus when he says that every 'thing,' "consists almost completely of holes, being a void for the greatest part, only interwoven by centers of energy" (pp. 226-227).

For centuries now scientists have been finding that the smallest structures they know of are largely 'voids interwoven by centers of energy' with even smaller structures moving within them. Exactly this reoccurred recently with the discovery of quarks as the constituents of protons and neutrons, otherwise known as quantum chromodynamics. Any concept of 'thing' that either Aristotle or Newton had was based on the idea of a singular entity, like the canonical billiard ball. It turns out that everything is divisible - every structure is actually a 'social' aggregate. Even more disconcerting, when these social entities are examined they are not solid like a rock. Solid things are largely voids. Things are empty. They are composed of energy relationships. Witness quantum chromodynamics. The result of this revelation is that the features of the universe we live in are best described in terms of relationships. It turns out that things are composed of relationships.

The difference between a relationship and a thing is the difference between dynamic and static. A relationship is a persistent process of regular transformation. A thing is a snapshot of the immediate state of a persistent process of regular transformation. The difference between dynamic relationships and static things will be further explored in the chapter "Processes and Structures." For now let us say that internally things are movement. Things are change, and it is the regularities in their movements, or changes, that define them. Thus, if everything is constantly moving then there are no special 'movers' or 'changers.' There is only movement and transformation at all levels of every 'thing' we currently know of. From this point of view, the history of the universe can be seen as a dynamic process with a history of selection for the accumulation of regularities, not as a frozen structure with smaller structures causing larger structures.

Matter-energy just after the Big Bang was interacting chaotically. Causality, in the form of the things that have appeared since then, has always been in the form of emergent relationships that limit chaotic interaction to regular interaction. In a fundamental way causality is linked to limiting interactive processes, not inciting interactive structures. Thus, entropy doesn't need to be explained, regularity does. Randomness does not need to be explained, regularity does, system existence does. Causality is about constraint accrual, not what caused what. The changer in this universe is not things, but the localized loss of degrees of freedom that initially random transformative states have undergone.

*When we imagine our universe we misrepresent it by saying that one thing causes another. This implies that the natural state of the universe is stasis. That there is no change until a changer comes. A truer representation is that everything has always been changing and that things can only be regular versions of that change.*

This brings us to the last point, infinite causality. Both Aristotle and Newton focused on causality in the form of proximate, or 'efficient,' causes. Movement is caused by another movement and change in movement is caused by a change in affecting movement. Both of them interpret the world in terms of linear causes and effects. The problem is that logically, this way of thinking requires a prime event, an unmoved mover. Nothing changes unless something else does. The world we know changes so either something started that change or it has always been changing. Aristotle's statement that infinite causality is impossible is just that, a statement. Another statement is that everything has always been interacting. You can either believe that matter-energy was created and was so created to interact (contradicting the first law of thermodynamics and presupposing some form of 'creator' or 'creation event') or you can believe that matter-energy has always existed, and always interacted.

Long before Aristotle, the Greek philosopher Heraclitus famously argued that, "nothing endures but change." More contemporary to Aristotle, but on the other side of the continent, people were practicing Vedanta (the source of Hinduism and Buddhism) which is based on cycles of existence. The metaphysicians of this belief system would have laughed at the linear concept of an unmoved mover. In Europe, Wilhelm Gottfried Leibniz, a contemporary and intellectual rival to Newton, would have also seen the problem with an unmoved mover. The concept that 'effects can be reduced to single causes' would have been untenable to him seeing as, "every simple substance has relations which express all the others" (Leibniz, 1962, p. 263). For Leibniz physical reality was also about relationships not things.

This view is not new; it is just unpopular in the current scientific paradigm. Most science as it is practiced now is in the business of contextualizing their questions in the form of experiments. Experiments construct a box within which any 'thing' included is isolated as a cause or an effect. Under such and such limiting conditions, what is the result? Not surprisingly, we find what we are looking for - isolated solutions to isolated problems in isolated contexts. This is a fantastic way to engineer things. Technology is very useful. However, it has proven insufficient to explain how the universe evolved - how we evolved as part of it. Our current scientific paradigm has taken us to the moon and extended our lives, but it has failed to illuminate our hopes, fears and desires. It has also failed to consider everything transforming simultaneously by acting on itself.

# Thought

*"'Reason' in language: oh what a deceptive old witch it has been! I fear we shall never be rid of God so long as we still believe in grammar."*

*Friedrich Wilhelm Nietzsche (2007, p. 20)*

*"Perception and experience categories need not mirror the 'real' world; they must, however, be isomorphic to it to such a degree as to allow orientation and survival."*

*Ludwig Von Bertalanffy (1968, p. 241)*

Rational thought is language and language is an inherited behavior. As humans we have inherited two different kinds of symbolic behavior. Ernst Cassirer (1965) calls these two the 'symbolism of number,' and the 'symbolism of speech,' math and whatever languages you speak. The former operates through logical transformations and the latter through metaphorical transformations. Both human languages are flawed for the same reason.

The languages and thoughts we have inherited are formulas for parsing our environment. A fundamental rule of these formulas is that one symbolic 'thing' must follow another. Both mathematical and metaphorical languages understand the universe through this process of separating the universe into representational symbols and then putting one symbol after another. This process digitizes and serializes everything. It separates things that are by nature simultaneously interrelated. Language has one thing act on another in serial lines of cause and effect. One word follows another - one thought follows another - one number follows another - one operation follows another.

The act of 'symbolizing' limits the kinds of thoughts we can have about the universe to a digital, serial code.

Linguists especially love to expound upon the wonderful, infinite qualities of semantic information. Simon Kirby (2007) demonstrates how taken for a given this presumption is when he mentions, "the unbounded semantic information that we can convey with linguistic expressions" (p. 10). Guy Deutscher (2005) presumes the same idea in the following quote, "[language] allows us to communicate unboundedly complex thoughts and ideas" (p. 8). The implication is that our powers of language are capable of harvesting and conveying unlimited types and amounts of information about the universe. The assumption is that there is nothing language cannot describe. This is profoundly untrue. Try to think of two separate events *simultaneously*! Now try to think of ten events *at once*! Now try to imagine how those ten events will transform as they interact to the various degrees to which they interrelate! Now ignore the billions of simultaneous transformations occurring around you and just try and imagine the billions of transformations that are occurring inside you, at this moment. Thought can only begin to represent what is really happening.

The languages we inherit, and by extension the thoughts we are capable of, are massively constrained by their digital and serial structures. With tremendous difficulty we have invented potentiated parallel processing computer networks, governments and other systems of complex simultaneous transformations, but these are highly limited and difficult undertakings. By in large, we are each individually handicapped to understand the universe not as it is transforming, but as how the structure of language parses that transformation for us.

Fortunately, the universe is not serial or digital. Unfortunately, we are limited to understanding it as such.

# Time

*"Einstein's picture of time was radically different... Nobody ever measures some absolute phenomenon called 'time'; what one measures is the rate of some physical change in the universe."*

*John D. Barrow (1994, chapter 6, para. 3-4)*

'Time' has been a problem for a long time. According to Ludwig von Bertalanffy (1968) "modern science applies whatever sort of space and time is most convenient and appropriate for describing the events in nature" (p. 226). Julian Barbour (1999) says, "space and time in their previous role as the stage of the world are redundant. There is no container. The world does not contain things, it is things" (p. 16).

The problem is that an objective account of the Newtonian idea of an absolute and ideal 'time' cannot be scientifically justified. The only consistent way to define time is by physical transformation. The two most exact sources of time in the world are atomic clocks, America's NIST-F1 and Switzerland's FOCS 1. Neither of these clocks is *time*, they are *clocks* that measure physical transformations. So in a very real sense our idea of what time is, is just a highly correlational coupling of regular physical transformation with words like 'time,' 'temps,' 'chronos,' 'kairos,' 'zeit,' 'tiempo,' etc. As a result, the traditional concept of absolute 'time' as existing apart from some real physical process will not be useful to this investigation. In its place will be the concept of 'change.' Change, as defined as 'transformation,' can at least be directly measured and defined as per the object or system that is unfolding.

For the purposes of measurement, time is change. Change is transformation. Transformation happens in two ways, either the thing or process changes its relation to itself or it changes its relation to other things or processes. Another way to say this is that systems in this universe are constituted by two kinds of relationships: internal relationships or external relationships. As we shall see, one important feature of universal interaction is whether positive, neutral or negative symmetry relations are formed in these interactions. Even more importantly, what are the results of those few areas of the universe where conditions differentially preserve positive symmetry interactions over long periods of transformation? This, however, is the subject of the rest of this investigation. For now, time will be seen as transformation.

Time is change in the structural relations of matter-energy systems.

# Symmetry

*"It is only slightly overstating the case to say that physics is the study of symmetry."*

*Nobel Laureate Phillip Anderson (1972, p. 393)*

The universe is interacting. A crucial concept for understanding how it interacts is symmetry. Mathematically, symmetry is, "*a way to rearrange things . . .* it is the rule you apply to get the rearrangement" (Stewart, 2007, p. 118). Mathematician Ian Stewart continues by observing, "there are three key words in the definition of symmetry; 'transformation,' 'structure,' and 'preserve.'" So symmetric things and symmetric processes are those that transform while preserving structure. Here is a version of the example Stewart gives:

**FIGURE 1: TRIANGLE SYMMETRY**

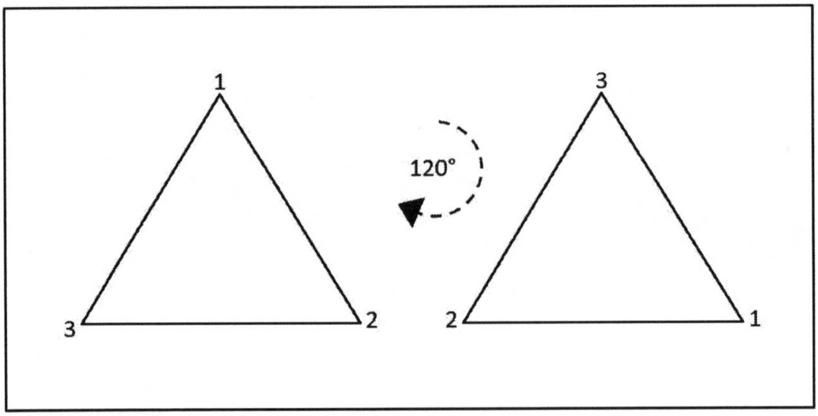


Source: author illustration

Rotate an equilateral triangle 120° and you get the same triangle. It has transformed and preserved its structure. This is mathematical symmetry. Symmetry in physics is much more complicated because it is not applied to idealized, two dimensional abstract structures. Real arrangements of matter-energy can have many measurable dimensions, such as chirality, charge, angular momentum, wave-function, etc. The rules of rearrangement in physics are not only highly multi-dimensional, but they also depend on notoriously 'messy' means of measure and observation as well as rather vague ideas of complexity and information.[7] However, as physics Nobel laureate Phillip Anderson notes in the quote at the beginning of this chapter, real physical relationships also seem to be overwhelmingly symmetric.

What Anderson means is that the ways that real matter-energy systems transform themselves and others are limited and defined by rules. As another Nobel laureate, Murray Gell-Mann, is fond of saying, these rules are just 'regularities.'[8] A physical regularity in this sense has two qualities. It changes and it repeats this change. What Both Anderson and Gell-Mann are implying is that nature has rules and these 'natural laws' are expressed as repetitive transformations. If it only happens once it is not a law. If it happens again and again, every time, then it is generalizable as a law.

In this investigation the terms 'law,' 'rule' and 'regularity' will be considered fungible in the sense that each term refers to symmetry. Each term refers to a class of transformations that preserve some degree of structure. As such, a *'regularity'* has a low degree of symmetry. Regularities are short-term transformations that preserve little structure.[9] By this kind of accounting, a *'rule'* is a class of transformations that have a medium degree of symmetry. It preserves more structure because it lasts longer. And, of course, a *'law'* has a maximum degree of symmetry. It refers to a class of transformations that preserve structure perfectly, in every case, forever. Gell-Mann calls laws regularities because that is what laws are, perfect and eternal degrees of regularity.

---

7 For more on complexity and information see Appendix 1.

8 This description by Murray Gell-Mann is reported by Stuart Kauffmann (2010) in his book *Reinventing the Sacred* (p. 292).

9 Regularities that preserve more structure (in a change case) last longer (in a time case). If time and change cases are themselves fungible, as previously suggested, then no distinction is necessary. Regularity is stability over change-time.

One way to imagine natural regularities, rules or laws is as differing degrees of symmetry. All of them exist only because they repeat themselves. What do they repeat? At a minimum they must repeat some kind of energy transformation. The cyclic repetition of energy transformation over time is what a rule, or a regularity, is. It is what physicists mean when they say that wave structures 'propagate.' Waves repeat the same structure again and again. From quantum wave function to chemical bonding to rotating galaxies to genetic reproduction, all rules or laws or regularities must constrain transformation to regular cycles that maintain some kind of measurable pattern through space-time. Only so do they persist. Existence in a ubiquitously and continuously transforming universe is regularity. It is the persistence of structure through regular transformation.

Symmetry is those rules that define a system's transformational regularities. In physics these rules are typically measured and described by mathematics. The formula $E=MC^2$ describes a basic symmetry of this universe: that energy can be transformed into mass, but only through the preservation of a regular relationship with each other, and the speed of light. The concept of symmetry is so fundamental to all interaction that imagining a universe without it is next to impossible. By way of example, describing a universe without symmetry would require imagining mathematics without an equals (=) sign.

This investigation will focus on processes not 'things.' Symmetry is transformation with preservation of structure where structure can be not only real physical things, but also the relationships these things form and that form them - their patterns of interaction, their rules of transformation. Nominally these measurements are included in traditional symmetry as the algorithms that describe the transformation, in the mathematics. However, these functional relationships are real. They exist and can be measured as information.

Claude Shannon helped define 'information' as 'the reduction of uncertainty.' Where natural selection is selection of a few surviving organisms from a range of many possible organisms, so is information the selection of a few signals from a range of many possible noises (Deacon, 2007a, 2007b). Both of these fundamental processes limit what can happen in the future. They select highly unlikely actual futures from vast spectrums of possible futures.

Although vast, these spectrums of possible futures are not infinite. They are always to some degree constrained by two variables: the current actual state of affairs and the regularities, rules and laws that constrain how that actual state of affairs can transform. Persistence under these conditions are those states of affairs that are sustainable, or repeatable. Another way to say this is that only relationships that are regular over time persist. Information in the physics sense implies high dimensional arrangements of regular matter-energy *relationships*. These relationships constrain all possible interactions to a determinably regular probability distribution - a compressible code. Thus, information measures constrained interaction - it measures the reduction of uncertainty of what can happen next.

In natural selection and information, what emerges are new symmetries. The advantage of using symmetry to unite emergent phenomena is that it is a mathematical abstraction designed to represent both structure and process. Symmetry is a 'process' of transformation that preserves a 'structure.' In natural selection, symmetry can represent reproductive processes like mitosis that maintain a class of biotic structures, like protists. Mitosis can be described as a transformational regularity that happens again and again, generating work cycles that both maintain and reproduce new organisms that, in turn constrain conditions to those that maintain and reproduce the same work cycles. Shannon information is founded on symmetry by its process of mathematizing systems energetically displaced from equilibrium. Systems described by Shannon information already include the entropic process by which their structure is defined (i.e. in the continuously measured statistical reduction of all possible states to an actual state).

The unification of process and structure under the mathematical generalization of symmetry allows us to see what emerges in terms of structure or process as the same thing. What emerges are new transformational regularities. In a phrase, the furniture of the universe can be defined as the regularities in the processes that maintain them. Through symmetry, structure inheres process and process inheres structure.

Information relationships have structure just like matter-energy systems have structure. Just like matter-energy systems, these relationships transform by rules - by 'preserving structure.' However, their structures are not always constrained by physical laws like the Pauli exclusion principle or the first law of thermodynamics.[10] Information structures can be created and destroyed. Two of them can exist in one place at one time.[11] This distinction between physical and informational structures is important because although matter-energy cannot be destroyed or created, the relationships it forms (or the information measured in its relationships) can.

In this view of the universe, physical symmetry is largely constrained by the first law of thermodynamics and the Pauli exclusion principle. Two different material structures cannot be in one place at the same time. Matter-energy structures cannot appear and disappear. However, informational symmetry is under no such constraints. Informational symmetry is the reduction of uncertainty in the transformations that preserve the system's structure. Informational symmetry relationships can be destroyed or remain the same, but most importantly informational symmetry can be created. When this investigation refers to 'symmetry' it will be referring to informational symmetry unless otherwise indicated.

[10] The Pauli exclusion principle says that *matter* structures, in the form of fermions like quarks and electrons, cannot occupy the same state. It does not apply to *force* structures, in the form of Higgs or gauge bosons. Physical symmetry applies to matter structures and informational symmetry applies to force structures (or transformational regularities). Because the fundamental relationship between bosons and fermions is currently unknown, the *physical* and *informational* categories of transformation must be viewed as loose representations even though they refer to current state of the art physics.

[11] Stuart Kauffman (2008) goes to great lengths in his book *Reinventing the Sacred* to show how two informational structures can exist in one place at the same time. His favorite example is the tractor engine block which initially performed one job, stabilizing the engine, and then came to perform a second job as a chassis mount for the rest of the tractor parts. This single engine block existing in one place at one time is maintaining two different informational symmetry structures, or jobs. It is simultaneously stabilizing the engine and mounting the parts.

# Synchrony

> *"But it is not possible that two things alone should be conjoined without a third; for there must needs be some intermediary bond to connect the two."*
>
> *Plato (n.d., p. 31)*

> *"I find that at least one further phenomenon seems to be identifiable and either universal or remarkably common, namely, ordering (regularity or periodicity) in the time dimension."*
>
> *Nobel Laureate Phillip Anderson (1972, p. 395)*

Symmetry is related to synchrony. For Strogatz (2003) synchrony means order through time. Since time is isometric to change, synchrony is also order through change. In this simple sense synchrony is informational symmetry. It is the reduction of uncertainty in the transformations that preserve the relationship between at least two systems, or two components of the same system, over time.[12] For a duration of transformation (i.e. time) two synchronous systems are 'coupled.' According to Strogatz they are coupled, "if some physical or chemical process allows them to

[12] The two different systems referred to here can also be interpreted as two different states of the same system. If a faucet is dripping water and that dripping contains some form of statistical regularity that can be represented as information then one can either call each drop a different system or a different state of the same system. This investigation will not address this issue, but it will consider the idea of different states of the same system as a symmetry through time and the idea of different systems as symmetry through space. Einstein made time and space fungible in a similarly functional sense.

influence one another" (p. 3).[13] Here are examples of nonsynchronous and synchronous systems:

**FIGURE 2: NON SYNCHRONOUS SYSTEMS**

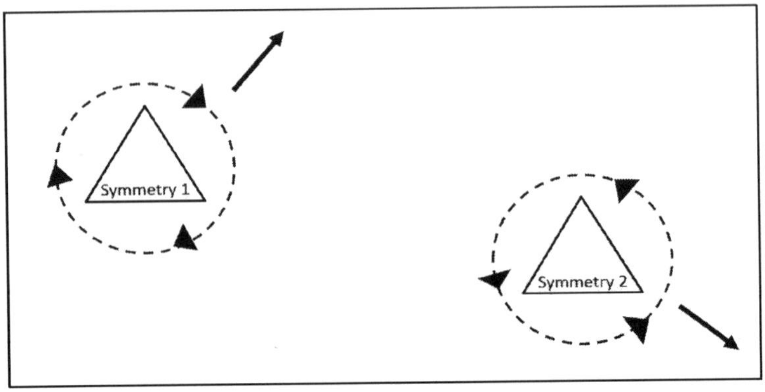


Source: author illustration

In Figure 2 two systems are transforming while preserving their structures. The algorithms that describe how each is maintaining its own transformation with preservation of structure are called 'Symmetry 1' and 'Symmetry 2.' In this example each system is independently transforming symmetrically (in this case rotating) and travelling away from the other. Each system is individually and independently synchronous, but together they represent no additional synchrony. Their relative futures are ever diverging chaotic paths in which uncertainty is constantly increasing as they further interact with their environments. Another way to say this is that they will each break their own symmetries in their separate futures.

13 Synchronous relationships need not be simple couplings. They can be very complex as in the case of strange attractors where the point of equilibrium follows a chaotic landscape. This is only to say that synchronous order need not necessarily be determinate.

**FIGURE 3: SYNCHRONOUS SYSTEMS**

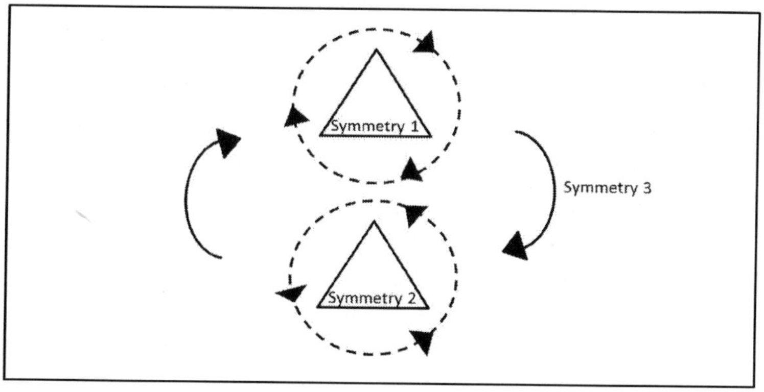


Source: author illustration

In Figure 3 these same two systems happen to meet and maintain a new regular relationship (imagine an electron and proton meeting or two hydrogen atoms meeting an oxygen atom). Together the two systems are coupled to form the synchronous relationship of the whole system - described by the algorithm 'Symmetry 3.' Even though no physical thing has been added to this example (e.g. the same amount of matter-energy is present in both examples) this example has added information in it now. Now, it not only contains the information that describes the transformation of each individual triangle, it has added the information that describes how each triangle now relates to the other. Their relative futures now include a regular relationship that can be described as information as they further interact. Another way to say this is that they will together break mutual symmetries in their coupled future. Their future can be described as a reduction of the uncertainty of what each can do in relation to the other. Stuart Kauffman (2008) calls this loss of 'degrees of freedom' a 'constraint.' Notice that no matter-energy has been added, but information has increased. The addition of information comes only in considering the environment of either triangle, namely, its new relationship with the other triangle and their unified effect on their shared future environment.

In this idealized case, an observer wishing to measure either triangle alone may find it transforming according to the same rules whether it is in or out of sync with the other (e.g. the triangles in Figure 2 are the same as the triangles in Figure 3 when considered alone). In real matter-energy interactions one or both of the systems may constrain to some degree the transformation of the other. In this way a planet changes how its sun 'transforms' just as that sun changes how that planet 'transforms.' Very importantly, however, information is not necessarily conserved in the transformation from nonsynchronous to synchronous systems. The amount of change the sun effects on the Earth is not related to the amount of change the two together can effect on the rest of the universe. Imagine the myriad effects that part of the Earth called 'humans' could have on the universe when space travel becomes possible.

In short, the description of Symmetry 3 does not entirely depend on Symmetries 1 and 2. Symmetry 3 also depends on its environment. Humans are part of symmetry 3 in the sun and Earth example. Humans exist only through the sun and Earth's synchronous relationship. However, what humans end up doing to the rest of universe (if anything) will be determined by what we find out there, by our cosmic environment. Symmetry 3 (human exploration) could have vast consequences, far beyond the relatively simple relationship the Earth maintains with the sun (i.e. orbit). The description of the human effect on the rest of the universe is not constrained by the Earth orbiting the sun. The Earth and sun are necessary but not sufficient descriptions of the human future. This concept is at the heart of a new way to understand emergence and will be fully explained in the "An Environmental Theory of Emergence" chapters at the end of the "Universe" section.

Canonically, 'symmetry' has been measured as a comparison in space. Psychologists map one side of a person's face and compare that map to the other side and the degree of similarity is typically described as 'facial symmetry,' order over space. Canonically, 'synchrony' has been measured as a comparison in time. Brainwaves, extinction periodicities and orbits are typically described in terms of synchrony, order over time. When all systems are reduced to the regularities that describe them, these regularities are transformational; they are dynamic, not static. What this means is that in a massively interactive universe where transformation is the natural state and regular versions of this transformation are only rarely selected for persistence, symmetry itself can be expressed as synchrony. All systems can be expressed as regularities in both time and space. Essentially, one result of treating space and time as a single fabric is that symmetry and synchrony are unified.

In this investigation synchrony will be explored in various different contexts. In the physics and astronomy contexts synchrony will often be referred to as 'positive symmetry.' Positive symmetry is the formation of relationships between abiotic systems that embody order through time (e.g. molecule formation, solar system formation). Another way to say this is that newly synced systems express an added level of information, added constraints or an added level of order through time. In the biological context synchrony will often be referred to as 'cooperation.' Cooperation is relationships between biotic systems that embody order through time. In this sense, cooperation is biological synchrony.

## Interaction

*"You are a complex system of activities that makes temporary use of various kinds of matter, but that matter is not you. You, and all other organisms, are continuous systems of material flux, of matter moving in, playing a role, and moving out."*

*George C. Williams (1997, p. 115)*

Interaction can be defined as the process by which the function of one 'thing' alters the functions of other 'things.'[14] Before the three results of interaction are introduced, the idea of 'things' must be addressed.

The 'things' that populate the universe we know are matter-energy arrangements where the structure of each arrangement is a product of the history of how that matter-energy has transformed according to the four fundamental forces (i.e. gravity, electromagnetism, strong nuclear force, weak nuclear force). Each of these things is characterized by some degree of symmetry. As Michael Field and Martin Golubitsky (1992) say, "Thinking of symmetry in this way, we see that every object has at least one trivial symmetry obtained by picking it up and putting it back in its original position" (p. 5). That is, every 'thing' transforms with some degree of preservation of structure. If it didn't, it would cease to exist.

All things are in a constant state of internal transformation. How they maintain structure through this transformation is what they are. Externally these same things can be seen as moving in space relative to other things. This also defines what they are. Maintaining external structure through relative movement in space and time is Field's and Golubitsky's "one trivial

[14] This definition is derived from E. O. Wilson's (1975) definition of cooperation in his book, *Sociobiology*.

symmetry." In both internal and external contexts some degree of symmetry, trivial or not, must be preserved for a thing to remain a thing. At a minimum, matter-energy itself must be conserved. Even in situations of maximum entropy, matter-energy cannot disappear, it must transform. In this universe, the transformation of matter-energy has rules. It is constrained by certain regularities.

From quanta up, all 'things' transform and the transformations have characteristic rules. Those rules are the four fundamental forces and their historical derivatives. Even quantum particles are subject to symmetry in their features as per charge conjugation (c-symmetry) and parity (p-symmetry).[15] In fact, if the thing has no symmetry at all it is not a thing. For a feature of the universe to be recognized by us as some 'thing,' it needs to be describable as transforming with some degree of preservation of structure. At the very least it needs to be capable of preserving the syntactic and metaphorical structure of language or the relational logic of mathematics. All persistent arrangements of matter and the forces are such because they consistently preserve some degree of relative structure.

Things are expressions of regularity. As such, the symmetry of any 'thing' is a matter of degree.

[15] The weak nuclear force is unique in this aspect because it works only on left handed (parity) particles.

# Systems

*Systems are oriented by their environment not just occasionally and adaptively, but structurally, and they cannot exist without an environment. They constitute and maintain themselves by creating and maintaining a difference from their environment, and they use their boundaries to regulate this difference.*

*Niklas Luhmann (1995, p. 17)*

The idea of systems has a long history. The word itself comes from the Greek 'syn' for 'together' and 'histanai' for 'cause to stand' (Harper, 2013). In the last century the word 'system' became the centerpiece for a new viewpoint on scientific unification. Ludwig von Bertalanffy (1968) was a founder of 'General Systems Theory' in which he tried to unite the physical, biological and social sciences by showing, "that the world (i.e. the total of observable phenomena) shows a structural uniformity, manifesting itself by isomorphic traces of order in its different levels or realms" (p. 87). Bertalanffy continues his thought by stating that the, "Unity of Science is granted, not by a utopian reduction of all sciences to physics and chemistry, but by the structural uniformities of the different levels of reality." He believed that science would not be unified by reduction to physical terms, but by the uniformity of the rules that appeared at each level of systemic organization. He often characterizes systems as "the whole is more than the sum of its parts" because "different levels of reality" like physics, biology and sociology share no causally reducible set of relationships. They are 'emergent.' Physics, biology and the social sciences are related, not by a reducible chain of physical logic, but by a processual logic that exists at each emergent level. In other words, different systems on different levels share the same process of existing even though they do not share the same structures or rules of existence.

Talcott Parsons and Niklas Luhmann developed Bertalanffy's 'systems' logic in the social sciences while Humberto Maturana and Francisco Varela developed it in the biological sciences. Bertalanffy (1968) defines systems rather vaguely as "a set of elements standing in interrelations" (p. 55). Parsons weighs in with: "When a set of interdependent phenomena shows sufficiently definite patterning and stability over time, then we can say that it has a 'structure' and that it is fruitful to treat it as a system" (p. 502). For Parsons (1951, pp. 5-7), social systems were composed of actors whose "needs-dispositions" caused them to organize into different action systems, like families or institutions, which worked together to build social systems. All of these action systems acted according to "needs-dispositions" maintained by relationship to the environment that made his whole theory like a nested hierarchy of interdependent relationships. Both Bertalanffy's and Parsons' definitions of systems tend to downplay environments and focus on the component relations of systems.

Luhmann is infamously obscure, but he is more inclusive than Parsons in defining a system. Where Parsons worked from individual actors up, from parts to whole in an engineering style of construction, Luhmann worked from the whole to the parts in an organic style. For Luhmann (1995),

> Systems are oriented by their environment not just occasionally and adaptively, but structurally, and they cannot exist without an environment. They constitute and maintain themselves by creating and maintaining a difference from their environment, and they use their boundaries to regulate this difference (p. 16-17).

Parsons seems to want to use how people interact to describe society while Luhmann claims that how people interact is society and how society functions is people. For Luhmann the important feature is the differentiation of the system from its environment. This differentiation is characterized by simplified and regular internal states negotiating for what they need to maintain themselves from a complex and chaotic environment through a boundary. For him, the system itself is defined not by its simplified internal relationships nor by its complex environment, but by the maintenance of the differential over time. Systems are open to their environments only to the extent that they successfully negotiate the conversion of various external relationships to feedback with limited internal relationships. By Luhmann's logic the system is neither the thing itself nor its environment, but the cycles of maintenance and reproduction that maintain both states through time and space. Luhmann does not use these exact words, but his thinking inspired the definition of systems that this investigation will rely on. This definition will be given presently.

Luhmann's social systems theory borrowed heavily from Maturana and Varela's idea of autopoiesis. Auto-poiesis is Greek for 'self'-'creation.' Maturana and Varela come from a biological background and thus envisioned autopoiesis primarily as biological systems - organisms. The backbone of their contribution was to focus on the need for any organismic system to maintain and reproduce itself. It must self-create to exist. While self-creation requires closing the cycles of maintenance and reproduction by forming regular relationships, they seem somewhat unclear, or torn, as to how to characterize this process. While autopoietic structures must close these cycles to outside interference to persist, they must also regularly include outside elements to persist. Maturana and Varela (1980) argue that,

> Living systems are units of interactions; they exist in an ambience. From a purely biological point of view they cannot be understood independently of that part of the ambience with which they interact: the niche; nor can the niche be defined independently of the living system that specifies it (p. 9).

Maturana and Varela seem to admit that systems need environments, but they spend most of their time characterizing the internal cycles that define the organism, not its relationship to its environment. The result is the lopsided impression that autopoiesis is closure of those *internal* cycles that maintain and reproduce the system. Luhmann does a better job of insisting on systems not only requiring some environment, but actually being part of, and defined by, their environment.

In fact, none of these brilliant thinkers seems to have been able to convey a consistently clear idea of what a system is. They all agree that environment plays a crucial role, but none of them illuminate that role. Bertalanffy believed that different social, biological and physical environments existed as unique and distinct levels of interaction. Parsons, Maturana and Varela claimed autopoietic and action systems needed environments, but failed to include them substantively in what their systems *were*. Only Luhmann explicitly includes environments in what systems are. He, however, fails to show us *how* systems depend on their environments.

All of these theoretical approaches contributed to the development of systems theory, but they all share a related weakness. Emergence is the unpredictable appearance of new systems and, as we shall see, new systems are just old systems locking into new relationships, new cycles of reproduction that maintain these new relationships. What this means is that emergent systems emerge because of, and from, the 'environments' of old systems. Emergent systems do not contain new components. They contain old components in new relationships. The first atoms did not invent electrons; they just began to include electrons in regular relationships with nuclei.

None of the systems that exist today existed at the Big Bang. All of the systems we will examine have emerged since the Big Bang. In this sense all systems are composed of their historical environments and are maintained and reproduced by their current environments. In fact, this investigation will propose that systems are just persistently energetically dense *parts* of their environments. If one imagines environments as encompassing an enormous spectrum of interaction ranging from almost no interaction to irregular interaction to semi-regular interaction to regular interaction, it becomes possible to see that the systems we know and study are just those very rare instances of regular interaction. The trick is in understanding that while irregular interaction has no effect on the future, regular interaction is the only thing that can exist in the future. The future is only ever populated by those very few interactions that are regular and only they exist to further interact. Existence itself is relegated to how those rare instances of regularity feedback with each other. Existence is regularity. Nonexistence is irregularity.

Essentially what this means is that systems are specially organized parts of environments and to consider them alone or outside of their environments is to limit what you can understand about them. From Bertalanffy to Luhmann there is a consistent realization that environments are important to systems, but they all consider the system as primary. The system is the primary object of investigation and the environment helps to explain the system. In truth though, it is the environment that is primary. It existed first and it defines what can be regular where, when and how. The system itself, whether it is an atom, a brain or a stock market is defined, in part, by the environment from which it emerged, in which it survives, on which it acts, and with which it feeds back. In fact, all physical systems emerged from the Big Bang environment. All biological systems emerged from an Earth-bound physical environment and all social systems emerged from a biological environment. Environments are not part of what a system definition is, systems are part of what an environment definition is. Bertalanffy, Parsons, Maturana, Varela and Luhmann all put the cart before the horse by focusing on what environments could tell us about systems. This investigation will put environments first. It will look at what systems can tell us about environments.

It is easy to laundry list what is inside a system and call that list the definition of the system. One just draws a line around the system and says everything within that line is 'system,' everything without is 'environment.' Much more difficult is to adequately describe what is the boundary and how does that boundary negotiate those interactions with a messy and undefined 'environment' so as to consistently maintain and reproduce the system.

This investigation looks at the relationship between systems and their environments and what emerges from such attempts is an increasing appreciation of the unreality of static structures as compared to dynamic processes. The closer one looks at individual systems with static definitions, the more they dissolve into increasingly fine-grained groups of relationships. You can't define a fish without the water it inhales, the food it eats and the symbiotic bacteria that make it their home. Like fish, our bodies do not contain us, they identify a location characterized by a web of relationships that constantly take external molecules and atoms and use them to replace internal molecules and atoms. Not only is our location always changing as we move around our environment, but our bodies themselves are always incorporating this environment just as our environment is always incorporating our bodies. What differentiates us from our environment is not the stuff that makes us, but the way that stuff is maintained as energetically different from its surroundings. Despite what we are in the habit of thinking, we are more process than thing.

As mentioned, there are at least two historical contexts, in which the concepts of systems as 'things' have been misconceived. The original context is the inherent fallacy of language that divides all experience into nouns and verbs, doers and deeds, numbers and operations, structures and processes. This investigation proposes that the division of structures (e.g. nouns/subjects) from processes (e.g. verbs/predicates) is an illusion created by how language parses reality for us. In reality 'things' embody actions and actions are the expressions of 'things.' The two are one. A system is a 'thing' only in the sense that it maintains certain regular actions over time. For now, though, it is important to consider that using the tools we were given (e.g. language, math and thought) the phenomenal universe can only be conceived as *either* 'things' *or* 'what things do' at any given moment. This fundamental misconception of 'things,' as distinct from their 'interactions,' eventually gave rise to the misconceived reification of 'things,' the search for the basic 'things' of this universe - the 'materia prima.' Much of the scientific community is currently struggling with the fact that these 'things' probably do not exist, that all 'things' are just relational regularities between energy states.

A basic tenet of this investigation is that all 'things' are interactive 'systems.' Additionally all systems are symmetric because, as mentioned in the previous chapter, systems are defined by transformations that preserve structure. In this investigation 'systems' will be defined in two parts, internal and external.

1) *Internally, systems are the differential connectivity within the historically unified and symmetrically persistent relations of distinct components (i.e. matter-energy states).*

2) *Externally, systems are their own differential connectivity with future external complexity mediated through a boundary that separates their energetically dense internal component relations from less energetically dense external environmental relations, yet also connects the internal to the external in symmetrically persistent relationships.*[16]

Systems are composed of distinct, but mutually affective, internal and external relationships. These relationships can be measured as information. Internal information describes how the component interaction is constrained (i.e. ruled). External information describes how the system itself interacts with other systems, how it is environmentally constrained (i.e. ruled). The internal information describes primarily the historical relationships the system is composed of. The external information describes how it is interacting right now and into the future. Thus this approach to understanding systems and how they change proposes an additional way to understand what they are. Systems are not just internal regularities, but external regularities as well. They interact to constitute their environments just as their own components interacted to constitute them. An important part of what maintains systems is their environment. Actually, systems are part of their environment. Systems are specially maintained concentrations of energy constraints within environments of less dense energy relationships. The external relationships of a system are as necessary to its existence as its internal relationships. Just as no man exists on an island, no 'thing' exists in a void.

This is just the introduction. The bulk of this investigation will be unpacking and illustrating how the different levels of internal and external interaction explain the emergence of those systems that populate our universe - atoms, molecules, solar systems, cells, animals and human minds. In particular, the external definition of systems will be an essential ingredient to establishing a functional theory of emergence. This theory will be presented at the end of Section 2.

The rub, however, is that those regular actions that define the system itself are both internal and external. A system can only be defined as a thing if you consider both its component rule-set and its environmental rule-set.

16 These definitions were inspired by Luhmann (1995) and his definitions of social systems.

Its component rule-set is the traditional definition of what a system is. It is the components and rules that have composed that system up until the time it was defined. However, there is another way to define that same system. This second way is not traditional and it depends on a new external perspective on what a system is. This new perspective is called the reactive face.

## Reactive Face

Systems have two environments, external and internal. Internally they have components; externally they are components. There are two guidelines that differentiate internal relationships from external relationships.

The first guideline is that internal relationships are more energetically dense than external relationships. On average, more energy per unit volume is symmetrically embodied within the system boundary than without. Humans preserve structure by transforming more energy per unit volume than their biosphere. The biosphere preserves structure by transforming more energy than its solar system. The solar system preserves structure by transforming more energy than its galaxy and etcetera. This result was discovered by physicist Eric Chaisson (2001). He calls it a measure of "power density." We will revisit this metric in more detail in chapters "Environment and Organism: external and internal" and "Toward a Universal Theory of Directionality" in the "Life" section.

The second guideline that differentiates the internal from the external is that the external predates the internal. Energetically dense internal relationships emerge from less energetically dense external relationships. When the universe was too hot and dense, protons and neutrons as persistent quark-gluon-quark relationships could not survive. Quark-gluon-quark relationships could only survive when their energetically dense relationship (strong nuclear force) was stronger (hotter) than their less dense (cooler) external ambient. When it is too hot outside, protons and neutrons cannot exist. Protons and neutrons (and possibly the strong nuclear force) needed to wait until the quark environment cooled down enough before they could emerge. This seems to be true of all emergent systems. They emerge from, and can only persist within, certain kinds of preexisting environments. First comes the environment, then comes the system.

What this means is that there are two different contexts that define each system: the internal system-*with*-components and the external system-*as*-component. The internal system-with-components is the traditional scientific definition of any system. It is the atom as defined by a nucleus with orbiting electrons. It is the human as defined by his cells. Up to now, science has done a wonderful job of taking apart the systems of our universe to understand level upon level of internal component relationships. This is the material reductionist tradition. This tradition has lifted our cultures up from murky mysticism to soaring technological heights, but in the process we have run into two seemingly related limits to our understanding. How do systems emerge and how do we predict what will emerge from them? Basically, once systems exist we know how to dissect them, but we do not know how they got there or what they will do in the future. As usual, not knowing annoys us.

This investigation proposes a new approach to solving this problem. The external system-as-component has not traditionally been considered as part of what a system is. Essentially, how a system acts in its ambient is part of what it is. That means the system itself is a regular feature of its environment. Whatever random arrangement of external interactions currently surrounds that system, the system itself preserves a structure as it transforms *within* that ambient. This external point of view reduces the internal complexity of any system to a single *reactive face*. To the environment, it is not functionally important what occurs within the system boundary, only that that system presents a localized, unified set of features to the rest of the environment's components. The reactive face is how a system interacts with other systems in its environment. It is the external definition of what a system is.

Consider a water molecule for example. A water molecule is composed of one oxygen atom and two hydrogen atoms. The oxygen atom has eight protons and eight neutrons and eight electrons, which means that it is composed of fifty-six elementary particles and some gluons. All of these parts are changing relative position to each other with relative predictability. The quarks and gluons are constrained to certain relationships within the protons and neutrons, the protons and neutrons are constrained to certain relationships within the nucleus, the nucleus is constrained within the electron halo, and the oxygen atom itself is constrained within the bonded water molecule. However, where any one component is, in comparison to the others, is always changing. Despite all this internal change, the water molecule acts completely predictably within certain external environments. In light gas environments like the Earth's surface, it bonds lightly with other water molecules to make bodies of water that have the emergent feature of surface tension. As components of an environment, every water molecule presents a constrained set of possible interactions with other features of that environment. The famous three phases (i.e. gas, liquid, solid) are perfect examples of the absolute regularity of water molecule behavior in certain water molecule environments, or conditions.

Now imagine each type of atom, molecule, solar system, bacteria, plant and animal that we know of. Each system not only is a system, but is part of a bigger system. Also, each system represents a functionally predictable location of interaction within its larger environment. Atoms can do a constrained number of things within molecules and molecules can do a constrained number of things within solar systems. Like quarks in atoms; bacteria, plants and animals are constrained to certain relationships within constrained areas of our solar system, planet and even our biosphere.

A reactive face is that constrained set of relationships that define the regularities of that system within its environment. It is what a system can regularly do externally, regardless of its internal changes. What is interesting about reactive faces is that they are independent of their internal substrate. No matter how the internal state of a water molecule is currently arranged, all water molecules behave in the same way when presented with certain external conditions. This behavior is impossible until the system itself exists. When the system comes to exist it is not its components that emerge, but the *behavior* of all of them together within that environment. The reactive face is the emergent informational symmetry that a novel system represents in its environment. It is a new potential for regular signals to be selected from a noisy environment. It reduces the uncertainty of what can happen at its location. Shannon would call this emergent behavior, new information.

As mentioned in the previous chapter, many systems theorists mention environments in their definitions of systems, but none seriously consider environments to be primary. One of the results of considering environments as a primary descriptive feature of systems is the need for an external definition for systems. What systems do in their environments is as much a part of what they are as what system components do inside them.

This new external point of view adds a new kind of time boundary to the consideration of what a system is. Relationships inside the system boundary share a history of regular transformation. All of these relationships have a history of how they came to be part of the system. Relationships outside the system boundary do not share this internal history of regularity. The reactive face samples an ever growing spectrum of interactions that leads the system into its own unique future. Most of these interactions are neutral or negative, leading to no new regular relationships. However, if the environmental conditions are right, the reactive face can find and maintain new positive relationships.

What this means is that the traditional internal definition of systems as what they contain is restricted to history. Any internal component or regular relationship can be physically traced back through time to when and how it became part of the system. On the other hand the reactive face is the external definition of systems and it is restricted to the future. Any emergent positive relationship that system can potentially find can only be accessed by the reactive face stepping into an unknowable, statistically determined future environment. Internal regularities are by definition historically knowable. External regularities are by definition unpredictable.

The reactive face is a unified historical regularity constantly stepping into an unknowable future ambient.

Environments predate systems. Environments naturally diversify because their component systems sample each other and thus expand the spectrum of interaction. Systems are those statistical regularities that emerge from these environmental systems sampling each other. These emergent systems are just historical regularities, but they also represent a new reactive face to the environment itself. They represent a novel and persistent feature within their environment that begins to resample the environment for further potential emergent regularities. Thus the system-with-components is historical, but the system-as-component is forever stepping into a statistical future, searching for emergent relationships to make a new historical system with a new reactive face.

By adding the reactive face as the external definition of systems, certain conditions within the universe can become generators of fractal pattern. Where conditions are critical, self-organization in the form of emergent systems can introduce new reactive faces to those conditions feeding back into the self-organization of further emergent systems.[17] All of this requires considering environments primary and systems as secondary. It requires an environmental point of view in which systems present a statistical future potential for synchronous behavior. This new feature of systems will be called the reactive face. The idea of a reactive face will be developed throughout this investigation and will be a cornerstone of the definition of emergence at the end of the next section.

Systems are not static structures. They are composed of relationships and it is the persistence of these relationships that determine them as systems. These dynamic systems have different rules for internal interaction and external interaction. However, not only does the one affect the other, but all instances of interaction are constrained to just three results.

17 The terms 'critical' and 'self-organization' both refer to Per Bak's (1996) work on self-organized criticality. For more on self-organized criticality see chapter "Cosmic Emergence: universal symmetry theory."

# Three Results of Interaction

There are three fundamental results of physical interaction in the universe.

*Result 1: POSITIVE SYMMETRY*

When two (or more) things interact and that interaction has some degree of symmetry, then the two together are synchronous for the duration of the interaction. The duration of the synchrony of the system is called positive symmetry. Positive symmetry is when new relationships are formed.

Positive symmetry is relational not summative. That is: in a positive symmetry relationship the algorithm that describes 'symmetry 1' (S1) when added to the algorithm that describes 'symmetry 2' (S2) does not equal the sum of the two algorithms. The sum of the addition of S1 to S2 is the sum of the two algorithms *plus* the algorithm that describes the regularities of how the two together will interact in the future (i.e. plus 'symmetry 3').

**FIGURE 2: NON SYNCHRONOUS SYSTEMS**

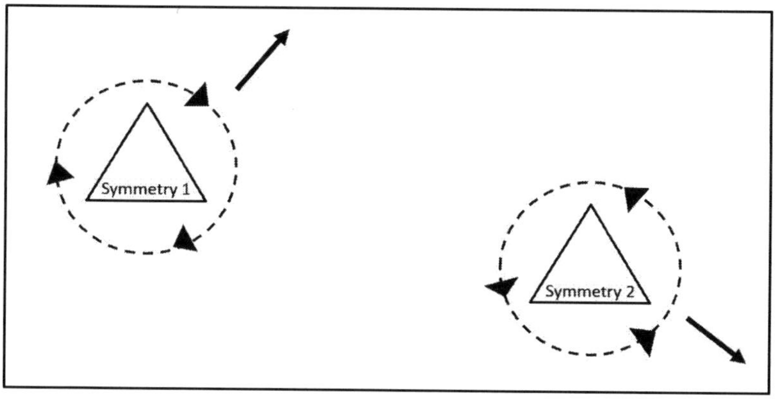


Source: author illustration

**FIGURE 3: SYNCHRONOUS SYSTEMS**

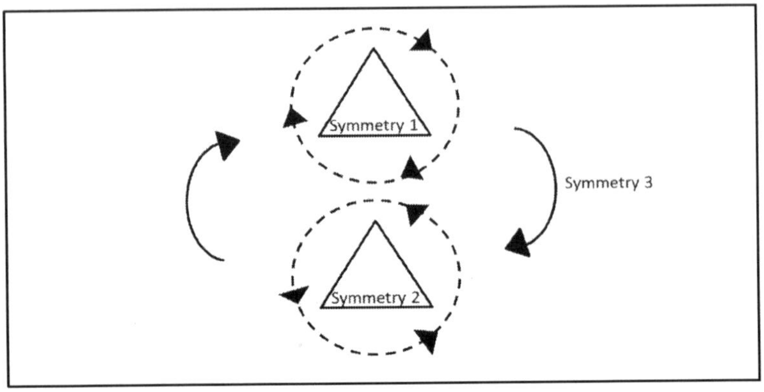


Source: author illustration

These are the same two examples from previous chapters. In Figure 2 there are two distinct systems that have no relationship to each other. They haven't met yet. In Figure 3 these systems meet and maintain a new relationship - as in the case of a sun capturing an asteroid. In Figure 3 no 'thing' has been added to the situation, there are still only those two systems in that area. However, Figure 3 does have added information. The added information is the new regularity that the two systems share in their new relationship. The mathematical description of Figure 2 is different than that of Figure 3 despite the fact that no 'thing' has been added, subtracted, multiplied or divided. The operation that describes the transformation of Figure 2 into Figure 3 includes a 'mysterious' addition of information.[18]

Positive symmetry is the 'mysterious' addition of information. Positive symmetry is characterized by emergent symmetries, new rules by which transformation preserves structure.[19] Positive symmetry is a gain of symmetric relationships between systems.

[18] The 'mysterious addition of information' may be compared to a well-known experiment. Thomas Young's double-slit light diffraction experiment is famous for showing that light is both a particle and a wave. The mysterious result of this experiment is that light from one source seems to act differently than light from two simultaneous sources (from a 'behind the screen point-of-view'). Photons appear to act in two different ways according to whether they are alone (one source) or synced (two 'slit' sources or a photon and a measuring device). Alone photons act one way, in systems they act differently.

[19] However, it may be possible that the emergent symmetry is so low dimensional, in the sense that it represents no new coupling potentials to the environment that the result of the

*Result 2) NEUTRAL SYMMETRY*

Neutral symmetry is the case where two things interact and no emergent symmetry arises. One example could be extremely brief interactions where no emergent symmetry could be formed yet both systems retain their individual symmetry. This is called neutral symmetry because no positive or negative aggregation of relationships emerges.

A substance in equilibrium is an example of neutral symmetry. In equilibrium there are no macro-state differentials that can maintain any kind of regular transformation. Everything interacts equally (at least statistically). Since a substance in equilibrium cannot lose or gain information, heat, energy or anything else, it is an example of neutral symmetry. Neutral symmetry is neither gain nor loss of symmetric relationships between systems.

*Result 3) NEGATIVE SYMMETRY (Entropy)*

When two things interact and that interaction causes one or both things to lose symmetry, to preserve less regularity, then the two together - as well as each one that lost symmetry - are interacting with negative symmetry. Negative symmetry in the classical domain is isometric to 'entropy.' Negative symmetry in the quantum domain is the loss of statistical relationships to interaction. The weak nuclear force can cause quantum systems to lose energy in the form of radiation, reducing the dimensionality of the system's statistical regularities. Negative symmetry is a loss of symmetric relationships between systems.

To avoid confusion it is important to recognize that the term 'symmetry' itself refers to the degree of positive symmetry that a system's transformations preserve. Remember, all things known to us have some degree of symmetry. Terms like 'entropy' and 'equilibrium' refer to the loss or lack of something, not the presence of it. In this sense, things that maintain no regular relationships (e.g. most quanta to each other) cannot lose symmetry they never had. Positive, neutral and negative symmetry all presume and refer to the familiar universe of 'things' that already

---

interaction appears to be just the sum of the two component symmetries.

maintain some positive degree of symmetry that can be lost, gained or maintained.

Just as negative symmetry represents a 'mysterious' or 'statistical' loss of information between systems so does positive symmetry represent a 'mysterious' or 'statistical' gain of information between systems. They are two extremes along the same spectrum of interactive possibilities. In this sense entropy (negative symmetry) itself implies positive symmetry just as one side of a coin implies the other.

## Universal Symmetry: Propagating Symmetry

*"This order is generic, and what we see in evolution may be primarily an emergence of states generic to the dynamics of living systems."*

*Brian Goodwin (1994, p. 186)*

Order exists. Order is not ours to deny, it is ours to explain. Currently we have two working theories of how order is lost (i.e. entropy and chaos), but no theory of how order is won.[20]

Universal Symmetry Theory proposes to show how the kind of generic order Brian Goodwin finds in biological systems (see quote above) can be a fractal subset of the generation of generic order in all universal systems. It proposes that there is a pattern to order, all order. That pattern is *the selection for symmetric persistence from expanding spectrums of interaction.* This investigation is an outline of how it works - of how order propagates.

The basic principle of Universal Symmetry Theory is that all systems that exist participate in cycles of maintenance and reproduction or they disappear. Both entropy and chaos are working to deconstruct all structure, all the time, so existence necessitates a construction process that, under some conditions, discovers, creates, maintains and reproduces structure in the form of regular relationships. Many creationists would love this construction process to be God. Most scientists would prefer this construction process to be a non-teleological, or non-intentional, process - a natural process. The most powerful and all-encompassing of such non-teleological processes is natural selection.

[20] See Appendix 1 for more on 'order.'

Natural selection proposes an immense amount of organismic interaction from which a few select organisms are chosen to exist in future descent lines. What chooses them? The environment in which they interact chooses them. The vast majority of all the extinction events that have so exquisitely tuned today's biosphere were the result of changing environmental conditions like those caused by asteroid impact, overpopulation, volcanism, atmospheric changes and glaciation. In fact, every organism is also surrounded by other organisms whose behavior also constitutes their environment. It is largely environmental constraints that determine which organisms reproduce and which die by differentiating local conditions. Death is the rule and life is reproduction. The organisms that persist do so by out-reproducing their deadly environments.

Universal Symmetry Theory proposes that the same holds true for all evolved structure since the Big Bang. It proposes that there was no structure at the beginning, just an interactive landscape of ever differentiating conditions. As the expanding universe grew, so did the spectrum of possible interactions. At first this interactive spectrum was highly chaotic as per the heat and density of early Big Bang interaction. However, as expansion continued the universe cooled into diverse local interactive landscapes. With this cooling and differentiation the interactive spectrum grew.[21] As the spectrum grew it became increasingly fine grained and eventually began to include very special regions of what is now called criticality. Criticality is those conditions of not too hot, not too cold interactions suspended between static order and dynamic chaos (Bak, 1996). Within these Goldilocks conditions, cycles of regular 'push' and 'pull,' or what Stuart Kauffman (2008) likes to call 'spontaneous' and 'nonspontaneous' work cycles, can be maintained. Within the boundary conditions of these critical regions, regular interactions could survive once sampled.

Within these critical boundary conditions, interaction stochastically sampled different types of relationships. Most of these relationships were irregular, however, the few relationships that were regular were eventually discovered and the first systems emerged. In order to persist, these relationships needed both push and pull, both hot and cold. They needed

[21] Harvard Physicist David Layzer (1975) supports a similar cosmogony in his 1975 article "The Arrow of Time."

areas where downwave was balanced by upwave, where equivalence was balanced by differentiation and where positive was balanced by negative. They needed not only energy differential to maintain the work cycle that constituted their existence, but they needed this energy differential to be reproduced on a regular basis. Entropy is a one way road to non-existence. Existence requires the regular reproduction of an energy differential and this reproduction is constrained to those areas of the interactive spectrum that are critical. In short, a universe with no initial systems needs to first grow the dimensionality of its interactive spectrum, especially critical regions which can then host regular reproductive relationships when they emerge. This process of selection for persistent synchronies from a background of ubiquitous and varied interaction is the key to Universal Symmetry Theory. It is the key to the over thirteen billion year growth of order that resulted, in our case, in the human mind.

Here is an example of how symmetry propagates through positive symmetry. As the universe cooled the strong nuclear force selected for protons where gluon/quark interaction was within certain conditional boundaries. These boundary conditions delineated localities where positive symmetry interactions out-produced negative symmetry interactions. In these areas the production of new protons outweighed the entropic or chaotic loss of newly formed protons.

The localities where symmetry systems were selected for and preserved continued to reproduce those systems until the conditions changed (e.g. too hot, too cold and/or not enough free quarks left to interact to create more protons and neutrons). Over change-time many localities passed through the boundary conditions within which quarks swept each other up in triplets. These localities became populated with new strong nuclear force systems (i.e. nucleons: protons and neutrons). These new systems, fed back with other initial systems, creating new boundary conditions from which ever newer systems could be selected for persistence. Thus, when protons began to appear, their interaction with other quanta produced just one new regular interaction, the regular interaction of protons with electrons to maintain atoms - electromagnetism.

In this case, the persistence of quark/gluon relationships in the form of 'protons' presented a new reactive face to the quantum environment, the 'proton' reactive face. As mentioned, a reactive face is the interactive regularities presented by the combined features of a system to its environment. It is the 'external' definition of a system. It is the regularities of that system's 'behavior' as per its interaction with its environment. In this case the reactive face of protons preserved just one new regular interaction from the quantum environment, the electromagnetic coupling of free electrons. This positive symmetry interaction selected for the light atoms (e.g. hydrogen, helium, lithium and beryllium). As more and more light atoms were produced and maintained they also went on to become part of the interactive background. This process where systems pass through boundary conditions that select for the preservation of positive symmetry relationships is called propagating symmetry.

The basic idea of 'propagating symmetry' is the maintenance of the conditions for positive symmetry interactions. Just as synchrony is the duration of increased symmetric interactions, propagating symmetry is the duration of increased positive symmetry interactions. The result of maintaining novel regular interactions is more variety in the spectrum of background interactions. A bigger spectrum means bigger critical regions and bigger critical regions can sample more relationships faster to find emergent structures which can then feedback into growing the size of the spectrum. Thus propagating symmetry is a compounding process as long as background conditions continue to maintain and re-produce substrate systems. The more different types of regularities that exist in constant interaction, the more novel interactions can be sampled, from which more novel regularities can be selected for persistence.

When discussing propagating symmetries, everything depends on the conditions that maintain the interactive background. If conditions heat up too much, proton/electron relationships dissolve and atoms come apart. If conditions cool too much, interaction becomes too infrequent for new regularities to be selected. The background becomes non-interactive. Nothing new can be selected for persistence. Fred Spier (2010) calls those 'critical' background conditions that propagate symmetries 'Goldilocks Conditions' for the obvious reason that they have to be 'just right' (e.g. not too hot and not too cold).

Propagating symmetry is a new way of imagining the growth of order in certain universal localities and under certain universal conditions. It requires only that localized differentiation increases; that expanding matter-energy samples a growing number of interactions. Within this growing spectrum of interactions it presumes that some tiny percentage of these interactions will be regular, or symmetric, for some duration. This presumption is justified by the appearance of the four fundamental forces, three of which are principally 'attractive,' or create positive symmetry relationships.[22]

Not only is the universe we find ourselves in ordered, it also seems to have the same rules everywhere. Astronomers refer to this as the homogeneity of the universe. We presume that gravity, electro-magnetism, the weak nuclear force and the strong nuclear force (as well as principles like the Pauli exclusion principle and constants like the speed of light) are all the same no matter where you are in the universe. This means that we expect to see the same kind of order in galaxies far, far away as we see in our own, and up to now, we do. All over the universe, at all scale levels, order is being won despite almost fourteen billion years of constant entropy and chaos.

Scientifically and mathematically we do not yet know how order is won. Universal Symmetry Theory is the theoretical beginning of an answer to how things emerge in this universe. By assuming a background of ubiquitous-simultaneous-and-varied-interaction, the 'fundamental forces,' and the 'three results of interaction' we will see that local areas of symmetric interaction can build up under the right conditions. This investigation is not a rigorous proof of how order is generated. What it is, is a beginning of how to imagine what, up until now, has been scientifically unimaginable. It is a way to imagine how order is won.

A few experimental possibilities will be suggested at the end of the next section that should confirm that order generation is generic to certain conditions of universal interaction.[23]

---

[22] This argument is circular, but productive. For more see chapter "Boundary Conditions" at the end of Section 2.

[23] Not incidentally, developmental biologist Brian Goodwin (1994) refers to a biological version of propagating symmetry when he talks about the generic biological structure created in morphogenetic fields by developing organisms.

# Section 2:
# Universe

This section outlines how symmetry can be seen as the principle by which the universe organizes itself.

# Cosmic Emergence

*"The idea of 'states' will have no place in the final theory, which will be framed around the idea of processes and the information conveyed between them and modified within them."*

*Lee Smolin (2001, p. 211)*

*"The power of natural selection theory to explain the origins of adaptive complexity derives from its replicative processes in the context of emergence."*

*Terrence Deacon (1997, p. 109)*

**FIGURE 4: BIG BANG**

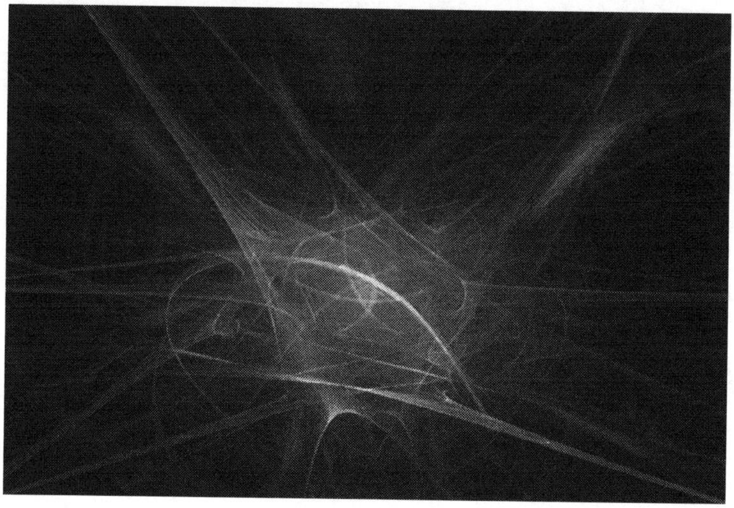

Source: Mashankar (n.d.)

Universal Symmetry Theory (just Symmetry Theory from now on) is the idea that there is a recognizable pattern to how the universe gains order. This pattern exists at all levels of interaction. Symmetry Theory is an extension of Darwinism to the whole fabric of the universe. Like natural selection is applicable to all levels of biology, Symmetry Theory should be applicable to all levels of interaction and existence. For this reason it would be unusual if such a 'theory of everything' had not been glimpsed 'through a glass darkly' before.

In 1996 Per Bak published a fascinating book on what he called 'self-organized criticality.' Self-organized criticality describes the conditions under which many biotic and abiotic systems organize themselves into characteristic, scale free functional regularities. Bak's book is interesting for two reasons. The first is the diversity of these systems and the second is the commonality between them. It would seem that from quantum systems up to astronomic systems, from structures in space to processes in time, there exists a universe-wide pattern of relationships. Bak is able to recognize the pattern everywhere, but he has no explanation for why. It is like Mendel finding the mathematical principles of genetic inheritance, but lacking an explanatory framework (i.e. lacking an understanding of DNA and meiosis) into which to fit it. For exactly this reason Mendel's idea found no home in the science of his time. For decades Mendel's work was forgotten. Per Bak caught a glimpse of something similar. He saw a ghost in the machine - a glimmer of pattern.

At the beginning of *How Nature Works* Per Bak (1996) asks a critical question:

> How can the universe start with a few types of elementary particles at the Big Bang, and end up with life, history, economics, and literature? The question is screaming out to be answered but it is seldom even asked. Why did the Big Bang not form a simple gas of particles, or condense into one big crystal? (p. 1).

What are the principles that arranged the universe into the organized structures of protons, atoms, planets and galaxies? Despite entropy, after thirteen billion years of matter-energy dispersal, the universe is highly patterned. Astronomer John D. Barrow (1994) says, "when we calculate the present entropy of the universe, we find it is staggeringly low; that is, we can conceive of distributing the forms of energy in the universe in ways that are far more disordered" (p. 27).

Currently there is no theory for the source of this ordered organization. This investigation is the beginning of such a theory. It is a step-by-step history of how the principle of symmetry brought the universe from the Big Bang to life, language and minds. This history is the story of how matter-energy arranged itself and it began almost fourteen billion years ago.

## One Hot Mess

> *In the beginning the universe was an inferno of radiation, too hot for any atoms to survive. In the first few minutes, it cooled enough for the nuclei of the lightest elements to form. Only millions of years later would the cosmos be cool enough for whole atoms to appear, followed soon by simple molecules, and after billions of years by the complex sequence of events that saw the condensation of material into stars and galaxies. Then with the appearance of stable planetary environments, the complicated products of biochemistry were nurtured, by processes we still do not understand.*
>
> *John D. Barrow (1994, p. xii)*

In the beginning, the universe was one hot mess. By best accounts the universe began in some kind of point inflation event. This event is referred to as the Big Bang and is metaphorically associated with an explosive expansion. Little is known about the initial state of the Big Bang because the size, heat, pressure and density of it are beyond experiment. Only hundreds of thousands of years later did the dense radiation created by the Big Bang become 'clear' enough to leave a visible 'record' of any evidence of our origin.[24]

[24] Photons were unable to travel through the early universe due to dense radiation. Eventually, the strong nuclear force and electromagnetism began to localize enough energy in the form of atoms that photons could travel through the empty spaces between and the universe became visible. The remnant of this initially opaque universe is called the cosmic background radiation and it is the historical boundary beyond which there is no archeological light record for cosmologist to interpret.

In the beginning, no known regularities existed.[25] Everything was interacting with everything else, but no regular relationships could be maintained in that environment. Within the first minute the universe cooled and local regularities became evident. These first regularities are often referred to as broken symmetries.[26] Broken symmetries are historical 'frozen accidents.' They happen once and then become a permanent feature of history. Another way to imagine a broken symmetry is when one of many possible results becomes actual. The details of that result can affect other results further along the chain of causality. For example, how a nucleus happens to become arranged constrains what kind of atom it can become and, ultimately, the molecular and chemical relationships it can maintain in its future. In a similar way, whether someone happens to go to college or not constrains certain events in their own future.

In a sense, a broken symmetry can be imagined as stepping into an adjacent possible.[27] Of all the possible next moves, the actual move taken makes all of the other possibilities irrelevant to future interaction. The move breaks the symmetry of many current adjacent possibles and steps into a new 'now,' with a new set of adjacent possibles. Until recently, traditional science has imagined 'what happens next' not as a selection from a field of possibilities, but as determined by fixed conditions. This is one of the reasons science has considered the universe to be deterministic, a Laplacian place where all actions are derivable from previous actions. A relatively new theory, chaos theory, has shaken this belief. It has shown that the future actually consists of many possibilities only one of which becomes actual. Chaos theory has confirmed that the future is fundamentally unpredictable.

In the expanding universe, the first regularities to be expressed as broken symmetries were between quarks and gluons. Regular relationships

---

[25] One could argue that the expansion process itself was a regularity, but since there is nothing outside it to measure that regularity by, this investigation will just treat regularities between universal components *within* the entire universe as 'regular.'

[26] The 'symmetry' in broken symmetry is different than the 'symmetry' in Symmetry Theory. 'Symmetry' in a broken symmetry refers to a space of equally possible results. 'Symmetry' in Symmetry Theory refers to actual regular relationships. The first is a reference to a statistical state of equipotentiality, a mathematically idealized abstraction. The second is a reference to a determinate state of affairs, a testable relationship.

[27] The term 'adjacent possible' was popularized by Stuart Kauffman (2000, 2008).

between these elementary particles formed the nuclei (i.e. protons and neutrons) of the light atoms. These would have been the nuclei of what would become hydrogen, helium, lithium and beryllium atoms (in order of appearance and increasing atomic weight). However, this is not exactly true. For those nuclei to form, the symmetry that describes the organizing principles of the strong nuclear force would need to emerge as the rule-set by which quarks, gluons, protons and neutrons attract and connect to each other. There is no known reason for why the strong nuclear force expresses quark and gluon regularities as it does, but it started to do so in the first instants of universal interaction.[28] The emergence of the strong nuclear force, therefore, could be considered the first broken symmetry to be expressed as a regularity. The products of strong nuclear force persistence are protons and neutrons.

The strong nuclear force is only one of the four fundamental forces. The other three are the weak nuclear force, electromagnetism and gravity. The four fundamental forces are distinct because they express regularities at different levels of interaction. Their emergence can be seen in two ways. Either they pre-existed in their current form, or they emerged as broken symmetries as the universe expanded and cooled. If the former is true it is uninteresting - it is beyond testing and it changes nothing. The latter implies that originally there was a singularity of forces. Ian Stewart (2007) describes it: "At first, all four forces acted in exactly the same way. But as the universe cooled, its symmetry broke and the forces split into individuals with distinguishable characteristics" (p. 242). As Stewart implies, the canonical view of the origins of these forces is to say that they split from an unknown singular state, "all four forces acted in exactly the same way." This way of expressing an unknown 'original state' presupposes that there is a law of everything, one unifying force, from which the four fundamental forces differentiated.

This investigation proposes a different approach; namely, that the 'original state' was a 'multiplicity.'[29] In this view the original state was totally

---

[28] In fact, the difference between what physicists call 'the strong nuclear force' and 'gluons' doesn't seem to be entirely clear. This may be another instance where structure (i.e. elementary particle) and process (i.e. strong nuclear force) are undifferentiable. For the purpose of clarity the two will be treated as separate concepts even if they are not yet entirely distinct.

[29] See Appendix 1 for more on the Big Bang and initial conditions.

differentiated and the four fundamental forces emerged as the regularities selected from interaction within this multiplicity. The traditional view is of *a unified regularity that mysteriously divides.* All subsequent events are related because they are parts of this unified origin. The view proposed in this analysis is of *a multiplicity that consistently selects for regularities according to a pattern.* All subsequent events are related because they share the same pattern of emergence even though what emerges in each case is just a broken symmetry, just one path selected from many possible paths. Symmetry Theory describes this pattern of emergence.

Many theorists imagine the original state to be a unified whole, like a rock, that is blown into pieces, a single equation. The universe we live in is made of those pieces, the derivatives of that original equation. They are searching for how those pieces can be reverse engineered to show us what the original whole looked like. This original reconstructed unit would be the elusive 'theory of everything.'

Now imagine a different scenario. Imagine the original state as having not one, but googols of different fine grained starting points. Stepping into the adjacent possible meant each starting point stepping into its own adjacent possible. This original Big Bang state is totally differentiated. It is at maximum entropy. It is defined by not being able to become more random. This original state is at the maximum degree of random interaction possible. Steven Jay Gould would say that this kind of universe was born at a left wall of minimum regularity.[30] From this point forward, only more-regular interactions were possible. Over time, the conditions in which these original states interact change and as they sample more and more adjacent possibles they expand the spectrum of kinds of interactions they can partake of. As this spectrum of interaction grows, so does its distribution tail of regular interactions. Some tiny number of these interactions is regular and that number grows in proportion to the growth of the overall spectrum.[31]

Regular interactions have the definitional character of existing over time. The result is that only regularities accrue over time because irregularities do

---

[30] See Gould (1996) for his argument against *progress* in the growth of natural complexity.

[31] This cosmological ontogeny has distinct parallels to Layzer's (1975) interpretation of the universe's initial state and subsequent evolution. One notable parallel is the presumption of an initial maximum entropy equilibrium state that can only vary in the direction of less entropy. Essentially, this is another way to say that interactive spectrums grow and with them grows the Goldilocks sweet spot of available energy from which emergent order can be selected.

not persist.[32] In this model, irregular interaction is the noise of the universe from which is consistently selected a tiny sample of regular interactions. Seen from the current human point of view, at the end of this process, very little of the universe's massive history of irregular interaction is evincible because irregular interaction leaves nothing over time to trace. We exist at the far right tail of the tiny slice of regular interaction that has persisted over the entire history of the universe. From our point of view the universe contained no regularities at its beginning and over time, as its interactive spectrum has grown, so have the total number of regularities grown with it. The universe has accumulated a tiny, but consistently growing trace of persistent, regular interactions.

From the traditional point of view the universe started out unified and has diversified since. From the Symmetry Theory point of view the universe started out diversified and from this diversity only unified regularities could be selected. What has accrued are regularities, not differences. This version of the growth of order will not be further argued here, it will be demonstrated throughout.

Symmetry is defined as transformation with preservation of structure. Where most of the initial universe transformed without preserving any kind of structure, some small part of it began to sample strong nuclear force symmetry and the structures of protons, neutrons and nuclei began to be preserved under certain conditions. The symmetry of the strong nuclear force is evidenced by the preservation of the highly ordered nuclei structures that organize the periodic table of elements.

So, the first regularity to preserve relationships was the strong force drawing the quarks and gluons together into characteristic arrangements of flavor and colour (i.e. quantum chromodynamics). Some of these arrangements would have been stable (i.e. persistent) and others not. The unstable products of quark/gluon interactions degraded, became lost or were rearranged. They remained irregular. Over time this selection process left more and more of the stable arrangements like protons and neutrons. As cooling progressed and interaction constantly sampled relationships

[32] If time is isomorphic to transformation then regular time, time as we know it, did not even exist until regular processes like quark-gluon relationships emerged to measure it.

for regularities, protons and neutrons met and the strong nuclear force maintained their relationships to form the aforementioned light element nuclei. These 'nuclear' arrangements are products of the symmetry of strong force interaction. As persistent quantum systems were maintained and re-created, and energy levels fell, a complex flux of interacting particles and forces began to shape varied environments of localized regularities in the expanding universe.

The fundamental forces are, by definition, the stereotypic interactions of elementary particles under different conditions.[33] They are regularities and as such they are symmetries embodied in synchronic interaction. These synchronic interactions feedback with the local irregularities of other broken symmetries such as angular momentum, density, wave functions and heat to provide an environment that differentiates the contexts and specializes the interactions forming in certain localities. Wills and Bada (2000) describe a very similar process on the surface of the primordial Earth where the first cells formed. The stereotypical affinities and features of molecules constitute a substrate of chemical symmetry upon which the cyclical geophysical regularities of tides, storms, filtering sands, waves, catalytic surfaces, volcanoes, and day-and-night operate to create a natural process of sorting and ordering (i.e. selection). Molecular arrangements of one sort are differentially maintained in one locality and come to interact regularly with different molecular arrangements from different historical and interactive conditions.

In the case of the first quantum systems, local interactive backgrounds of broken symmetries were the substrate upon which the rules of quantum mechanics sampled interactions for persistence (i.e. synchrony).[34] The result was the sorting and ordering of a few localities in the universe into aggregations of highly ordered matter-energy systems interacting according to highly symmetric sets of rules. Quark and gluon interaction preserved protons and neutrons whose further interaction preserved nuclei whose further interaction with electrons preserved atoms whose molecular affiliations and local irregularities caused massive gas clouds that condensed into massive gas stars that exploded to form local gas clouds that sorted themselves by

33 An exception is gravity, which is special in many ways.

34 Examples of such backgrounds of broken symmetries would be broken symmetries that resulted in fluctuating radiation, heat, density, electromagnetic fields, etc.

their own chemical/gravitational attractions into the giant clusters of spiral galaxies that, if you frame shift down, further collapsed into individual spiral galaxies and solar systems and the handful of other recognizable cosmic structures. During all this time the vast majority of everything else continued to interact without preserving any new structures.

Of course, the universe need not appear to be arranged so symmetrically. Looking at it now, much of it appears to be nothing at all, which reinforces the question of whether we could recognize non-symmetry arranged parts of the universe as any 'thing' at all. This raises an intriguing cosmological sidebar. Symmetry theory is essentially a theory of selection. Like Darwinian natural selection it necessitates and presupposes a vast amount of waste. For every animal whose lineage persists, millions must die. Likewise, every emergent regularity (i.e. symmetry) needs a vast spectrum of irregular interaction. Only by sampling billions of different irregular interactions can those few regular interactions be discovered. Therefore, the existence of any persistent structure presumes a massive historical background of unstructured interaction. If the universe operates by symmetry theory principles, by far the largest part of it should be asymmetric, should preserve no structure when it transforms. Just recently cosmologists have been stunned by the discovery of dark matter and dark energy. Although this discovery is little understood, the fact that it is unexpected in the current framework and necessary in the symmetry framework does seem suggestive.[35] Symmetry theory assumes that the existence of anything requires a vast amount of 'nothing' - and we seem to have just found a vast amount of nothing.

Another thing should be added to the concept of 'rules' before moving on to the emergence of the systems themselves. As mentioned, rules are a class of transformations that preserve some degree of structure. How this class of transformations is preserved however, depends on the environmental conditions within which these rules persist. So rules (and regularities and

---

[35] Dark matter may simply be the matter-energy that was never caught up in positive symmetry relationships. It may be what is left of the interactive background from which all symmetry systems have been selected. Dark matter may be hard to detect for exactly this reason. It displays no interactive regularities under current conditions. How do you test for something that for fourteen billion years has failed to show any persistent positive interaction potential? Dark matter may simply preserve no structure when it transforms. Under current conditions, it is *asymmetric*.

laws) can be thought of as a class of transformations that only exist under specific conditions. This investigation takes the environmental or external definition of causality seriously, which means that rules depend on those external boundary conditions within which the select interactions that define those rules persist. The investigation refers to 'symmetry rules,' or 'rule-sets,' as those groups of rules that maintain and reproduce certain classes of structural regularities *under certain conditions* in this universe. For example, quantum structures are defined by the environment within which the rule-set of quantum mechanics is maintained. The boundaries of these conditions are most of the known universe.[36] Molecular structures are additionally maintained by the rule-set of chemistry, located within the already established boundaries of quantum environments. Astronomic structures are additionally maintained by the rule-set of astrophysics, located within the already established boundaries of quantum and chemical environments. So ultimately, rule-sets can also be seen as environments or conditions.

For now what is important to remember about rules is that; 1) they apply to just a small percentage of regular interactions that persist in a much larger spectrum of irregular interaction and 2) they depend on the maintenance of environmental conditions, beyond the boundaries of which they fail to maintain those regularities. For more on rules and boundary conditions see the chapter "Boundary Conditions."

In this model it is the local conditions, or 'environments,' that initially differentiated in this universe. After the Big Bang, cooler and less dense conditions consistently selected for strong nuclear interactions which left more and more protons, neutrons and nuclei. However, the specific conditions at each locality varied, producing more or less of each of these products in unique dispositional arrangements. Although we don't have an exact picture of what did happen, we do know that of those original differentiations only protons, neutrons and their nuclei have survived subsequent evolution as unified systems. They are the only *regular* strong force interaction that commonly persists in nature today. This means

36 Examples of candidate environments that exclude the quantum rule-set would be the locations we call black holes, the tiny string dimensions and perhaps even the vastness of space itself, depending on how we decide to count space as 'nothing.'

that either all the protons, neutrons and nuclei that will ever exist were produced after the Big Bang and are slowly, one by one, disappearing as they interact over time, or they are indestructible 'perpetual machines' or they are consistently re-created when the conditions (e.g. not too hot and not too cold) are right - and the conditions in some places are still right. The Symmetry Theory model proposes the latter. The conditions are still right for both the maintenance and the re-creation of protons, neutrons and nuclei. This means that the fundamental building blocks of all matter, strong force products like neutrons and protons, exist conditionally. They only exist in environments where the conditions for both their maintenance and, eventually, their re-creation are statistically favorable. This argument will be extended to atoms, molecules, galaxies, cells, humans and words over the course of this investigation. Atoms, molecules, galaxies, cells and words also fail to exist if they are not supported by environmental cycles of maintenance and re-creation.

Ultimately, this investigation seeks to describe the history of order in the universe. It presumes that order did not exist at the beginning. As such, this investigation will describe the three basic sources for the formation of order that caused the symmetry rules of quantum mechanics to add chemistry rules and later add astronomy rule-sets under certain conditions at certain localities. The first source was the ubiquitous and varied interaction of everything. The second was the sampling within that interactive environment for synchronous (i.e. regular or symmetric) relationships. Some small part of the spectrum of interaction was synchronous and by definition only those relationships could become regularities over time. Only they could persist as systems. The third and final source of the current universal order was the re-creation of these synchronous relationships. Only those regularities that were consistently reproduced ended up populating their localities thereby creating environments where their structures fed back into the interactive environment from which new synchronous relationships could be sampled. Only those localities that 'reproduced' their regularities propagated.

## Quantum Symmetry

**FIGURE 5: ATOMS**

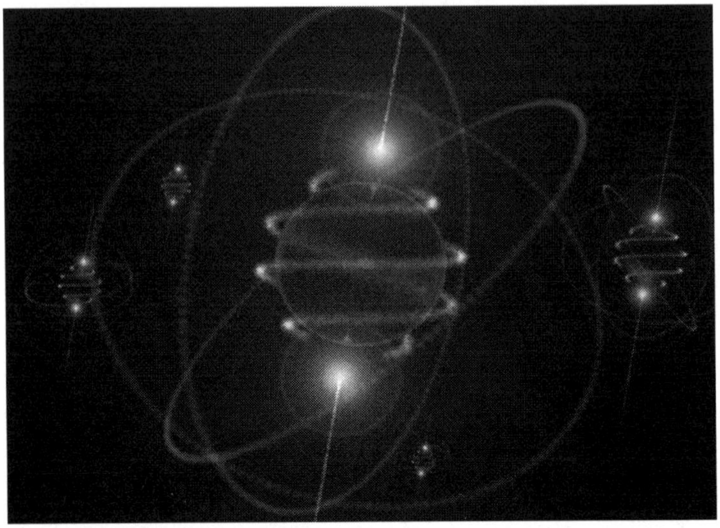

Source: Triggs (2012)

Symmetry is transformation that preserves structure. It is the rule-set by which persistent matter-energy structures transform themselves and each other. Only recently, since Heisenberg and Schrödinger, have we come to a limited understanding of the rule-set of quantum interaction. These rules are overwhelmingly symmetric.

Every quantum particle has a charge compound; that is, it is either positive or negative. Another way to put it is every particle has an antiparticle. Every particle also has parity, or 'handedness.' For example, 'right handed' particles interact distinctly from 'left handed' particles. Matter-energy arrangements, as we know them, from quarks to galaxy clusters to neural nets, are those arrangements that are stable enough to persist and are composed of elements that are either positive or negative (c-symmetry) and left or right handed (p-symmetry).[37] Everything preserves some basic structure when it transforms.

37 Photons may be an exception to this rule seeing as they have no c-symmetry.

Of course, these features of quantum 'particles' are human abstractions. No one has seen a neutrino. Quanta do not have 'hands.' What these features do is symbolize how these particles interact with each other in stereotypical (i.e. regular) ways. In quantum chromodynamics quarks have red, green, or blue characteristics attributed to them by the gluons that bind them. A red quark transforms into a blue quark by emitting a red-blue gluon. This happens invariably. It is a rule, a regularity. If a red quark loses a red-blue gluon it transforms into a blue quark every time. Quantum mechanics is the rule-set for the transformations that preserve quantum structure. In the quantum mechanical world a scant sixty to sixty one quanta (i.e. elementary structures or processes) are combined to describe quantum transformation (Gell-Mann, 1995, pp. 183-187).

Sixty to sixty one elementary particles are far too many for unified theory physicists and astronomically too few for the imaginable universes. The rules for these transformations are embodied in the rules of quantum mechanics. Likewise, the rules of quantum mechanics are the result of the aforementioned symmetry building process. Massive interaction sampled the spectrum of quantum interaction, eventually selecting for those few regularities, like chromodynamics, that can persist under broad universal conditions.

Later in this investigation we will see a similar symmetry building process in the evolution of language. Terrence Deacon (1998) describes how language structure was built:

> The rules of combination that are implicit in this structure are discovered as novel combinations are progressively sampled. As a result, new rules may be discovered to be emergent requirements of encountering novel combinatorial problems, in much the same way as new mathematical laws are discovered to be implicit in novel manipulations of known operations (p. 100).

As the sixty or so elementary quanta interacted to varying degrees, under varying conditions, they were expanding the spectrum of what could occur between them. Luckily for us, they hit upon the first regularity of the conditions our universe was experiencing, the strong force. The strong force preserved the building blocks of material systems, protons and neutrons. Protons and neutrons were new entities. They opened up new opportunities for even more regularities to be sampled under the right conditions. "As a result, new rules," like ionic bonding, autocatalysis, mitosis and altruism were, "discovered to be emergent requirements of encountering novel combinatorial problems, in much the same way as new mathematical laws are discovered to be implicit in novel manipulations of known operations." The rule-set of quantum mechanics is as much a fossil of how elementary particles sampled each other as the rule-set of mathematics is a fossil of how humans sampled and combined Pythagoras' original precepts. This same symmetry process will lead us to conclude that language itself is a fossil of how our ancestors originally sampled and combined cooperative behaviors.

The ordered parts of the universe began as very few kinds of particles operating on each other from a highly restricted set of rules describing the limited transformations that could be maintained and reproduced within a growing spectrum of environmental conditions. We assume that the rules for these transformations are, and have been, invariant. We assume that these rules have always preserved the same structures in the same ways. The first set of rules, upon which all further pattern depends on and was selected from, is quantum mechanics. It has survived over thirteen billion years of constant transformation. The only structures that could have survived the maelstrom of subsequent interaction would necessarily have been selected from persistent quantum regularities like neutrons and protons. As such, quantum symmetry is embodied in the strong nuclear force that differentially maintains those neutrons and protons. The quantum transformations that preserve quantum structures are the embryonic structure of pattern in our universe. They are the first regularities. Quantum symmetry was the first level of order to be selected to persist.

## Chemical Symmetry

**FIGURE 6: MOLECULES**

Source: Google Images (n.d.). *Molecules and atoms*

Chemical symmetry refers to the entire range of electromagnetic processes and structures. It presumes a substrate of preexistent nucleons composed of quanta constrained by the rules of quantum mechanics. Quantum symmetry exists, but its origins are largely unknown. This is less so with chemical symmetry.

Shortly after the Big Bang simple atoms and molecules began to become selected for persistence from the quantum environment.[38] This is the post-bang point where nuclei began to capture their appropriate number

[38] The quantum environment is the sixty or so quanta plus the emergent strong nuclear force products like protons and neutrons (i.e. nucleons).

of electrons and these fully operational atoms began to interact with each other en-masse in a few areas of the universe. Atoms interact, much like quanta, in a variety of stereotypical ways according to a small number of rules. However, they arrange themselves into 'chemical' groups (i.e. molecules) through two kinds of electromagnetic bonds, ionic and covalent. An ionic bond is created by the electromagnetic attraction between two atoms that have exchanged electrons. A covalent bond is created between two atoms that electromagnetically share electrons. In both bonds a physical symmetry is maintained because the sum of the parts of the two atoms remains the same.[39] In combining the two atoms new matter is neither created nor destroyed. However, new information is created. Not only does the bond maintain the historically established physical sum of the component atoms, it adds information in the form of the emergent properties the bonded system presents to its future environment. This is *positive symmetry.*

The compound formed in the bonding includes the sum of its physical symmetries, plus a new communal informational symmetry that exists only in relation to its environment. This emergent informational symmetry is how the unified system's reactive face can interact in a novel way with its environment. Of course, molecular emergence involves the adjustment of both internal and external relationships. Internally, the statistical distribution of electrons within their halos becomes more highly constrained (e.g. p-orbital shifts). Externally, the molecule's new geometric logic of molecular bonding deconstrains those relationships that molecule, as a whole, can now sample, maintain and reproduce. That is, the emergent reactive face deconstrains the molecule's future relationships to its environment.

In the chemical domain there are three general rules for interaction: chemical bonding, chemical equilibrium and chemical decomposition. When molecules meet, they either increase symmetric interaction, exhibit no change in symmetric interaction or lose symmetric interaction. These general rules are not only true for chemical interaction, but reflect the "three results of interaction." In quantum interaction some elementary particles

[39] Matter-energy by-products may be produced in the form of lost particles or energy, but these by-products equal exactly the loss of energy/matter from the bond, as per the law of conservation of energy. Physical symmetry is maintained and matter-energy is neither created nor destroyed.

maintain persistent relationships (i.e. bonds) under certain conditions - like quarks, gluons, neutrons and protons. Many elementary particles exhibit no change when they meet and under certain extreme environmental conditions (e.g. large hadron colliders) some have been shown to lose symmetry when smashed hard enough. These three categories of results can be generalized to all levels of interaction as *negative symmetry, neutral symmetry* and *positive symmetry*. Suns can explode, remain stable or draw cosmic material from their environment to maintain more planets and moons (or even larger core reactivity patterns). Organisms also evolve a similar spectrum of interactions (e.g. predatory, neutral or symbiotic relationships).

As the universe expanded it sampled more and more interactions and the products of this sampling were emergent systems. Emergent systems were the result of those very few positive symmetry interactions that persisted under existing environmental conditions. These emergent systems were relationships that had to be re-created within their environments in order to become persistent. Once they became persistent they changed that environment itself, sampling more and different spectrums of interaction and thus finding new positive symmetry relationships. This is how a universe with just quanta and nucleons discovered atoms and how a universe of atoms discovered molecules. This is how the universe has discovered level after level, domain after domain, of emergent systems. The chemical domain is a rule-set for those relationships that persist from electromagnetic interaction conditions.

Negative symmetry has always existed as both entropy and chaos. For exactly this reason only those systems that are both maintained and re-created can persist. All systems experience both chaos and entropy. Only those systems that feedback positively with their environments, in the form of being able to maintain themselves in the short term and reproduce themselves in the long term, are able to persist despite constant chaos and entropy (i.e. despite negative symmetry). Protons and electrons feedback positively through their electromagnetic bond. They persist in the forms of atoms and molecules.

Both positive and negative interactions step into the future statistically. The disordered universe that emerged after the Big Bang was almost maximally entropic. It contained no ordered or regular systems that could lose their order through negative symmetry. The first step towards the balanced universe that we currently live in required the discovery of a few regular systems like protons, nuclei and atoms. These systems were unified by regular relationships, which only then could be disrupted by entropy and chaos. As that old chestnut goes, you cannot lose what you never had.

Once positive symmetry sampled for a few regular systems, then negative symmetry, in the form of entropy and chaos, could statistically act on these systems causing them to go extinct over time. What survives this process of statistical sampling are only those systems whose statistical gain of cycles of maintenance and reproduction (i.e. positive symmetry) are greater than their statistical loss of said cycles. Everything statistically steps into its future. This has long been recognized in the forms of entropy and chaos. What has not been recognized is the balancing force of positive symmetry. All known systems are necessarily descended from selection from those few, but persistent, positive symmetry cycles that our universe has both maintained and re-created over the last thirteen point eight billion years.

Quantum symmetry is not chemical symmetry. Chemical symmetry largely ignores the strong nuclear force. That is, electromagnetic bonding synchronizes different systems in different ways than the strong nuclear force does. Also, chemistry is based on how communal quantum systems (e.g. nucleons and electrons) interact, not on how individual quanta interact (e.g. quarks, gluons, etc.). In fact, the rules of chemistry were well known centuries before quarks and gluons were even discovered. The rule-set of chemistry emerges from interactive environments pre-organized by the rule-set of quantum mechanics, but each domain describes different systems constrained by different rule-sets on different environmental levels of interaction. In fact, the chemical domain is a subdomain of the quantum domain. It exists only in quantum environments where the heat and density conditions are favorable to maintaining and reproducing quantum symmetry products like nucleons. Chemistry requires the preexistence of a much larger quantum domain within which certain localities of regular 'chemical' interaction are maintained. Chemical symmetry is embodied in the electromagnetic force that differentially maintains atoms and molecules. Ultimately, chemistry is a rule set that constrains the interactions of synchronized 'groups' of quanta. Chemistry is how groups of quanta 'socialize' under certain conditions.

## Astronomic Symmetry

FIGURE 7: GALAXIES

Source: NASA, ESA and the Hubble Heritage Team (2011)

At the point in universal expansion where the rules of astronomical symmetry began to be selected for, the universe was a sea of quanta, nucleons, atoms and molecules. Density irregularities had become more pronounced as the sorting processes of the four fundamental forces, combined with a history of amplified broken symmetries and the cooling of the universe, together produced a differentiated arrangement of expanding cosmic matter-energy. The universe was still hot and dense, but it was beginning to be structured by the symmetries that, to this day, constrain its interactions. That is to say, it was beginning to be constrained by the transformations that, to this day, preserve its structures.

By now the range of interaction was greatly expanded, starting from Planck length quantum interactions and extending on out to the presumed infinite interaction limit of gravity.[40] At denser localities, gravitational and mechanical forces (such as angular momentum) added more constraints and shaped the emerging cosmic patterns to differing degrees. Massive, light element suns collapsed under gravitational pressure until the nuclear reactions at their cores resisted further collapse with radiant energy pressure. When these suns eventually exploded, they dispersed their newly made heavier elements throughout these localities, adding new elements and angular momentums to other elements from other exploding suns and further differentiating the expansion pattern.[41]

Harold Morowitz (2002) comments on this process: "So, as the result of the stellar emergences, the universe consisted of a great variety of stars, galaxies, occasional novae, pulsars, quasars, black holes, and an array of known and unknown entities generated by the laws of gravity, mechanics, thermodynamics, and nuclear synthesis" (p. 52). These relatively few types of symmetry products are selected from those few elementary particles that maintain persistent relationships (e.g. quarks, gluons and electrons) from which in turn are selected those relatively few simple atoms (e.g. hydrogen, helium, lithium and beryllium), from which those relatively few molecules that collapsed to form the first stars were selected, from the deaths of which are constantly reproduced the entire spectrum of atoms in the periodic table of elements.

This feedback process is an excellent example of one of the first cycles of reproduction on which our current universe is founded. The process is called nucleosynthesis. It is the constant production and reproduction of most of the heavy elements in the universe by nuclear fusion in solar cores. Almost every physical thing known to humans is dependent on this gravitationally driven cycle of stardust condensation and dispersal. The Earth and us Earthlings are all part of this cyclical process. Not only would our body's complex mix of elements be impossible without this

40 Whether gravity itself was newly emergent or preexistent is at this point irrelevant.

41 This mix of angular momentums, along with the gravitational 'toilet bowl' effect, is the cause of solar system and galaxy angular momentums that give them the characteristic rotating disc structure.

cycle of reproduction, but most molecules and most heavenly bodies as we know them would be impossible. The maintenance and reproduction of the astronomical domain is not only a necessary precursor to life-on-Earth, but it also feeds back into the chemical and quantum domains by reproducing those molecules, atoms and quanta that fill those domain's current spectrum of interaction. It could be additionally argued that the gravitational cycle of reproduction that maintains the astronomical domain also maintains a range of heat and energy differentials that are equally necessary for conditions of emergence. One of these energy differential ranges is the warm sun bathing our cool planet surface - human emergence conditions.

Quantum symmetry rules describe and constrain how elementary particles may be synchronously arranged. The most consistent, stable and ubiquitous arrangements of quantum symmetry are the light atom nuclei of the periodic table of elements. Chemical symmetry rules describe and constrain how these atoms may be synchronously arranged into molecules to form the familiar phases of matter: gas, liquid, solid and other less familiar phase states like dense plasma. As these chemical aggregates gain mass they also begin to be significantly affected by gravitational symmetry rules. That is, they attract each other in proportion to their mass and distance. As atoms interacted to make molecules, molecules interacted to make molecular aggregates that, as they grew, attracted each other more and more. These chemical and gravitational aggregates evolved into the massive heavenly bodies that pattern our universe. The laws that describe and constrain these bodies are Morowitz's aforementioned laws of mechanics, thermodynamics and nuclear synthesis operating in the context of gravity - the rule-set of astronomic symmetry.

At each stage in the evolution of the universe a limited number of product structures are preserved by the rule-set that maintains interaction at that level, under those boundary conditions. As a result, at the quantum level, strong nuclear force structures differentially survive. At the chemical level, electromagnetic structures differentially survive and at the astronomic level, gravitational structures differentially survive. Each level also represents new rule-sets for persistence. Covalent bonding does not exist at quantum levels and dense plasma thermodynamics does not exist at chemical levels. In one sense there is a complexification of matter-energy systems as selection for persistence samples new products that interact by new rule-sets. However, even though these product systems are more complex than their predecessors, they represent a fraction of the imaginable systems - the fraction that results from maintainable component interaction. These products may appear to be more complex, but they are in fact composed of simplicities - of constraints. Of all the things that matter-energy could be doing at that location, what it is doing is massively simplified, regular and constrained. What it is doing is adding information to the universe by reducing the uncertainty of what its components can do next. It is actually reducing the complexity of its description by constraining action to regular, rule based interactions.[42] For more on the problem of defining complexity see Appendix 1.

Ultimately, emergent rule-sets constrain the relationships system components can have to each other while at the same time they de-constrain the unified system's potential for finding additional synchronic relationships within its future environment. The example given in the previous chapter was of the emergent molecule whose internal nucleons and electrons are more highly strong force and electromagnetically constrained, but its external reactive face is gravitationally deconstrained. Emergent systems can find new relationships in the old universe they just changed by being born into it.

Astronomic symmetry is embodied in the gravitational force that differentially maintains galactic and solar systems. It is the additional rule-set by which heavenly bodies interact to maintain and reproduce the stereotypical structures of the cosmos (e.g. stars, planets, galaxies, black holes, etc.). Astronomic symmetry emerges from the interactive background (i.e.

[42] For a more in-depth discussion on this theme, see chapter "Environment and Organism: external and internal." Also, Terrence Deacon has recently written of "constitutive absence" as that which defines both systems capable of work (machines and animals) and information. His idea is an extension of Michael Polanyi's idea of boundary conditions and constraints. Both sources cast information and living systems as regular relationships highly constrained by external conditions (environment). In this book the ideas of 'selection for persistence' and 'internal vs. external definitions of complexity' are highly congruous with these perspectives. I highly recommend Deacon (2007a, 2007b) and Polanyi (1968) to those who are curious.

environment) of chemical symmetry just as chemical symmetry emerges from the interactive background of quantum symmetry. Positive symmetry is the addition of regular relationships over time. This is accomplished by the discovery of emergent synchronous relationships. In fact, positive symmetry is emergent synchrony. In this universe, from quarks to stars, synchronous interaction selected the maintainable and reproducible structures of our universe.

# Discourses on the Universe

*"A 'natural law' is a compact description beforehand of the regularities of a process."*

*Murray Gell-Mann (as cited in Kauffman, 2008, p. 5)*

## Selection and Persistence

Did the recognizable universe begin as a few types of quanta acting on each other in stereotypical ways? At the origin of the Big Bang many different arrangements of matter-energy may have been created. In the universe we live in, thirteen billion years later, very little of this initial plethora may have survived.

That means the conditions of the early universe, particularly its heat and density, may have created matter-energy arrangements that could not persist as the universe expanded and cooled. So far, physicists simulating high-energy interactions in particle accelerators have discovered hundreds of different strong nuclear force compounds. While these unusual compounds do express the fundamental symmetry of strong nuclear force interactions (e.g. they do appear to be made of the same symmetries inherent in quark/gluon type relationships) they are not nearly as common as protons and neutrons (Gell-Mann 1995).[43] This may be for at least three reasons. First, the unusual compounds may need the high-energy environment that no longer exists in the universe. Second, they may degrade rapidly due to weak force radiation or a lack of persistent binding relations (e.g. nucleus-electron binding

[43] In another example, unusual particles like muonic and tauonic leptons do express the fundamental symmetry of quantum mechanical interactions, but they are not nearly as common as electrons because they rapidly decay into electrons under current conditions.

relations). Or third, they may simply have been sorted out of our area of the universe by a process we do not yet understand (or one that we do[44]).

All of these reasons highlight the importance of selection as a process of differentiating and specializing matter-energy arrangements. Specific matter-energy systems are defined by the physical symmetries that arranged them and by how they interact with other systems, their reactive face. In the case of the original systems, the symmetry rule-set is called quantum mechanics. These are the elementary ways that structure has been preserved during transformation in our universe. They are elementary because they appeared shortly after the Big Bang and continue, unchanged, up until now. These symmetries have persisted because they have been maintained within certain conditional boundaries. They maintain themselves by maintaining those conditions that encourage the reproduction of their own product structures (the conditions of propagating symmetry).[45] If atoms were only created once, and then never re-created, the universe would slowly lose its atomic environment as they degraded (e.g. weak nuclear force) or interacted negatively (i.e. as they lost their electrons due to negative symmetry interactions). Entropy and its disruptive forces would break the emerging order of the quantum mechanical universe. The positive symmetry cycles of maintenance and reproduction (like the re-creation of atoms in stellar nucleosynthesis) defy this kind of 'disordering' of the universe.

The universe we are studying more than thirteen billion years after its inception is highly ordered at the quantum level, sixty to sixty-one elementary particles and four fundamental forces. The main reason protons and neutrons have survived this transformation, and many other possible, but not common particles haven't, is because the uncommon arrangements are not currently maintained or reproduced by universal conditions and the quantum symmetry rule-set itself.

---

[44] It's not uncommon for cosmologists to propose one way processes where once a resource is used up, nothing else can emerge. Cosmologists often propose that shortly after the big bang quarks 'swept up' all the gluons into proton/neutron relationships and shortly after that, nuclei swept up all the electrons into atomic relationships.

[45] Quantum systems support chemical system (atomic/molecular systems) emergence and chemical systems support astronomic system emergence. In turn, Astronomic systems (most notably suns) re-create a spectrum of quanta and atoms supporting a feedback cycle. This will be further elaborated in the "Maintenance, Feedback and Reproduction" chapter.

Over the last thirteen billion years those particles not created by the cycles of transformation included within the symmetry set of quantum mechanics and its historical derivatives (chemical and astronomic rule-sets) have either ceased to interact (lost in space) or have interacted and have been thus captured and included in the cycles of quantum interactions (see chapter on "Inclusion" in the "Life" section).

One way to think of persistence is in terms of selection. Only those systems that persist within the context of universal interaction survive. In this sense persistent systems are selected by conditions of interaction. As systems emerge and populate those areas where conditions favour their selection they interact more and more. Only persistent, or selected, interactions are available for further interaction under those conditions. All other interaction is trivial because it is not preserved.

A key feature of symmetry is persistence. Persistence is characterized by maintenance and, especially once entropy is involved, reproduction. Those symmetries that do not persist, either through maintenance or reproduction, are, over long periods of transformation, not symmetries. Through time (i.e. transformation) their structures are not preserved. They do not survive as 'things' and they do not go on to become the interactive substrates of further 'things.' They are the random results of broken symmetries. They are important because they extend the spectrum of the interactive background from which persistent synchronies are selected, but they cannot be recognized as systems because they preserve no regularities in their interactions. They present no novel reactive faces to the spectrum of interaction.

The conclusion is that the post-bang universe may have created a massive variety of matter-energy arrangements of which, over the last thirteen billion years, the majority has gone extinct, leaving the universe as we find it now. The universe as we find it now is populated by matter-energy arrangements that are stable products of quantum symmetry transformations and their historical derivations.

All systems in this universe require both selection and persistence. Selection requires maintenance within an environment. Persistence requires both maintenance and re-creation due to constant entropy. In a nutshell one could summarize this state of affairs as: *anything that can happen does happen, but only those things that happen regularly persist.*[46]

## Process and Structure

> *"The contrast between structure and process breaks down in the atom as well as in the living organism whose structure is at the same time the expression and the bearer of a continuous flow of matter and energy."*
>
> *Ludwig Von Bertalanffy (1968, p. 248)*

> *"Both [structure and process] mutually presuppose each other, because under increasingly demanding conditions (i.e., those not determined purely by chance), structuring is a process, and processes have structure. They differ through their relation to time."*
>
> *Niklas Luhmann (1995, p. 44)*

Symmetry is transformation with preservation of structure. It is a process. However, symmetry can be described in two distinct contexts: forces (processes) and their matter-energy products (structures). Forces are simply consistent ways of relating. They are transformations that always maintain certain structures. They are regular relationships. The oldest ones are the fundamental four that were not generated by any structure we know of. All further forces of nature are results of historical relationships between matter-energy structures that have persisted long enough to significantly affect the phylogeny (i.e. descent line) of matter-energy transformation. The structures of the universe are the matter-energy 'products' of the descent lines of regular interaction.

These forces and products are the "processes" and "structures" referred to by Bertalanffy and Luhmann above. This division is elemental to our understanding

[46] The phrase, "anything that can happen, does happen" was inspired by Cox & Forshaw (2011).

of the universe. However, our understanding is limited by how symbolic language and thought parse reality for us. We feel the need to separate the forces from the products because we symbolically separate everything into actions and things, operations and numbers, verbs and nouns, deeds and doers. This is the nature of the language tool we have inherited to extend our sensitivity to our environment. Obviously there would be no forces if there were no things and vice versa. Reality is matter-energy unfolding itself. It is a unified event.

Even so, to speak of this unified unfolding we must reduce it to thinkable terms, thus the convenient convention of forces and products. The first notable side effect of the artificial separation of forces and their products is the necessary division of measures of state symmetry from measures of transformational symmetry. Transformational symmetries measure the preservation of the structure of the rule-sets that constrain the transformations of certain kinds of matter-energy systems. Transformational symmetry is measured as informational symmetry in that it is defined as a relationship between what does exist and what could exist. For example, the collapse of intra-galactic particle clouds into characteristic disc solar systems with massive sun-like centers and a series of receding elliptically orbiting planets evidence the preservation of the rule-set of astronomic symmetry. That same matter-energy could be doing any number of things and the relationship of what it is doing compared to what it could be doing tells us how its transformations are constrained. This relationship gives us information. That is, transformational symmetries describe the constraints on the ways certain matter-energy arrangements can change over time. They describe dynamic relations - processes.

State symmetries, on the other hand, measure the preservation of the physical arrangement of the structure itself at any one moment. State symmetry is measured as a physical symmetry in that it is defined as a comparison of two static physical states. For example, the geometric form of a planet, when compared from any angle at any one time, is an almost perfect symmetrical circle. That is, one half has the same geometric form as the other half. One half preserves the physical structure of the other when they are compared (i.e. transformed in our minds). State symmetries describe how certain physical structures are arranged at any one time. They describe static relations - structures.

It should be remembered, however, that the universe constantly and ubiquitously transforms. State symmetry is a consequence of humans symbolizing their environment. Plato's idea that things have true, unchanging and ultimate forms is a feat of imagination - it is not necessarily a true and ultimate representation of things in this universe. The more we measure and learn about this universe the more evidence accrues that supports a dynamic, not static view. Ultimately, *processes* constitute structure - *not material.* Material passes through systems. It is the canalized routes of transformation, the symmetry process through which the material passes, that give structure its 'structure.' Things are made of relations, not things. As Bertalanffy (1968) says, "matter is largely voids interwoven by centers of energy" (p. 227).[47]

So the question is, why and how do transformational symmetries that preserve rule-sets so often produce high degrees of state symmetry in their products? Why do quantum symmetry rule-sets produce beautiful state symmetries in the form of the distribution halos of electrons orbiting atoms for example? Part of the answer must be that it is the physical arrangements of the emergent structures themselves that constrain the rule-sets the emergent structures will abide by. It is the actual structure - the state of the relationships at any instant - that constrains how its transformation will proceed in the following instants. Another way to say this is that the physical arrangement of a system's component relationships and reactive face at any one instant constrains its future relationships.[48]

In the case of atoms, it is the physical structures of the atoms themselves, their characteristic disposition of quarks, gluons and electrons that result in the chemical symmetry rules. This is a chicken or egg situation that can only be solved by unifying the two categories. State symmetries are snapshots of transformational symmetries. One causes the other and the other causes the one because they are two ways of measuring the same thing. Even though state symmetry may

[47] Perhaps a more exact way to say "centers of energy" would be "displacements of regular energetic relations." These "centers of energy" are just constraints on the expression of differentiated energy relationships.

[48] This is not an argument for deterministic reductionism. Rather, it casts the future as an adjacent possible, whose dimensional fine graining is codetermined by both system components and system environment. This adjacent possible is both constrained by deterministic historical components and, at the same time, deconstrained by stochastic future environments. This makes stepping into the adjacent possible both deterministic and chaotic. This is an argument for deterministic chaos.

not exist as a physical reality, it can exist in an idealized framework such as mathematics. State symmetry can exist for us as an important measurement concept for describing reality. The universe and its 'things' are an unfolding process that we choose to measure in at least these two different ways.

Einstein himself took steps towards formalizing the unity of structure and process for us. He showed us that matter is energy and thus space is time. By equating the *structure* of matter in space with the *process* of energy in time Einstein made matter transformable into energy. Not only does $E=MC^2$, but physicists now refer to 'matter-energy' as often as they refer to 'space-time.' When Einstein unified process and structure he opened the door to understanding structure as transformation. Structure inheres process and vice versa.

Perhaps a good metaphor to imagine the false dichotomy between process and structure is to compare them to art. Music is fluid. Its beauty is measured by an aesthetic judgment of balance in its transformations through time. Sculpture is static. Its beauty is also measured by an aesthetic judgment of balance in its transformations through space. To an aesthetic mind a sculpture can sing and a piece of music can be monolithic. They are two different measures of the same aesthetic sense. The music of the universe is transformational symmetry. The sculptures of the universe are static symmetries. Fundamentally however, the two are expressions of the same artifice.[49]

> *"Beauty is truth."*
>
> *John Keats (1795-1821)*

## Boundary Conditions

This investigation will repeatedly refer to 'interactive environments' and 'interactive backgrounds' from which persistent relationships can be selected. It will also refer to 'conditions' that are necessary for selection for persistence to occur. These concepts are related.

---

[49] Kenneth Clark's (1969) book *Civilization: a personal view* draws many compelling cases of how philosophy, architecture, politics, music, painting and sculpture all embody each other as they transform through the ages. There is also a BBC TV series by the same title.

An interactive environment is the phase space of all possible interactive components sampling interaction and breaking symmetries over time (i.e. change). This is not a Newtonian definition of an area of space that contains 'things' or 'systems.' Interactive environments are the systems themselves as they transform through interaction. Interactive environments can be called interactive backgrounds when specific systems or interactions are considered in the foreground. Thus specific systems or events can have interactive backgrounds and all systems have, and are part of their own, interactive environments. This is an opportunity for further work, but in this investigation the two concepts will be considered fungible.

In this universe interactive environments grow by constantly sampling their adjacent possible and presenting more variation for selection. In these cases the interactive environment can be characterized as an *expanding spectrum of interaction.*

'Conditions' are related to what is often called 'boundary conditions' in mathematics. Boundary conditions are those sets of values for which the relationships described by the mathematical equations are true. In a similar sense, for any system to exist (i.e. be true) it needs certain conditions. These conditions are usually ranges of measurements of transformation like heat, density or wave functions. Perhaps the most important range of measurements is the amount of interaction. As Per Bak and others have noted, too much interaction (e.g. gasses, plasmas) prevent synchronies from stabilizing. Too little interaction (e.g. solids, voids) precludes new synchronies from ever forming. Both of these conditions are traditionally related with the too hot and too cold measurements of Fred Spier's goldilocks conditions (Spier, 2010). Both theories identify limited ranges of 'push-pull' interaction that are defined by separating hot from cold, chaotic from static. These spectrums are the conditions that are necessary for the maintenance of those systems that persist within them. In fact, systems are defined as those critical locations in the spectrum where cycles of chaos and stasis, push and pull, life and death can exist as symmetric work cycles. Within boundary conditions certain regular transformations can and do exist.

There was a time when the conditions of interaction did not maintain any regularity that we know of. This post bang interactive environment was close to what is colloquially referred to as chaotic, or unpredictable, or un-ruled.

Each elementary quanta broke symmetries and moved chaotically into its own adjacent possible with unmeasurable degrees of freedom, maintaining no durable synchronous interaction. As interactive sampling through time progressed and conditions changed (e.g. heat, density, angular momentums) then quantum regularities began to be maintained under certain conditions. One result was populations of emergent systems that created energetically differentiated (i.e. varied) areas of interaction. This means, that in the beginning, quanta interactively sampling each other began to present more regularities to each other in some places than others. For example protons and neutrons were not perfectly evenly distributed. This differentiation of environments was also a differentiation of conditions from which, eventually, different combinations of further regularities could be selected. The result was that although the same kinds of regularities were preserved throughout the universe, the specific states preserved at any one instant or in any one phylogeny (i.e. descent line) of regularities varied locally. The homogeneously un-ruled universe began to maintain heterogeneous localities of regular structures. It began to have differentiated environments that preserved specialized systems.

In a Leibnizian universe where interactive environments are the systems themselves (as opposed to an abstract Newtonian space with systems 'in' it) then rule-sets are boundary conditions and boundary conditions are interactive environments and interactive environments are rule-sets. The boundary conditions for the persistence of symmetry products are their interactive environments. Their interactive environments are those localities that maintain those symmetry products. Obviously, this definition is circular and we will return to it shortly in the context of Darwinian natural selection.

The picture that emerges from this way of dividing the universe is reminiscent of the Russian matryoshka doll. The biggest doll is the interactive environment of the universe. Some small part inside that doll describes those areas of the universe that maintain quantum regularities (e.g. protons and neutrons - strong nuclear force rules). Some small part of that doll (interactive environment) contains those conditions that maintain chemical systems (electromagnetic rules) and some small part of that doll contains the conditions that maintain astronomic systems (gravitational rules), and so on.[50] In each reduction a certain percentage of the interaction maintained goes on to further maintain a smaller percentage of even more complex regularities. However, these extended environments of complex regularities are maintained (continually selected for persistence) by the nested environments they were selected from. Figure 8 below represents the nesting of system domains, i.e. system environments.

**FIGURE 8: SYSTEM DOMAINS**

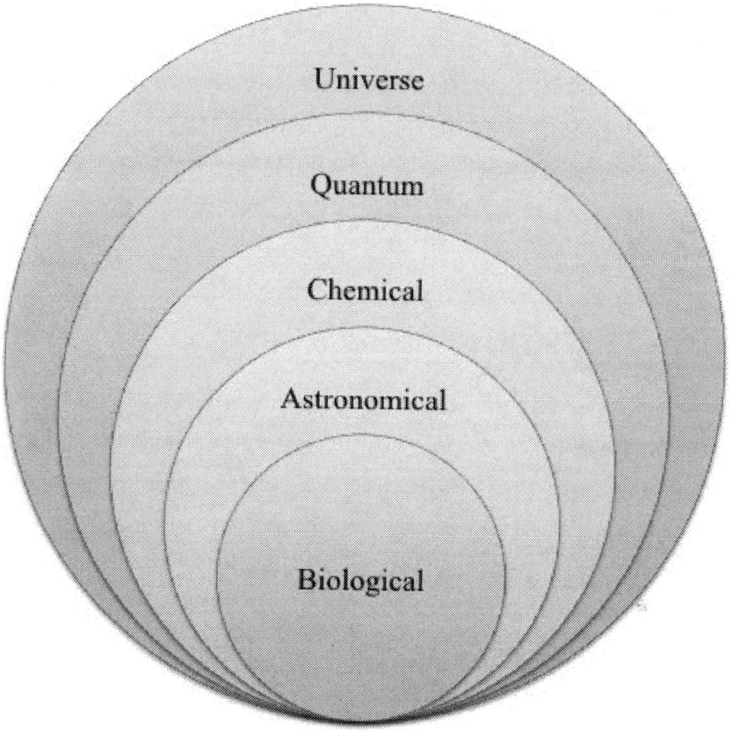


Source: author illustration

[50] This nested doll metaphor oversimplifies the case. It would be unlikely that one single unified part of the overall interactive background would represent all quantum regularities. More likely, the universe broke into different 'pockets' (i.e. perhaps galaxy clusters or galaxies) of quantum regularities and would be better represented as the biggest doll (the whole universe) having several similar dolls within it (i.e. several pockets of quantum regularities). It is also doubtful that the boundaries of these pockets of regularities are as definite as the doll boundary implies. More likely these pockets extend like fractals as persistent symmetries feedback and further pockets emerge.

This nesting of smaller, more highly constrained domains of emergent order is not as uniform as either the Russian doll metaphor or the above illustration simplistically imply. A truer representation would be a fractal pattern where each level of order can be 'zoomed in on' to reveal ever more detailed levels of the same pattern unfolding. Although a fractal process of emergent pattern generation is perhaps a 'truer' mental model, it is also difficult for our minds to maintain and use. For this reason this investigation will continue to refer to the more static, hierarchical representations like those presented above (for more on metaphorical representation see chapter "Social Behavior and the Delta Metaphor").

At this point such a fractal, or nested, point of view is highly theoretical and undeveloped. However, something like it needs to be clarified if this approach to universal evolution is to become predictively productive. The following are just two examples of the kind of predictive bounty this new point of view could provide.

The 'Anthropic Principle' has long been a thorn in the side of astrophysicists and cosmologists. One of their fundamental questions is why the universe's fundamental forces and constants (e.g. speed of light, Planck's constant, elementary particle mass ratios, etc.) are so perfectly tuned to maintain life? Any small deviation from the values that exist and nothing like us could exist. The Anthropic Principle gives scientists an answer that they rarely love. It states that the universe has such constants because we exist. Or to put it another way, if we didn't exist, neither would such a universe, but we do, so it does.

Such tautological arguments are rarely espoused in science. What scientists want is a reason that not only can be tested, but that has the power to explain the most phenomena in the shortest most elegant way. This principle is commonly known as Occam's razor. What scientists currently lack is a succinct, testable way to explain why the universe so conveniently supports life.

Science has made at least one enormous exception to its tautological anathema - natural selection. Herbert Spencer described natural selection as 'survival of the fittest.' Since then, the idea of 'fitness' has been integral to natural selection. Unfortunately, fitness is typically defined as 'those who survive.' The result is the classic Darwinian tautology, 'natural selection is the survival of those who survive.' Despite this obviously circular logic, natural selection is arguably the most successful scientific theory of the last millennium - why?

Natural selection has succeeded because it is elegant, powerful and testable. Symmetry Theory is simply an expansion of Darwin's Theory of Natural Selection beyond the scale levels of biology. 'Selection for symmetric persistence' is a recasting of 'survival of the fittest.' Only those systems that maintain synchrony persist. Maintaining synchrony is persistence. Only those organisms that are fit, survive. Maintaining fitness is survival. Although Symmetry Theory carries the same tautological flaw, it also carries natural selection's greatest strengths - elegance, power of explanation and testability.

The Anthropic Principle does not explain anything 'productively' and it is not testable. Symmetry Theory does and it is. The universe supports life because some part of what the universe is doing is always supportable. The universe represents a vast range of interaction of which only vanishingly small parts of this range maintain positive symmetry relationships. The evidence of positive symmetry relationships is emergent systems. The interaction of these emergent systems throughout the universe represents a new range of possible interactions from which new emergent systems can be selected for persistence. Only what is maintained can be further selected. We are here because we are supportable - we exist because we are descended from what can exist.

The four fundamental forces are only those regular relationships that last. Whatever complexity results from the continued interaction of the derivatives of the fundamental forces, exists because it represents further relationships that are maintainable. What scientists want is a succinct, testable connection between the fundamental forces and us. While the Symmetry Theory explanation is succinct it has yet to be proven. The only way to prove it is to learn how to test environments as well as 'things.' This means a massive augmentation of the current scientific method and mind-set. It is, however, necessary. The connection between the fundamental constants and us is not explicable by testing how things in test-vacuums can be made to affect other things in test-vacuums. It is explicable only by learning how to identify the positive symmetry range of the overall spectrum of interaction. That is, it is explicable only by including 'noisy' environments. The modern scientific paradigm looks for 'causes.' It needs to start looking for 'contexts.' In this sense, humans and all other universal systems are just highly ordered localities within their environmental contexts.

In the end, something like the Anthropic Principle tautology will probably be embraced because it is a natural extension of the most powerful and elegant way we know to explain the forms and functions that pattern existence (i.e. Darwin's Theory of Natural Selection). Unfortunately, natural selection is tautological. However, Symmetry Theory takes the wholly untestable and 'magical' Anthropic Principle and places it squarely in the scientific realm. We are finely tuned to this universe because we are descended from what has survived.

Here is a second example of the possible predictive bounty of Symmetry Theory. There must be a range of positive to negative interactions for selection to sample for persistent relationships (i.e. propagating symmetry). Regular coupling and decoupling is how matter-energy finds partners it can persistently sync to (i.e. 'cooperate' with). This kind of positive/negative, coupling/decoupling, not too hot/not too cold interaction would appear to provide for an environment of all kinds of 'push-pull' relationships. Per Bak (1996) describes just such an environment when he describes self-organized criticality.

In a nutshell, self-organized criticality is the conditions from which multitudinous types of natural order spontaneously emerge. Self-organized criticality is the family of regular fractal and logarithmic relationships that naturally emerge from different kinds of phenomena interacting at the edge of chaos. The edge of chaos is that part of the spectrum of interaction that divides static order (too little change) from random chaos (too much change), e.g. 'order for free' as per Stuart Kauffman. The tension of the push-pull relationships that characterize select environments within this universe could presumably provide the kind of 'criticality' Per Bak finds self-organized systems emerging from. In fact, energy itself is defined as the ability to do work. This basic constituent of our universe is itself defined by such a stable push-pull tension relationship.[51] In a sense, push-pull relationships create a 'harmonic' environment that then becomes a candidate substrate that is sampled for further constraints, further differentiation and specialization.

Even more interestingly, the kind of push-pull environments resulting from massive amounts of positive to negative symmetry interactions would set up the fundamental conditions for oscillator and wave transformations. We find exactly these kinds of relationships at the bottom of all physical causality. Light and the elementary particles appear to be characterized by wave functions, field vibrations and positive/negative charges. In fact, it not only appears to be at the bottom of symmetry, but at the bottom of all persistent interaction. Starting with quantum field theory and string theory; cycles, oscillators and wave patterns seem to characterize all levels of persistent interaction from quantum to chemical to astronomic to biological to social to symbolic. When we get to the "Life" section, we will see that the basic definition of life also depends on the stable reproduction of work cycles - metabolism.

From this point of view, cyclic, oscillatory and wavelike push-pull relationships should be seen as results of environmental interaction not as features of things. Systems contain and maintain specially bounded environments of push-pull relationships because they were selected for persistence from push-pull environments. They are themselves, specialized push-pull environments.

In conceptualizing push-pull environments the idea of 'constraint' comes to the forefront. Push-pull symmetry seems to exist at the bottom of universal causality. Electrons do it with their nuclei and quarks probably do it with their chromodynamic partners. This low level universal 'vibration' exists as the environmental background for the emergence of all further systems. It is the original source of matter-energy difference and potential. It is the source of all work. This symmetry 'harmonic' may characterize certain regions of the spectrum of interaction - like the mathematical division of a musical string on a stringed instrument (e.g. note). These differentiated harmonic regions then become the environments for further selection, further harmonic differentiation and specialization within that region.

[51] See Appendix 1 for more on energy.

Metaphorically, the universe we are able to see and feel could be imagined like a guitar string whose original note(s) was selected from the spectrum of all interaction. This original note(s) has continually been constrained into higher and higher dimensional numbers of small notes that when they sound together maintain emergent harmonics that can be further interacted with and selected for further emergent notes.[52] This perspective seems naturally congruent with quantum field theory. In fact, string theory is also based on proposing a family of harmonic interactions that exist in extra micro-scale dimensions.

What emergent systems become selected for is novel ways to constrain harmonic relationships. These vibrating (synchronized) areas of interaction are the only original and stable sources of regular transformation. This changes the idea of causality from 'what' to 'how.' Physical reality is no longer exclusively explainable by what causes what, but by how new interactions constrain what was already occurring. We could stop considering causality in terms of the things that cause certain reactions and start considering it in terms of how preexisting interactions are further constrained. In fact, all systems can be seen as differentiated and specialized levels of constraints on those fundamental quantum, chemical and astronomic patterns of matter-energy expression.

52 This would have implications for the many universes hypotheses. It suggests that although the universe could express order in many ways, it actually expresses it this way. Two analogous metaphors are of a guitar that could play many tunes but is actually occupied with only one at the moment or an ocean surface that could express many wave states but is actually only expressing one at the moment.

# An Environmental Theory of Emergence

*"The central task of theoretical physics in our time is no longer to write down the ultimate equations but rather to catalogue and understand emergent behavior in its many guises, including potentially life itself."*

*Robert Laughlin and David Pines (2000, p. 30)*

## Creating Information

Before frame shifting to the case of the Earth and biological systems it will be helpful to take a more in-depth look at what characterizes universal systems and how these systems create information. All things are systems. All systems are made of components in regular relationships.[53] All components are parts of systems. All things in this universe were selected for persistence from an interactive background of previous things.

The question is how do things create different things? How do systems create new systems? The universe is composed of a historical hierarchy of systems. These systems were built by the confluence of the following factors:

1) conditions and constraints extant in the interactive background,
2) selection for persistence,
3) formal emergence of structures, and
4) reproduction and radiation of those structures to feedback into the interactive background.

[53] This is true with the possible exception of the 60-61 elementary particles, but see Appendix 1 for a fuller discussion of the probability that these 'elementary' particles actually turn out to be elementary.

This feedback fuels the cycle with novelty (see Figure 9). Broken symmetries propagating throughout this process give the structures their specific aspects, but the conditions, rules and selection process keep each domain organized stereotypically. The question remains, how did this process build the historical hierarchy of systems that birthed us?

FIGURE 9: CYCLE OF EMERGENCE

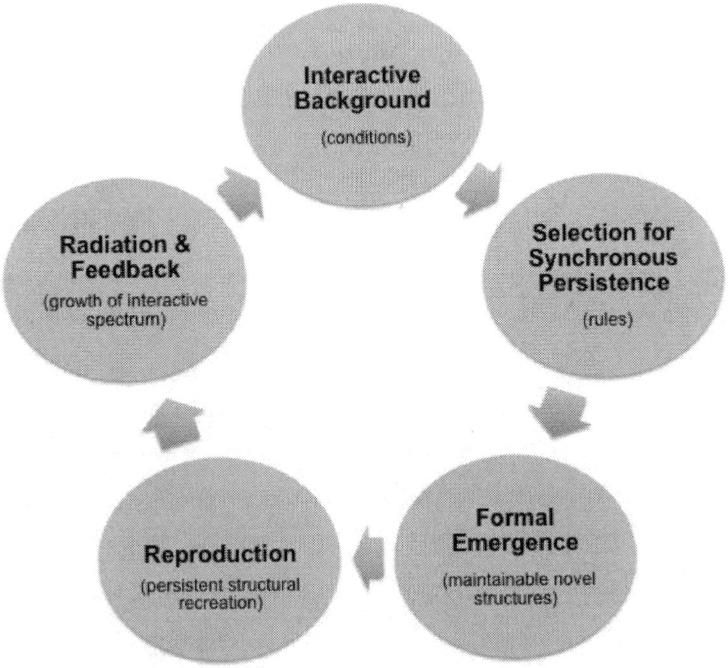


Source: author illustration

Components maintain a system by interacting in accordance with certain rules on a persistent basis. These rules can be seen as information in the sense that they describe a finite, compressible code of interaction that is persistent enough to maintain the system structure - a rule-set. Of all the things that the components could be doing (e.g. noise or entropy), what they are doing is a tiny subset (e.g. information or symmetry). The system is defined and constrained by the relationships that its component relations maintain, not by the material of which it is constructed.[54] This means that an atom behaves the same with any electron in orbit. Exchanging one electron for another does not change the internal or external kinds of relationships that constitute the atom, only its specific momentary state. Informational symmetry is the measurable spectrum of possible state symmetries within which the system can maintain itself or exist. Exchanging one electron for another changes the system's state symmetry, but not its transformational symmetry. On another level, a heart or kidney replacement changes the state of the individual, but (if successful) does not change that individual's maintenance of homeostasis.

If you assume certain conditions of the expanding distribution of matter-energy (i.e. the Big Bang) and a few basic rules (i.e. the strong nuclear force and electromagnetism) then you get basic systems like atoms. These systems are selected and maintained by the regularities described by quantum mechanics and electromagnetism under certain conditions in an expanding, diversifying, and cooling universe.

When the environment begins to include enough atoms interacting in enough different ways, some meet and bind into molecular systems. Molecular systems are built of quantum systems - nucleons and electrons. They include quantum information, but also generate the rule-set of molecular chemistry. Molecules present novel reactive faces to their environments. They present different opportunities for interaction than single nucleons or electrons. Thus molecules embody two rule-sets: internal and external, quantum rules and chemical rules, strong force rules and electromagnetic rules. Electromagnetic 'chemical' rules are a super-ordinate rule-set that describes how synchronized groups of quarks and electrons interact with their environments. Chemical rules also include as subordinate the information of how quanta interact (i.e. quantum rules). However, chemical rules are the external expression by group quanta interaction, not individual quanta interaction. This means that as long as disruption at the quantum level does not disrupt atomic structure, the molecule remains

[54] Another way to say this is that the material they are made of *is* the component relationships they maintain or, in even other terms; the material components that compose the system are themselves just energetic relationships. These relationships are measurable as information or code.

stable. Ammonia is a good example of this. Protons in Ammonia nuclei exhibit an unusual feature where they 'tunnel' through the nuclei rapidly from one side to the other like someone flipping a switch. Despite this disturbance at the quantum level, ammonia couples stably with other atoms like chlorine to make stable molecular compounds like salt.[55]

What is interesting in this process is the creation of information despite the lack of creation of matter-energy. What is created in the super-ordinate rule-set is new information that constrains the information in the subordinate rule-set. It is the subordinate rule-set plus new constraints. In the case of atoms and quanta, new molecular rules can additionally constrain the ways in which an atom's nucleons and electrons can relate to each other. This is seen when the electron halos of component atoms are redistributed to a restricted set to reflect the geometry of new ionic and covalent relationships in molecules. However, nowhere in the rules of the subordinate rule-set (i.e. the strong force of quantum mechanics) is there a description of the super-ordinate rule-set (i.e. the electromagnetism of molecular chemistry).

What creates the super-ordinate rule-set? Nothing does. It is not created. It is selected from the interactive background as those few relationships that persist. It is not the components that create the super-ordinate relationships, but the environment that preserves them.

Order in the universe started with the emergence of fundamental regularities. The strong force preserved symmetric quantum structures like nucleons. Electromagnetism preserved symmetric chemical structures like atoms and molecules. Molecules depend at one level on nucleons held together by the strong force. At a second level they depend on nucleons and electrons held together by electromagnetism. Before molecules appeared, the rules of molecular chemistry didn't exist. Once molecules emerged, they interacted according to rules generated by their own structure. In this way structures embody rules and rules maintain structures.

As previously mentioned it seems very likely that the human tradition of separating 'rules' from 'structures' conceptually is a misrepresentation of natural process. It is likely that natural process makes no distinction between

55 Ammonium chloride (i.e. sal ammoniac) is in bread yeasts, salty licorice and Koskenkorva vodka.

rules and structures. Structures are rules embodied and rules are structures en-ruled. New structures inhere new rules. Once there was one in the universe, there was the other. New interactions find new rules that select for new structures and vice versa.[56]

Like *On the Origin of Species* this investigation does not seek the origin of structure in the universe, but only to trace its development as we find it. It will be a success if this process of development can be given some form of continuity that can be sharpened by future efforts. The greatest obstacle to tracing this process is the lack of a transformational logic that describes how subordinate rule-sets *cause* those super-ordinate rule-sets that accompany the appearance of novel systems. Typically, this process is called emergence.

## Formal Emergence

> *"Relationship to the environment is constitutive in system formation."*
>
> *Niklas Luhmann (1995, p. 176)*

Protons, neutrons and their cousins are the smallest and first universal systems we know of. They are systems because they are unified in two ways. Internally, systems are unified by the persistent and constrained relations of their components (Luhmann, 1995, p. 22). Externally, they are unified by the constraints of their own relations to external features. Thus, protons are protons because they are composed of quarks and gluons in chromodynamically maintained relationships. Equally, protons are protons because they behave in certain predictable ways when they interact with other matter-energy structures (e.g. with other protons, neutrons or electrons).

These quantum systems are the building blocks of our universe. From these select arrangements, all further systems were derived. The problem comes in describing how the specific relations of components determine the behavior of the system as a whole. Niklas Luhmann (1995) states:

[56] This is true as long as certain interactive conditions such as number of constituents, density and heat are maintained within the boundary conditions for the maintenance of substrate systems.

> Whether the unity of an element should be explained as emergence 'from below' or as constitution 'from above' seems to be a matter of theoretical dispute. We opt decisively for the latter. Elements are elements only for the system that employs them as units and they are such only through this system . . . This also means that the complexity that is necessary or sufficient to a system is not predetermined 'materially,' but rather can be determined anew for every level of system formation with regard to the relevant environment. Thus emergence is not simply an accumulation of complexity, but rather an interruption and new beginning in the constitution of complexity (p. 22-23).

Luhmann wants to know if the components determine the system or the system determines the components, and he, "opt(s) decisively for the latter." A typical scientist would argue against this decision. For most scientists, 'things' are what composes them. Water is $H_2O$ molecules - it is not hydrodynamics. Hydrodynamics is what water (groups of $H_2O$ molecules) does under certain conditions, but it is never considered *to be* water. Luhmann implies that water *is* hydrodynamics. In his case, what the system is (e.g. hydrodynamics) determines what the components are (e.g. $H_2O$ molecules). For Luhmann, the regularities of a thing's functional interaction with its environment are what it is. Fortunately, there is no reason to see these two options as mutually exclusive. Why cannot water be $H_2O$ and hydrodynamics?

In the history of systems, components exist before systems do. In fact, components *are* systems at preexisting levels of interaction. As such, components are 'pre-systems' and there should be historical steps that document how these pre-systems began interacting persistently in a rule based way. At some point, the persistent interaction of these pre-systems defines the unity of a new system. Quarks and gluons came together under the rule constraints of chromodynamics to create protons and neutrons. This is emergence.

Obviously, the problem with emergence is not the internal logic of component relations. These are determined by the logical transformations of rule-sets like quantum physics. The problem is the lack of a transformational logic that describes how the components create and constrain the behavior of the unified system. Luhmann (1995) claims that components ("elements") are only components when "the system that employs them as units and they are such only through this system" (p. 22). It seems that internally, systems can be logically reduced to interacting components. However, how the unified system interacts with external features is not logically derivable from how the components interact. Luhmann also states: "This also means that the complexity that is necessary or sufficient to a system is not predetermined 'materially,' but rather can be determined anew for every level of system formation with regard to the relevant environment" (p. 23).

Emergent systems can be understood in two distinct contexts. The first is the historic precursors which at a lower level of causality, step-by-step led up to the new, or 'emergent,' unity. The second is the context in which the emergent unity functions within its now and future environment. In this context there is an unexplainable qualitative change in the way the emergent system interacts with its environment (i.e. how it 'behaves'). Ultimately, there appears to be an historical, but not a predictive continuity between component behavior and system behavior.

Symmetry Theory proposes a solution to this puzzling logical discontinuity. The question is: where does the information that is created come from if not from the components of the system? The answer: the information comes from the environment itself. Specifically, it comes from how environmental structures react to the emergent system. All systems interact all the time. If a new system becomes maintained it presents new opportunities for reactivity to the matter-energy structures in its environment. These opportunities for novel interaction are what generate the emergent information in novel systems. They are the background interaction from which some percentage of the new interactions sampled will be regular, persistent and over time propagative.

This creation of information is invisible in traditional scientific testing because testing is predicated on isolating the system and thereby eliminating environmental reactivity. Testing takes pre-evolved, pre-ordered states (i.e. systems) and watches for entropy to select for disorder. It does not take disordered states and watch for symmetry to select for order.[57] Who would test vast random soups? Before computers, who could test them? Testing *makes* causes *make* effects. It does not check for what is selected for persistence from backgrounds of ubiquitous and varied interactivity. Novelty is generated in external relationships, not internal relationships. The inexplicable "emergent" features of natural systems are not due to their obvious components, but due to their vast environments.

This seems obvious. Novel, seemingly inexplicable features of systems, like the colour of gold or the surface tension of water, are only inexplicable if you presume that systems can only be defined by their components. Once you allow systems to be made by their environments, by the regularities in how they interact with other systems, the problem disappears. In fact, the universe is arranging itself in this way. Atoms were not created by quarks alone. We know this because there is nothing in chromodynamics that predicts the atom. Atoms were built by the relationships quarks formed with other features of their environments, namely gluons and electrons. It is relationships that emerge - not things. Historically, you could say everything was made by *what* it is composed of. Predictively, you could also say everything was made by *how* it related to other things.

A difference between historical emergence and predictive emergence becomes evident from this point of view. Historically, emergence is based on component regularities. Predictively, emergence is based on environmental regularities. The same kind of comparison will be made when we discuss genetic and phenotypic points-of-view. Genotypes embody the history of the organism, they were selected for what worked historically. Phenotypes embody the organism's now and future functionality, they are what will or will not work now, and in future generations. Each organism's genetic code is fixed, but which organism's pass on their code is an open negotiation with their environment. The past is determined and the future is statistical. The future is emergent.[58] The result of the addition of an environmental point of view to the reductionist point of view is a framework within which how

---

[57] The emerging complexity sciences have started to occasionally test vast interactive backgrounds as per Kauffman's order for free, Bak's self-organized criticality, Conway's game of life, etc.

[58] For more see chapter "The Direction of Biological Novelty."

systems came together historically compliments how they come together in the future.

Understanding this form of emergence requires an adjustment to our stance on causality. Here is the adjustment. All adjacent possible states that the universe can access from any one 'here-and-now' state constitute a limit to what can happen in the next step. In this sense, all adjacent possibles are constrained. Not just anything can happen. The future state is always limited by the present state. However, the present state steps into the future stochastically. In fact, all interaction is to some degree 'stochastic' in the chaos theory sense that no direct measurement of the universe around us is perfect. This inherent inexactness is an ontological feature of the fine graining of our universe not an epistemological feature of our capacity to know or measure it. Causality is stepping into a constrained future stochastically.

Emergence, then, is just positive symmetry to a persistent degree. Where entropy and chaos describe the nature of how systems statistically step into their adjacent possibles in a negative way, emergence describes the nature of how systems statistically step into their adjacent possibles in a positive way. From this point of view emergence is really only the counterpart of entropy and chaos. A universal spectrum of interaction maintains both positive and negative relationships. Existence is stepping into a statistical future, the question is really only whether that future will maintain those positive relationships that define the system, or not. *To be in sync or not to be, that has always been the question.*

Emergence makes sense from this dualistic point of view. Emergent systems are the positive relationships sampled from a full spectrum of possible interactions. The nature of these positive relationships is as inherently unknowable, unpredictable and statistical as entropy and chaos are because they are two sides of the same coin. The past is deterministic and reducible. The future is emergent and irreducible because not only are systems constituted by their components, but they are also constituted by their relationships with their environments. They are constituted by what happened in their past, but are always stepping into a statistically (not deterministically) defined adjacent possible, an emergent future. Ultimately, it is their relationships with their constantly changing environments that create the next systems. Novel systems emerge from environments, from interactive backgrounds. In fact, their own components are really only canalized ancestral relationships that were at one time selected for persistence from previous historical environments.

## System Interpenetration and Chaos Damping

Symmetry Theory implies that as systems couple to form new systems, each emergent level of interaction produces super-ordinate rules that, to some degree, causally isolated it from the regularities that generated it. Environmental interaction for emergent systems is different from component interaction because upon unification component interaction is locked into a mutually constrained phase space. On the other hand, the unified system itself is free to explore the novel phase space of its reactive face interacting with an environment that its own existence is altering. The components are constrained to historical repetition while the unified system is free to explore the future. In a sense the novel reactive face presented by the unified system, isolates its component's reactive faces. Although this isolation is part of what defines the system as emergent, it is not complete. In fact, proximate levels of interaction can interpenetrate each other significantly. However, as systems beget further and further systems, the reactive faces of the lower level components become causally insignificant. Additionally, there is the possibility that this isolation stabilizes emerging structures by dampening chaotic propagation.

First of all, each emergent or super-ordinate level of rules and structures represents a jump in type of causal relationships. This jump is the subordination of the internal rules to a novel set of external rules. Before sync, each component broke its own symmetries in its own future. After sync, the components communally break symmetries in a mutual future. This emergence process reduces the dimensionality of each component's future interactive potential while at the same time expands the dimensionality of the group's future interactive potential. This constraining of component potential divides boundary conditions (i.e. rule-sets) into 'before' and 'after' contexts. The internal rules maintained components *up to* this moment and the emergent external rules maintain the unified group *from* this moment.

The emergent unified system enables the creation of new interactions with other existing and emerging structures *from* this moment because the emergent system constrains environmental interaction differently than its individual components did. Ammonia has a different reactive face than *sal ammoniac* salt. Try ingesting one, and then the other. Ultimately, this 'jump' in causal relationships is a recasting of possible interactions. In this recasting, the old component rules still exist, but are subjected to the further constraints of the new group rules. If there are no new rules or the new rules do not constrain the old (i.e. are not super-ordinate) the system is not 'emergent.' In this case the new system is simply an extension of its own domain (e.g. if atoms interact in groups too small to be ordered by the gravitational rules that characterize astronomic symmetry, they remain just extended expressions of chemical symmetry).

What this means is that jumps (i.e. emergent systems) superimpose new rules over old rules. The emergent system constrains its components physically by changing their reactive face to the environment. The degree to which the new rules override the old statistically varies. This causes a spectrum of novelty. Some new rules only blot out a fraction of the old ones making them basically extensions of the old rules. Some are so elaborate as to occlude the old rules, basically making the component substrate invisible to external interaction. This is what makes salt safe and ammonia deadly.

In the case of astronomic symmetries, the rules of chemical affiliation still affect them under certain conditions. Chemical transformations in nebulae help form the composition and behavior of the heavenly bodies that they eventually collapse to compose. This is the case with the different kinds of planets that constitute our solar system. Some planets have dense atmospheres or rings; others have life. Chemical symmetries are partially responsible for each of these astronomic features.

Even though interaction between super- and sub- ordinate levels is possible within the system, it is increasingly unlikely as the levels distance themselves. While quarks and gluons may subtly affect atomic interaction, they are more or less invisible to molecular interaction and almost totally invisible to astronomic interaction. This happens because the jumps between levels represent statistical losses of component reactivity rules. These rules disappear under the statistical superposition of the new rules.

As such, the order those previously mentioned collapsing nebulae eventually impose on their environments (e.g. the organization of the planets) is mostly created by gravitationally regulated astronomic rules like classical mechanics. Very little is the result of the same chemical rules that ordered the nebulae. Likewise, a molecule may to some degree affect individual cell behavior, but the whole multicellular organism functionally ignores it. Interpenetration exists and in some cases can be maintained by positive selective feedback, but it becomes increasingly unlikely as systems complexify and levels of constrained release of energy are added.

An interesting result of this discontinuity of rule-sets, or statistical jump between emergent levels, is the braking of chaotic propagation. One of the most shocking developments of the last century was the development of Chaos Theory. Chaos Theory basically says that small irregularities in any ordered state become amplified as that state transforms.[59] Error propagates through systems. The obvious question chaos brings to mind is, 'Then how can ordered systems persist?' If chaos and its cousin entropy both inevitably propagate disorder in ordered systems, then how was order ever maintained in systems?

The canonical answer to this question is that order is not maintained. Entropy is final. The universe marches from perfect order at the Big Bang to perfect disorder at its heat death (Carroll, 2010). All of these structures that appear to have appeared are really only different states of degeneration. Among other things, the linear finality of this current model is unsatisfactory. Dissipative processes do not persist and to put us in one requires we lean on a particularly strong version of the anthropic principle in which we just happen

[59] "Chaos in space" (the study of fractals) will be ignored here because although it is very real in the abstract mathematical world, it is paradoxical in the real world. If the universe can never be perfectly finely grained, as Chaos Theory implies, then what is the point of calling a mathematical freeze frame picture of it real? If the universe is more continuous than we can measure, then the choice of any one spatial representation is nothing more than a human abstraction of measurement that has no privileged reality. Space changes. Space is time. There is no privileged point-of-view from which to stop it to study it.

to be lucky enough to exist, as an anomaly, in an eternal sea of nonexistence. It's a species of the Boltzmann's brain argument.[60] Instead, we should seek to explain the universe in terms of what we find in the universe, namely waves, cycles, differentiation and emergence. To explain the existence of these structures we must go back to the question; if chaos and its cousin entropy both inevitably propagate disorder in ordered systems, then how was order ever maintained in systems?

Part of the answer could be related to the discontinuity of rule-sets between structures. If interactions at the quantum, or even molecular levels were directly linked to interactions at the organismic or planetary level, then wave-particle duality, uncertainty, superposition and chaotic error at the lower levels would constantly propagate up through this link destabilizing the structures.[61] However, if each level of organization was isolated from its lower levels by interrupted levels of algorithmic transformation, then each super-ordinate level of organization would effectively dampen chaotic propagation from below. In a sense, each emergent rule-set operates like a wave break, stopping disruption from moving on to the next rule-set.

This makes sense from a persistence point of view. Components that interact over long transformations and form emergent systems need to retain their own maintenance mechanisms. To be functional (i.e. to exist) these component mechanisms must have been able to deal with the chaos and entropy within their own systems and environments to have survived. Like the Russian matryoshka doll, each level of emergence exists on its own. Its independent persistence serves to brake chaotic and entropic propagation so that further levels of cooperative organization can emerge with its products as stable components.

Interestingly, just as chaos and quantum uncertainty may be systematically braked as they try to propagate from within systems, entropy may be similarly inhibited as it tries to propagate from the environment into systems.

---

60 The Boltzmann's brain argument comes from Boltzmann imagining the universe as a large phase space of randomly interacting particles which could, given infinite time, accidentally assemble a brain. This brain could then falsely perceive its own astronomically unlikely state of affairs as the way the universe works. A temporarily ordered state of affairs is mistook for a universe that is essentially and eternally chaotic.

61 Wave-particle duality, uncertainty and superposition are all aspects of the same odd 'statistical' nature of quantum interactions.

This conjecture assumes that disorder is represented by environmental conditions and order by the system itself. Under these conditions, interaction with external features that might produce reactions capable of chaotic amplification into the system would also be broken by the discontinuity of the system's information levels. Both of these possibilities are arguments for the necessity of separated information levels in persistent integrated physical systems.

Systems are composed of different levels of component interaction. Although all levels combine to make the whole system, all levels do not affect each other equally. Statistically, those closest to each other affect each other the most while levels more separated affect each other less.

Structures are systems maintained by rule-sets. Not all structures are maintainable just like not all rule-sets are maintainable. A prerequisite for emergent structures is that they remain stable, that they persist. To remain stable they cannot allow too much chaos to propagate up through their levels of organization. This is done by differential selection for those emergent structures whose relationships are causally disconnected from their component relationships, and whose component relationships are causally disconnected from their component relationships. Thus, quantum uncertainty, chaos and possibly entropy, do not propagate freely through the system because each level of constitution is preserved by different informational rule-sets, different kinds of transformations preserving different levels of structure.

This process should be compared to the well-known process of 'summing over' applied to quantum systems within classical systems.

# Conclusions: Universe

The matter-energy included in the Big Bang expansion is continuously and ubiquitously interacting with itself. The ensuing universe is the disposition of all matter and energy as it transforms constrained by the three results of interaction (e.g. positive, neutral and negative symmetry). Positive symmetry is largely responsible for the emergent four fundamental forces plus their accumulated structural phylogenies. This makes the history of the universe the history of maintainable matter-energy relations. The reason our universe is highly ordered is not because it was more highly ordered at the Big Bang and is slowly falling apart, but because order is the only structure that persists during constant and varied interaction.

Interaction in our universe is subject to at least three domains of order: quantum, chemical and astronomic symmetry. These domains are conditions under which certain rule-sets act as pruning algorithms selecting for certain regularities (i.e. certain persistent structures). The longer regular interaction persists within those conditions, the more matter-energy is sorted by these rule-sets, leaving more and more product structures. These structures persist as long as the rule-sets maintain and re-create them. They also feedback into the environment by populating it and thus changing the interactive background into that from which new regularities, new structures, can be selected.

The largest and first domain is the quantum symmetry domain. It is characterized by the strong nuclear force. Constant and varied quantum interaction at this level leaves mostly the structures that comprise nucleon systems (i.e. proton, neutron and nuclei systems). Start with enough of the sixty quanta in a space where they interact enough (i.e. the right conditions) and eventually you will end up with quantum order (and no more).

The middle and second domain is the chemical symmetry domain. It is characterized by electromagnetism. The addition of constant and varied nucleon interaction at this level leaves mostly the structures that comprise molecular systems (i.e. atoms). Start with enough nucleons and elementary particles in a space where they interact enough (i.e. the right conditions) and eventually you will end up with chemical order (including quantum order).

The smallest and last domain is the astronomic symmetry domain. It is characterized by gravity. The addition of constant and varied atomic/molecular interaction at this level leaves mostly the structures that comprise astronomic systems. Start with enough molecules in a space where they interact enough (i.e. the right conditions) and eventually you will end up with astronomic order (including chemical and quantum order).

Essentially, the result of a cooling environment of quarks and gluons is strong nuclear force bonding, which leaves more and more nucleons. The result of cooling environments of quanta and nucleons is electromagnetic bonding, which leaves more and more atoms and molecules. The result of cooling environments of quanta, nucleons, atoms and molecules is gravity, which leaves more and more astronomic structures. At each level, the emergent order maintained is generic to that interactive background.

As noted, each of the later domains includes ordered systems from the earlier domains. They do this by superimposing rules. These superimposed rules are just additional levels of structurally constrained release of energy (in the form of stereotypical interaction). These rules are also evidenced by the survival of a limited set of emergent communal organizations. This limited set of high order structures is constructed of lower order components. Collected components are to some degree constrained by the higher-level unified organization. They are less free to make novel relationships. The collection and conversion of free systems into unified systems both establishes an environment of higher level systems and limits the further reactivity of free systems to create further innovations (e.g. each level may self-stabilize, self-regulate).

Note that each level of higher order structures not only includes lower order structures, but also continues to interact with lower order structures. This implies that when only quantum systems are present, gravity does not affect interaction at the quantum level. However, once chemical systems large enough to be influenced by gravity exist, they can "pull" quantum systems. The same feedback should apply to chemical systems under astronomic forces. This should be testable. As well, there is no obvious reason why higher order systems feeding back into lower order systems couldn't produce novelty (i.e. inter-domain selective interaction). This too should be testable, particularly as per the emergence of life, which appears to be a result of feedback between the astronomical and chemical domains.

A further interesting result of this arrangement is that each domain seems to have a size threshold below which it remains sub critical, meaning that it will not self-organize under those conditions (e.g. in the Per Bak 'self-organized criticality' sense). Simply put: if the distance is too great and the matter-energy too little then there is not enough regular interactivity from which persistent structures can be selected for maintenance or re-creation. As a result, each of the domains needs a minimal threshold of density, heat and number of constituents to create them. These are parts of the *conditions* previously referred to. Presumably, the total matter-energy threshold for quantum order is many orders of magnitude lower than the threshold for chemical order and the threshold for chemical order is many orders of magnitude lower than astronomic order.

It would be interesting to test for these thresholds, at least at the quantum and chemical levels. It would be even more interesting if the threshold conditions were found to have similar values. This would be the case where the density, heat and number of elementary particles necessary to evoke nucleon order in an experimental space were either similar or statistically related to the density, heat and number of nucleons and quanta necessary to create atomic order in an experimental space. As mentioned at the end of Section 1, both of these ideas should be testable.

What chromodynamic interaction explains is proton and neutron structure - nothing further. What proton, neutron and electron interaction explains is atomic/molecular structure - nothing further. The parts can explain the composition, but not the behavior of the whole. The behavior of the whole is determined by the novel interactions its novel structure is now capable of within the environment it has changed by becoming part of it. So the answer to Luhmann's question, "are components determined by the system or do components determine the system?" is that even though components did determine the system historically, it is the system's environmental reactivity, its 'reactive face' that explains the logic of how the system will behave now and in the future.

The history of universal systems can be seen from the internal, component based point of view. This is what science currently does very well - trace the history of systems. However, the ontogeny, or logic of emergence, cannot be seen from just the historical, internal, component based point of view. What emerges in the future must be seen from the point of view of the interplay between emergent structures and their environments. This point of view includes the external.

Novel structures embody novel relationships with their environments. The only way for these novel relationships to persist in an entropic and chaotic universe is through re-creation. These novel relationships must be re-created to persist, to become universal regularities. Through re-creation they become significant parts of their own environments and feedback into the emergence of further novel unions.

It appears that the physical reality of universal transformation is stepped. Like system deterioration (entropy, chaos, negative symmetry), system creation (positive symmetry) is statistically stepping into the unknown. Each time systems get together to create more complex systems the behavior of the emergent systems can be as unpredictable as their environments. More to the point, the rule-set by which the new system interacts with its environment is unpredictable. As Luhmann (1995) noted,

> The complexity that is necessary or sufficient to a system is not predetermined 'materially' but rather can be determined anew for every level of system formation with regard to the relevant environment. Thus emergence is not simply an accumulation of complexity, but rather an interruption and new beginning in the constitution of complexity (p. 23).

What all this means is simply that universal transformation does not appear to be entirely deterministic. Things that emerge explicably can behave inexplicably, not because of what constitutes them, but because of what surrounds them. The concept of emergence has been abused for exactly this reason. It has two meanings. Most scientists use it to mean that new systems come into being through the deterministic interaction of previous systems. Others use it to mean that the behavior of these new systems can be un-derivable from previous systems - that something mysterious is happening. Both seem to be correct, but emergence is no more mysterious than entropy.

This investigation recasts the 'mysterious' nature of emergent system behavior in informational terms. Simply put, this universe started out with matter-energy and sorting rules that could only be realized as it cooled. As it cooled the sorting rules constrained the relationships matter-energy could take (their structures). Some of these structures were maintained by the sorting rules, others weren't. Those that persisted populated their own environments. These structures were novel. Their interaction with their interactive background and each other formed new relationships. Some of these new relationships lasted longer than others, further constraining the old structures for that duration. Novel structures and novel constraints created a new environment of interaction for both new and old structures. The new environment provided new opportunities for interaction. It provided new rules for the sorting of persistent relationships. System maintenance and environmental maintenance feed back into each other, making system maintenance recursive and creative.

In this universe, relational logics (i.e. information) can be added. Matter-energy cannot be destroyed, but it can be reorganized such that its relationships can change - information can be created. Organization is maintained by the four fundamental forces and those historical structures and conditions that feedback into themselves. Thus, emergence is simply information recast, organization recast, structure recast. A cooling universe with minimal rules and variable density (interaction) will produce emergent structures at the interstices where conditions for stable push-pull relationships exist. These new rule-sets maintain new structures that will vary further interaction and thereby continue to sample for further emergent structures, and so on.

This feedback process is a kind of descent with heritable variation. Structures are being selected for by rules. Some of these structures and their attendant rule-sets will maintain (re-create) themselves differentially. Now the universe has heritable variation and selection processes. All it needs is transformation (time) for those structures that are maintained and reproduced to increase their populational representation. Interaction will always continue. The popular emergent structures will feedback into the environment generating novel interactions that feedback into novel rule-sets that create new emergent structures. The only thing that changes is relationships and only the relationships that maintain certain classes of changes (symmetric transformations) survive. That is how the universe built and is still building order.

# Section 3:
# Life

This section is about the evolution of life-on-Earth and it shows two things. First, it shows how the abiotic and the biotic systems are tied, how the principles of one are related to the principles of the other, how the pattern of life-on-Earth is a fractal zoom in on the fractal fabric of the universe. Second, it shows how biotic systems have transformed over the history of life in the hopes that understanding this pattern of emerging systems will help us understand human origins.

# Life Emergence

*"Now it appears likely that similar transitions, from groups of organisms to groups as organisms, have occurred throughout the history of life, right down to the origin of life itself as social groups of cooperating molecular reactions."*

*David Sloan Wilson (2002, p. 17)*

**FIGURE 10: PLANET EARTH (WITH MAGNETIC FIELD)**

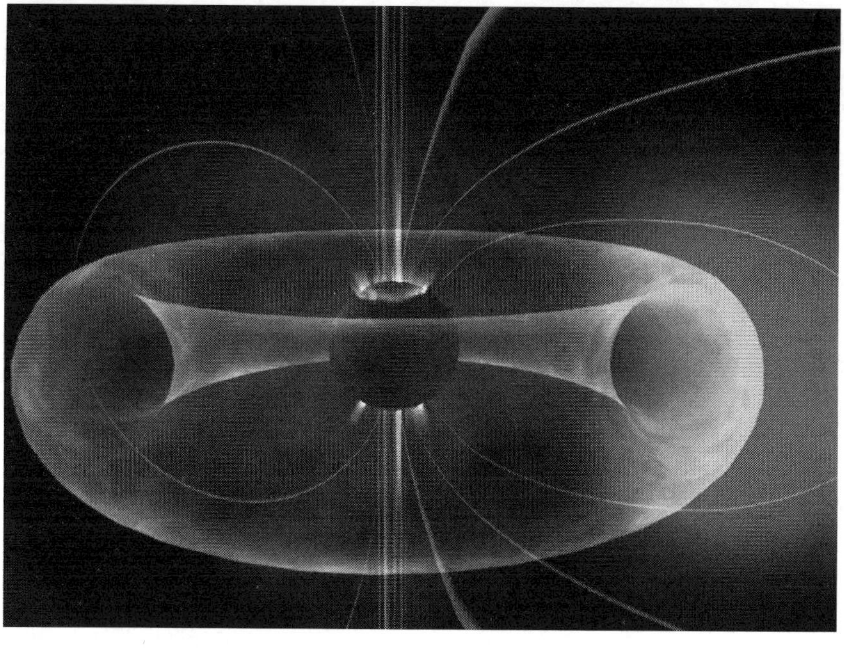

Source: O'Callaghan (2012)

Life-on-Earth persists on the thin shell of a small planet in the middle of an average solar system on the outskirts of a typical galaxy that is just one of many in this cluster of galaxies that is constantly distancing itself from unknown multitudes of other galaxy clusters in this universe we call home. As such, this section spans the last four billion years of a total of over thirteen billion years since the Big Bang. We are talking about a miniscule part of the universe over the last third of universal transformation.

One of the dense localities in the expanding and cooling universe became the galaxy cluster that contains the Milky Way. The forces of astronomic symmetry constrained the Milky Way to the characteristic rotating disc with increasing density towards the center. Likewise, many of the dense localities within the Milky Way, subject to the same rule-set, formed similar cyclic structures with dense centers. These structures are stellar systems. One of these stellar systems is our solar system. Within our solar system there are other variations of this same cosmic pattern in the form of a series of rotating spherical structures, each with increasing density towards their centers. Planet Earth is one of these structures.

The structure of the Earth is determined by its history and the rule-sets that describe gravity, thermodynamics, chemistry and mechanics (Morowitz, 2002). This structure is a series of less and less dense shells moving out from a heavy metal core. The last three, least dense shells are the layers most important to the formation of life. They are the lithosphere (the rock crust), the hydrosphere (the oceans and water systems), and the atmosphere (the gaseous outer ring of the Earth). As background conditions, the potential for ubiquitous and varied chemical interactions (i.e. the kind that result in push-pull environments) in these zones is great. It is within this vast spectrum of interaction, extending from Earth's molten iron core out to the almost infinite absolute zero of space, that our miniscule 'slice of life' conditions are buried. It was here, in this tiny sliver of the overall spectrum, where life was first sampled, maintained and reproduced.

## What is Life?

> *It appears that the chart of intermediary metabolism arises for a collection of atoms of the periodic table interacting by the rules of organic chemistry under the right conditions. The ultimate emergence of metabolism seems embedded in the laws of chemistry, but the reactions are a tiny subset of all possible organic reactions. We must search for the pruning algorithm.*
>
> *Harold Morowitz (2002, p. 76)*

Scientific definitions of life depend on two concepts, metabolism and reproduction. If it doesn´t have a metabolism to preserve itself and doesn´t have a reproductive system to preserve its kind, it isn´t considered alive.[62]

Metabolism is a specific biochemical Earth-bound version of autopoiesis. Autopoiesis means self-reference, self-maintenance, self-reproduction or self-organization - depending on whom you ask (Luhmann, 1995, p. 1; Margulis & Sagan, 1986, p. 256; Maturana & Varela, 1980). It is inclusion within a cycle of maintenance. As we have seen, many structures in this universe maintain themselves. From nuclei to galaxies, some select parts of the universe contain the right conditions for not only the creation of, but also the maintenance of, rule constrained structure. Metabolism is a chemical subset of this universal process of maintenance. Essentially, this subset is based on what Harold Morowitz calls the 'chart of intermediary metabolism.'

The chart of intermediary metabolism describes the hundreds of chemical interactions that characterize all known life. It is to biological structure what the periodic table of elements is to chemical structure, a set of rules delimiting those interactions that are maintained and re-created under certain environmental conditions (i.e. universal conditions for chemical structure and life-on-Earth conditions for biological structure). Living systems are like all systems. They are defined by their dynamic structure, by the relationships that both internally and externally maintain them. In the same sense that moons are part of the cycles of life and death of solar

[62] See Kauffman (2000) for an excellent discussion of autonomous agents.

systems, so are ATP molecules part of the cycles of life and death of animals. Metabolism is internal transformation with preservation of structure. It is the internal symmetry rule-set of biological systems.

A vital feature of metabolism is the distinction of internal from external. The chart of intermediary metabolism describes the internal regularities, but these regularities only persist because they mediate regular interaction with external regularities (i.e. external conditions). The basic structure of life is the cell. Contrary to intuition, the cell is not *different* from its environment. The cell is a sealed off *part* of the environment in which highly regularized maintenance and reproduction transformations are conserved. It is composed of the same stuff as the external environment, but in an energetically dense and improbable mixture of relationships.[63] Like the thin layer of biosphere sandwiched between space and the Earth's core, cells are a highly specialized part of the spectrum of the overall interaction of their environment.

While varied in their specific structure, cells all have one basic common feature: they differentiate their highly ordered internal milieu from the disorder of their environments by maintaining a boundary. What is inside this boundary is a complex steady state of transformation. That is, the inside of the cell maintains a constant state of order that can produce work cycles. Internally, the cell is an open thermodynamic system that captures energy from its ambient and uses this energy to perform the work cycles necessary to maintain a thermodynamic steady state far from true equilibrium (Bertalanffy, 1968, p. 142). In other words, one thing life must do to persist is to protect itself from loss of order. Loss of order - be it entropy or chaos - means death. The cell avoids death by maintaining a boundary separating the internally conserved biochemical process of the chart of intermediary metabolism from the slings and arrows of random ambient events.

The separation of internal order from external disorder is crucial to the maintenance of life. However, this separation has another effect. By

[63] These "energetically dense" relationships refer to the concept of "energy rate density" by Chaisson (2001). Energy rate density is a measurement of the amount of energy that passes through a set volume of system. In a sample of universal systems, solar systems process more energy per unit volume than galaxies, planets process more than solar systems, biospheres more than planets and animals more than plants. One surprising result of the logarithmic increase in power densities of systems as they emerge is that brains process orders of magnitude more energy than suns do, per unit volume.

encapsulating a highly organized chemical state the cell creates its own demise. The degree to which the cell wall isolates the internal system is also the degree to which the thermodynamic system becomes closed. Closed systems are entropic. Life isn´t possible without closing the system off to some degree, but death becomes inevitable when this happens. So how does life persist in the face of inevitable death?

Life persists for two reasons. The first reason life persists is because it was selected from what persisted before it emerged. Living systems persist because protons, neutrons, atoms, molecules and solar systems persist. Organisms are selected from a very specially organized and miniscule slice of the overall spectrum of interaction of these previously reproduced and reproducible systems. Living systems are those emergent relationships that are re-creatable or reproducible within this solar system's spectrum of conditions. As such, life emerged as one tiny fractal repetition of a much larger fractal pattern, a pattern that has repeated itself from quantum to chemical to astronomic domains over billions of years. As part of this process, life has created and re-created this same pattern within its boundaries again and again, ever descending into more detail of the same process of selection for persistence. Like all persistent systems, life is repeating only what is repeatable.

The second reason life persists is reproduction. The death of any specific classical system is as inevitable as entropy and chaos are ubiquitous. For systems to persist they must either maintain themselves indefinitely (i.e. the perpetual motion of electrons orbiting a nucleus) or they must be continually re-created by the rule-sets that govern them. Suns and their planets die in novae only to be re-created when that matter re-condenses at another locality. Life is no different. It maintains its specific incarnations as long as possible, but it also re-creates those incarnations as often as biological rule-sets allow. In part it is this re-creation, in the form of 're-production,' that defines biological systems as 'alive.'

Remember, structure itself is just regular transformation. Structure is a certain kind of statistically unlikely process, the regular or repetitive kind. However, even though regular transformation (structure) is highly statistically unlikely, it is the only kind of transformation that remains persistent. Symmetric transformation, in the form of structure, is the universe's only memory - and this memory, these persistent structures, are the only possible building blocks of any future imaginable. Thus, biological reproduction is a generational form of transformation with preservation of structure. It is a special Earth-history bound case of chemical symmetry. In general, reproduction is re-creation and it is a certain kind of structural maintenance necessitated by entropy. When entropy is a factor (e.g. when the structure is larger than a quantum-level one), reproduction is one of the key features of structural maintenance (i.e. persistence). This same concept was captured in the layman's definition of symmetry in the introduction: "Everything that can happen does happen, but only things that happen regularly persist."

In the abiotic symmetry domains, re-production is the re-creation of the familiar systems that persist in each domain (e.g. atoms, molecules, solar systems). These structures are created again and again by the conditions and rules of interaction that characterize their respective domains. These rules *seem* to exist outside the structures they create. That is, atoms don´t internally create further atoms. Likewise, atomic groups under certain conditions tend to create and re-create some of the simple universal molecules like $H_2O$. The creation and re-creation of water molecules all over the universe is due to these conditions, not to the water molecules themselves. However, once molecular groups have been established by local conditions, they can begin to interact with each other. Some of these common interactions have further interactions that can affect the original molecular populations. This process is called feedback. The process of feedback as it relates to processes of inclusion, maintenance and reproduction will be examined in the upcoming "Discourses on Life" chapters.

## Chemical Autocatalysis to Prokaryotes

> *"Life has been, since inception and with no discontinuity, chemically connected to its past."*
>
> *Lynn Margulis (1998, p. 79)*

**FIGURE 11: BACTERIA**

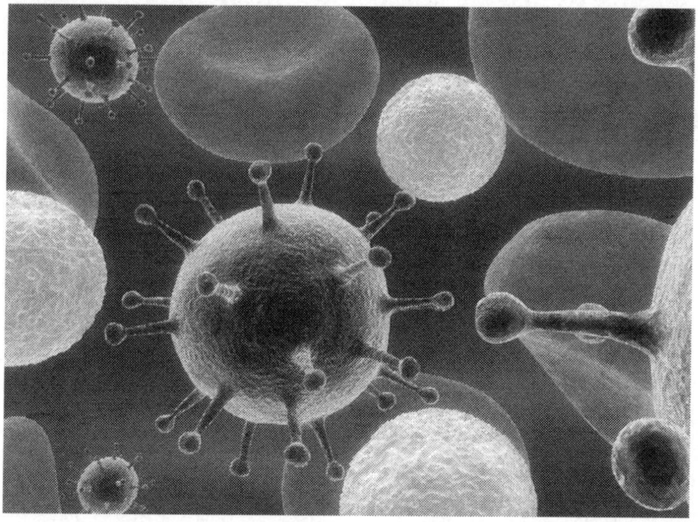

Source: Swain (2009)

How life-on-Earth began is subject to a lot of speculation. In general, the biochemical route to the first cell is usually described as chemical autocatalysis somehow becoming encapsulated to form protocells that eventually evolve into the DNA based cells we call prokaryotes, or bacteria.[64] But the question remains, how did this happen?

The structure of life-on-Earth emerged very much like atomic structures emerged from the laws of quantum symmetry, molecular structures emerged from the laws of chemical symmetry and cosmic structures emerged from the laws of astronomic symmetry. First there was a sorting process. The highly chemically interactive conditions on the surface of the Earth were first of all constrained by the quantum, chemical and astronomic laws specific to the Earth's cooling surface. Of particular importance were the laws of chemical interaction. The sorting of what will go with what, in which ways, was further constrained by the geophysical

[64] There are many different theories, but most of them include autocatalytic cycles and boundary formation. For general overviews see Margulis (1998) and Margulis & Sagan (1986). For more detailed information see Kauffman (2000), Smith & Szathmáry (1995) and Wills & Bada (2000).

synchronies of the Earth's surface like volcanic patterns, tidal flux, night and day, storm patterns, erosion patterns, wind and water wave-structures, sand filtering patterns etc. The growth of an increasing spectrum of localized boundary conditions supported an increasing number of complex molecules that interacted within an increasing number of microclimates throughout the outermost spheres of the Earth. As the spectrum of molecular interaction grew, the tiny Goldilocks parts of the spectrum that represented emergent autocatalytic cycles grew within it. The vast majority of interactions were non-autocatalytic; they didn't persist long enough to change the interactive backgrounds at their localities. The few autocatalytic interactions that did persist presented emergent reactive faces to their localities. Two of the proposed autocatalytic cycles necessary for the emergence of life were lipid membrane boundary formation and the emergence of RNA/DNA.

As the chemist Harold Urey was fond of pointing out, the laws of chemical symmetry, combined with some very basic historical conditions, can produce some very interesting life-like biochemistry. Just as in a universe with a few localities of patterned, self-maintaining matter-energy aggregations, a very few of these life-like bio-chemical aggregates were maintained in a very few unique locations. These molecular aggregates persisted in the face of constant entropy because the conditions that reproduced them were constantly maintained by the processes of quantum, chemical and geophysical symmetry specific to their locality. These regularities allowed a few molecular aggregates to persist long enough within these novel interactive backgrounds to sample new rules of interaction from which, eventually, new systems (e.g. protocells) could be selected for persistence.

This very process is currently being investigated under the name of the "theory of collectively autocatalytic sets." Stuart Kauffman (2008) describes this theory thus:

> The intuitive idea of the theory of the spontaneous emergence of autocatalytic sets is that given a set of molecules, reactions among the molecules, and distribution of catalysis

> of those reactions by the very same molecules, a diversity of molecular species will be reached, at which time collectively autocatalytic sets spontaneously form in the system (p. 66).

This approach should sound familiar. The theory of collectively autocatalytic sets is another version of the idea of emergence as described in the previous section. There is a chemically interactive background establishing the boundary conditions within which constant catalysis between molecules presents an emergent "diversity of molecular species." These emergent molecular species have novel reactive faces. As these novel species further sample their interactive environment they create a spectrum of negative (i.e. non-autocatalytic) to positive (i.e. autocatalytic) interactions. Only positive (i.e. autocatalytic) interactions are preserved within that specific environment because they are the ones that maintain and reproduce themselves - *auto*-catalytically.

The theory of autocatalytic sets has not yet been proven, but Kauffman (2008) says that, "recently, others have demonstrated mathematically that this phase transition to collectively autocatalytic set is guaranteed to occur" (p. 67). Kauffman goes on to say that experimental results should be forthcoming. Experimental confirmation of some version of the theory of collectively autocatalytic sets would additionally support Symmetry Theory, at least in the context of the origin of life. In fact, the experiment previously mentioned looks a lot like the experiments proposed in the "Conclusions: Universe" chapter that would support the kind of environmental emergence implied by Symmetry Theory throughout the universe. The same kind of experiment should confirm the emergence of generic order in the abiotic, as well as biotic, domains.

Interestingly, Kauffman mentions another implication of this kind of environmental emergence, the possibility of a 'general biology.' By this Kauffman refers to the fact that such a theory does not pinpoint the specific route collective chemical autocatalysis took to persistence, only a probability that it will take some route if an appropriate spectrum of push-pull (or in this case cleavage and bonding) conditions are maintained. This is exactly the same problem we faced when examining the emergences of the quantum, chemical and astronomic domains. Although the exact route quantum, chemical and astronomic symmetry products took to attain persistence is unknown, they did seem to emerge handily when conditions were appropriate.

It may be impossible to verify the exact route of broken symmetries that chemically led to the emergence of the first living cells. Although discovering the exact route would be interesting it may not be terribly informative to a "general biology" approach. A general biology approach would merely indicate that given the right conditions, some route to chemical autopoiesis would eventually be discovered. In this case, chemical autocatalytic cycles like RNA, DNA and lipid membranes are just three candidate structures within a potentially much larger interactive context for emergent life. The degree to which the specific structures that do constitute life-on-Earth are constrained to the structures they took (e.g. the degree to which *this* kind of life is predetermined) is as yet unknown.

In a miniscule corner of a miniscule tidal pool on the miniscule surface of a miniscule planet circling in a miniscule solar system of a miniscule galaxy in a miniscule galaxy cluster in an inconceivably tiny corner of the universe, life started. That is to say that the conditions that birthed life are by no means common. In fact, each of the above steps towards life represents the ever decreasing ranges (i.e. spectrums) of interactions within those symmetry domains that host positive interaction. The maintenance of positive symmetry is the interactive background for propagating symmetry. Life was born in a very specialized space of compounded universal symmetries.

This space can be imagined as a phase space, or as physicists would call it, a configuration space, "a wonderfully clear way to picture, all at once, everything that can possibly be" (Barbour, 1999, p. 43). Typically the phase space of life's origin is thought of as describing chemical possibilities and constraints, but in reality it should include the possibilities and constraints of both the quantum and the astronomic domains, although probably to lesser degrees. Molecules and molecular compounds all over the world were regularly sorted by gravity, mixed by storms, heated by radioactive decay and lapped by water waves, magnetic fields and photons. Somewhere in this vast spectrum of flux and reflux, new structures were born from the extreme end of the tiny range of persistent organized interaction.

Somewhere, perhaps in a tidal pool, organic molecules began to cooperate to produce autocatalytic cycles that helped maintain some degree of the system´s cyclic structure. These structures produced new rule-sets, but only those structures whose rule-sets best maintained and reproduced their structures persisted. Eventually, the system that emerged as the most persistent was the familiar prokaryotic cell (e.g. bacteria) based on genetic binary fission and a basic, stripped down rule-set of the chart of intermediary metabolism. The process was a familiar one. The first living cells were created by selection for persistence of chemical interactions that created the conditions for characteristic rule-based molecular interaction - that created the conditions for canalized production and reproduction of cellular structure. This is a universal process where conditions of ubiquitous and varied *interaction* are sampled for *synchronous* systems whose further interaction is selected for regular *re-creation*. This process turned the Big Bang universe into a quantum universe, the quantum universe into the molecular universe and the molecular universe into the astronomical universe. In Earth-bound biological terms, this same process could be called *communication* (i.e. interaction) selected for *cooperation* (i.e. synchrony) that is further selected for *reproduction* (i.e. re-creation). This frame shift, from terms traditionally associated with inanimate systems to terms traditionally associated with living systems, will be elaborated shortly.[65]

> *Whatever the mechanism by which the first replication was accomplished in the prebiotic world, it must surely have been a cooperative (emphasis added) enterprise involving more than one type of molecule, accompanied by substantial flows of energy from one molecule to another.*
>
> *Christopher Wills & Jeffrey Bada (2000, p. 130)*

[65] See chapter "Communication, Cooperation and Reproduction" for more on this lexical frame shift.

## Prokaryotes to Eukaryotes

> *A typical photosynthetic cryptomonad has a set of genes on its chromosomes and a second set - descended from bacteria - in its mitochondria. It has a third set, much damaged and inactivated, inside the degenerated nucleus of the red or green alga that it has swallowed. The alga in turn has its own set of mitochondria, contributing to a fourth set of genes. And finally, there is a fifth set contained in the chloroplast inside the alga!*
>
> *Christopher Wills and Jeffrey Bada (2000, p. 196)*

**FIGURE 12: PROTISTS - VOLVOX CARTERI (DARKFIELD)**

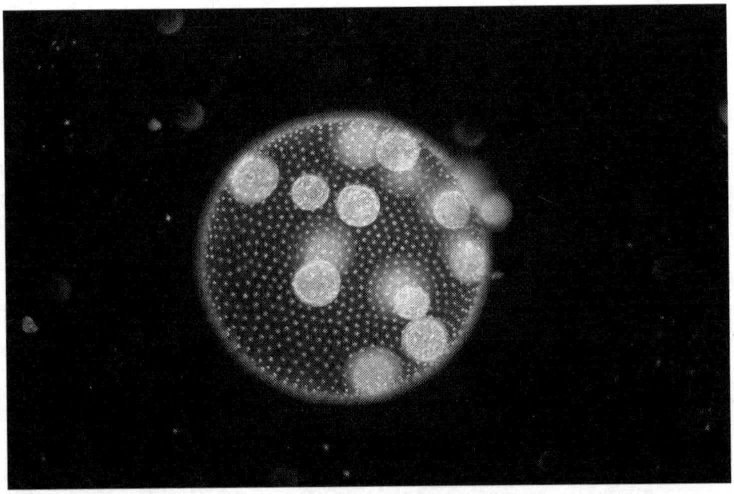

Source: Goldstein (2006)

For two billion years life-on-Earth consisted of single celled prokaryotes (e.g. bacteria). These prokaryotes showed little propensity to diversify morphologically, but developed a remarkably wide range of metabolic processes. To this day, most of the chart of intermediary metabolism has been developed by bacteria. These, "slow evolving generalists," as Gerhart & Kirschner (1997, p. 9) call them, did however interact extensively. For one, they invented sex. They developed a process for passing genetic material from one cell to others that greatly enabled adaptability. For another, they created and helped maintain the oxygen rich atmosphere that enabled the high energy metabolic systems that are necessary to support the larger, more complex eukaryotic cells that eventually evolved.

For close to two billion years prokaryotic cell interaction (i.e. communication) was widespread and complex. As species and populations increased, the spectrum of interaction increased. As with certain locations of the post-bang universe, some of these interaction localities began to maintain synergistic patterns of communication. The one favoured by Lynn Margulis is where one cell engulfs and attempts to eat another. In the end, the engulfed cell becomes a symbiotic partner with the engulfing cell. This partnership can be called symbiotic, synchronous or cooperative. In any case, the two operating together created a new cooperative relationship capable not only of self-maintenance through mitosis, but also of a novel set of behaviors that revolutionized the surface of the Earth.[66]

This story of emergence should be familiar by now. It is interactive prokaryotic *communication* selected for synchronic *cooperation* selected for an emergent *reproductive* system. The new cell is called a eukaryotic cell and it is the basis of almost all macroscopic life. Margulis´ Serial Endosymbiosis Theory posits that eukaryotic cells were created by mergers of cooperative prokaryotes into novel social unities. Only recently has "Serial Endosymbiosis Theory" become widely accepted by the scientific community.

---

[66] Margulis (1998, p. 31) has also proposed that more than two cells were symbiotically involved in the genesis of the eukaryotic cell.

## Eukaryotes to Multicellular Organisms

FIGURE 13: MULTICELULAR ORGANISMS

Source: Google Images (n.d.). *Colorful sea sponges*

The emergence of multicellular organisms (e.g. animals, plants and fungi) in the fossil record seven hundred million years ago is a landmark in this investigation. In the grand scheme of things far more important landmarks have already been passed, such as the origins of the three main symmetry domains, and origins of life and eukaryotic cells. However, the emergence of multicellular organisms is important for this investigation because it marks the emergence of structures like us. From here on out this investigation is working less from the Big Bang and more towards the human mind.

First, however, it must be noted that multicellular organisms emerged, much like all novel universal structures, through synchronous interaction resulting in novel rule-sets producing structures selected to maintain and reproduce themselves. Once again, this is interaction (i.e. communication) selecting for synchrony (i.e. cooperation) selecting for propagating symmetry (i.e. reproduction).

Multicellular organisms are collections of cooperating cells. The difference between the cooperation of prokaryotic cells to form eukaryotic cells and eukaryotic cells to form multicellular organisms is a difference of internal vs. external cooperation. Prokaryotes have a cell wall that seriously restricts communication with other cells. This constraint required the eventual internalization of other cells before a symbiotic relationship could be established. As should be expected, restricted communication restricts new cooperative arrangements which restrict new reproductive processes which ultimately results in restricted emergence. Remember, it took almost two billion years, more than half the history of life, for prokaryotic cells to start cooperating as unified organisms (e.g. eukaryotic cells).

Multicellular organisms had a different kind of breakthrough with the same cooperative result. The evolutionary breakthrough multicellular organisms had was the loss of the cell wall barrier and the biochemical creation of a system of cell adhesion.[67] Multicellular cells were novel because they were externally synchronized. One of the advantages of this arrangement is the loss of the size constraint that limited single celled organisms.[68]

Protists (i.e. single celled eukaryotes) are limited in their complexity by their size. Their size is limited, as is bacteria size, by the mechanical and chemical constraints of their component arrangements. When an unlimited number of cells can be connected (e.g. multicellular organisms) a new phase space of component arrangement becomes available for cooperative interaction. *Intra*-cellular constraints are now subordinated to new superordinate *inter*-cellular constraints. The language of inter-cellular communication is called signal transduction. As it increased between the cells, so did multicellular organisms′ ability to mix and match cooperative arrangements. Terrence Deacon (1997) says, "with each level of organization (molecular, cellular, structural, functional) in a complex multicellular body there is an *exponential increase* in the amount of information necessary to control its construction and operation" (p. 107). What he is saying here is that cooperation ("information necessary to control") between cells must increase

[67] Of course, there were also other adaptations.

[68] It must be mentioned that size and complexity constraints weren't lost as much as newly constrained at a much higher level. The word 'loss' is a simplification meant to indicate the massive expansion of cooperative phase space that cell wall loss enabled.

exponentially as organisms get larger and more complex. The hundreds of different kinds of mechanical, chemical and electrical signal transduction 'languages' that multicellular organisms embody are evidence of this exponential increase in information, or cooperation.

In short, what makes multicellular organisms so morphologically plastic, so rapidly evolving, so functionally specializing, and so often going extinct (Gerhart & Kirschner, 1997) is their whole new platform of communication. Novel cell structure enabled a novel class of interactions to be selected. Some of these interactions were maintained longer than others, and, over time, the most persistent ones were selected to cooperate, eventually reproducing new relationships.

Multicellularity is another example of novel rule-sets and structures emerging via cooperative aggregation (i.e. positive symmetry). Cosmic structure is maintained and reproduced by astronomic symmetry, but astronomic symmetry is built upon the structural and processual foundations of chemical and quantum symmetry. Multicellularity is reproduced by meiosis, but meiosis depends upon the biochemical processes and structures, specifically binary fission and mitosis, that were discovered by eukaryotes and prokaryotes. These processes include the autocatalytic cycles and the chart of intermediary metabolism that both prokaryotes and eukaryotes embody.

# Discourses on Life

## Maintenance, Feedback and Reproduction

Feedback is when the products of transformations alter the probability of the function of the transformations themselves. It is a kind of cyclical communication, where communication (i.e. interaction) has been previously defined as the process by which the function of one 'thing' alters the functions of other 'things.' In the case of feedback, those 'things' are processes; they are transformations. Classic feedback is described as being positive if it increases the transformations that produce more products. It is negative if it decreases those transformations.

In a universal context, if there are no atoms at some point, but the combination of strong nuclear force and electromagnetism start to produce enough of them that atomic interactions become common enough to produce molecules, then the gravitation between groups of those molecules will start to produce suns and the cycles of nucleosynthesis will proceed. Even though the universe started with just light elements, the interaction of these light elements selects for heavy elements and the combination of the two in astronomic systems leads to the creation, destruction and re-creation of a full range of elements. Atoms make molecules, molecules make suns and suns make more atoms. It's a circle of positive feedback.

In a biological context, metabolism is a classic case of positive feedback. It creates work cycles that maintain and reproduce organisms that in turn reproduce more organisms and create more work cycles. Stuart Kauffman (2000) describes this feedback as a, "vast web of constraint construction and constrained energy release used to construct yet more constraints" (p. 5). Probably the most ancient and fundamental metabolic feedback cycle necessary for life is reproduction.[69] Reproduction is another classic case of positive feedback. It maintains the conditions and rule-sets that create more organisms, even though the death of any one organism is inevitable.

Each of the three universal symmetry domains is characterized by the creation and re-creation of characteristic structures. However, in each of these domains the structures seem to be created and maintained by conditions that exist outside the structures themselves. Long periods of gravity at the molecular level are what draw solar systems into their characteristic forms. The strong nuclear force and electromagnetism at the quark, gluon and electron level are what re-creates atoms. For this kind of re-creation it seems that preexistent forces and conditions are enough. What seems strange and unique about biological symmetry is that the rules that re-create its structures do not seem to be dependent on external conditions. Biological systems, organisms, seem to have internalized the rules for their own reproduction. They seem different somehow. Is this true?

It is not. All universal structure is built and maintained on substrate systems. Planets are built and maintained by molecules interacting under certain conditions and molecules by atoms and atoms by quanta. Organisms are no different in this respect. They depend on stable quantum, chemical and astronomic substrates - namely life-on-Earth conditions. Like all universal structures, organisms just add an extra level of constraints to local combinations of preexisting constraints. Other universal systems also depend on rules from other levels to create and maintain the substrate upon which their rules can create and maintain their own structures. Without a broader context of electromagnetic conditions within which atoms, molecules and eventually stars can be selected for persistence, there is no astronomic symmetry domain. What appears to be a unique feature of life, namely the internalization of re-creation rules, really does happen at other levels too.

Re-creation is inclusion within a cycle. Only re-created systems can persist and any system that is re-created is inside the re-creation cycle. Every 'thing' that we know of (living or dead) is either in a re-creation cycle or disappearing.

Just as quantum order needs a minimum number of elementary particles, in a restricted heat range, at a minimum density to create the conditions

69 Reproduction is actually a subset of metabolism but it is such an ancient, distinct and specialized metabolic transformation that for the rest of this investigation I will often refer to them separately.

of interaction from which atomic order can be selected, so does biological order need a framework of boundary conditions. This framework is the boundary within which organisms can metabolically and reproductively maintain themselves. Quantum, chemical and astronomic structures also maintain the conditions for their own re-creation. They can only persist within their own cycles of re-creation. What appears to be a unique feature of the biological domain is in fact true of all domains. They all maintain and re-create their structures within their own self-contained, self-maintaining frameworks. All persistent systems are autopoietic.

In short, we are biological systems within a biological framework. From inside it seems like the rules that re-create us are internal and this is true. However, from a universal point of view, all levels of structure maintain and re-create themselves within their own cyclic frameworks (i.e. bounded interactive backgrounds).

Essentially, the concept of positive feedback boils down to processes of maintenance and re-creation (or 'reproduction' in biological parlance). Internal positive feedback, in the form of inclusion within a cycle, is not the exclusive property of the biological domain. Each substrate domain contains its own mechanisms for its own maintenance and re-creation. In fact, without those positive feedback mechanisms those domains would cease to exist. An elementary particle universe will create and re-create protons and neutrons as long as it is not too hot, not too cold, not too dense and not too dispersed. No matter how many protons and neutrons fall apart, there will be a continuous cycle of re-creation until those conditions change. And conditions always change. In fact, protons and neutrons themselves changed conditions by making first nuclei, then light atoms re-creatable, enabling cycles of molecular re-creation which ultimately changed gravitational and density conditions at some locations to create and re-create the full spectrum of atoms in nucleosynthesis. The whole time that the universe has been getting colder and less dense, certain feedback cycles at certain locations with the right conditions have been adding interactive variety and thus growing the spectrum of interaction from which synchronous cycles of maintenance can be selected for persistence.

Perhaps there is a measure by which biological feedback, in the form of genetic reproduction, is unique to the degree (rate? connectivity? specificity?) it re-creates structures within its own framework. However, it must be remembered that all domains are characterized by unique feedback mechanisms in the form of their own rule-sets, rates and conditions of transformation. Gravity appears to be meaningless in quantum systems. The strong nuclear force, electromagnetism, gravity and cellular autocatalysis all maintain unique product structures by feeding back into the conditions that recreate those structures.

Conditions for reproduction are an essential feature of structural maintenance at all domain levels in this universe. Positive feedback is the process that drives this reproduction. It is partly based on a historically persistent substrate, the components, and partly based on maintaining the right conditions in the interactive background, the environment. Just because organisms were selected to invert that interactive background from external to internal does not make them special. Upon close inspection the three universal symmetry domains also help conserve the interactive background that maintains their characteristic structures by defining restricted areas of interactive space in which their rule-sets maintain and reproduce the structures that create those conditions. They too invert the interactive background from external to internal. They too persist at different scale levels. They too contribute to the maintenance of those conditions within which they maintain and reproduce each other. They too conserve themselves through inclusion in cycles of positive feedback.

On a universal level this positive feedback can be seen as positive symmetry because it is always rule based, these transformations must preserve structure. Matter-energy cannot be destroyed or created and more importantly, aggregates of it only exist in regular forms under regular rules of interaction. Feedback is the engine for conditions of propagating symmetry like those we have here on Earth. Propagation of the symmetries of life is called reproduction. Propagation of the symmetries of the universe is called positive symmetry - and life is just a tiny fractal slice of it.

## Environment and Organism: external and internal

> *How can the events in space and time which take place within the spatial boundary of a living organism be accounted for by*

*physics and chemistry? ... The obvious inability of present-day physics and chemistry to account for such events is no reason at all for doubting that they can be accounted for by those sciences.*

*Erwin Schrödinger (1944, chapter 1)*

*"This distinction between everything on the inside of a closed boundary and everything in the external world - is at the heart of all biological processes."*

*Daniel Dennett (1991, p. 174)*

Science occurs within the shadow of what Thomas Kuhn called 'paradigms,' or worldviews (Kuhn, 1970). One of the fundamental tenets of our current scientific paradigm is that things are exclusively and only what they consist of, what is inside them. Nobel Laureate Erwin Schrödinger was convinced that life could be explained by what occurred, "within the spatial boundary of a living organism." As he mentions above, he believed that physics and chemistry could account for life. More than fifty years later, gravity, aperiodic crystals, and covalent bonding have contributed next to nothing to our understanding of jealousy, inspiration, hobbies or morality. It may be time to start looking outside the boundary of living organisms, outside the traditional purview of physics and chemistry, outside the inside, for answers. To do this we need an understanding of what 'outside' is and what our relationship as biological systems is to it.

Like all systems, organisms are regularities within their environments. The stuff that constitutes the environment is the same stuff that constitutes the systems within it. Systems are loci of persistently constrained environmental regularities. One result of selection for persistence from an interactive background is the selection of novel synchronic structures (e.g. positive symmetry, emergence). Another is the increased representation of these emergent structures in the environment (e.g. propagating symmetry). More novel structures lead to increased variety of the types of communication (i.e.

interaction) they can experience, both with each other and with the environments they are expanding into. If they are to exist they must maintain themselves - they must be re-created. Some of their environmental interaction must feedback positively with their own systems. This allows them to out-reproduce entropy, to survive over generations. This positive feedback can canalize communication pathways. It can find and maintain new interactive regularities.

Daniel Dennett (2003) gives an example where a butterfly lays eggs on the wrong plant. When the eggs hatch the larvae imprint on that plant (i.e. learn to recognize it as 'the right plant on which to lay their own eggs'). In a few cases this may create a happy accident. This new and maintainable (i.e. imprintable) information is now passed down to their offspring, not genetically but epigenetically, by learning. In this way a new "informational path" is being sampled for persistence:

> Once the informational path of transmission is established, and becomes 'relied upon' *by the genes* to do some of the carrying, it becomes subject to design improvements of its own, just like the myriad refinements that have beefed up the processes of DNA coding, replication, editing, and transmission over the eons. Genetic changes that tend to prolong parent-offspring contact and interaction, for instance, can raise the reliability of these pathways of social learning by giving them more time to operate, and then attentional biases (watch Mom!) can evolve to tune the transmission further. The path becomes a road becomes a highway, an informational channel *designed* by natural selection to enhance R&D in the lineages that rely on it. (p. 172).

What happened here is novel interaction reinforced by positive feedback. The mother butterfly makes a mistake (i.e. makes a novel interaction), but that mistake actually helps her offspring maintain and reproduce themselves (i.e. creates positive feedback). As interaction continues her offspring can sample more of their environment with their novel reactive faces and they can find more and more ways to maintain and reproduce themselves by canalizing those positive symmetry (i.e. positive feedback) relationships that maintain them. "The path becomes a road becomes a highway, an informational channel designed by natural selection to enhance R&D (i.e. sampling for novelty) in lineages that rely on it." Dennett is essentially describing an instance of how emergent structure becomes selected for persistence from an interactive environment. In fact, he is using the growth of emergent behavioral information relationships in the butterfly example to inform both genetic (DNA) and cultural (social learning) processes. New information is created when new positive symmetry relationships are formed, and as that information maintains and reproduces itself, it further samples the environment for further positive relationships.

In the case of the origin of life-on-Earth, the conditions of terrestrial interaction produced a few localities where molecules capable of autocatalysis came into regular communication with molecules that formed membranes. The novel informational symmetry that measured how these two systems began to maintain each other was the emergence of protocells. These protocells were formed by selecting ever more maintainable and more reproducible chemical interactions from the host of more or less maintained molecular relationships in their environments. Molecular information relationships were sampled and then selected for persistence and etcetera, etcetera. An information pathway becomes a road becomes a highway - the highway of life.

An environment is a slippery concept. Generally, it could be defined as a subjective human measure of matter-energy transformation to a certain level of coarse graining at a certain locality of the universe. Which kinds of transformations, and thus systems, are chosen to be included in any definition of an environment usually depends on a discipline's interest. Typically social science definitions of environment do not include fine-grained quantum or chemical transformations, nor do they include coarse-grained cosmological transformations like dense plasma thermodynamics. This is probably one of the greatest failures, and opportunities, of scientific integration. Each discipline chooses what to count and what to ignore, what is signal and what is noise.

In a scale-free universe, chasing an absolute definition of environment would lead to the same kind of pit of pendulating snakes and ladders that Bertalanffy, Maturana, Varela and Luhmann ended up in.[70] It is an important job, for the future. Today we need a practical definition. Practically, environments are those external systems that contain the internal systems we study. Usually the systems we study were created within the environment we find them in. As such, internal and external systems are causally interrelated in webs of maintenance and reproduction. What differentiates the internal from the external system? Internal systems include more symmetry. They transform by preserving more structure. They contain more Shannon information per unit volume. They reduce uncertainty and increase regularity. Systems are special organizations of the environment that maintains them. What makes systems special is the degree to which they constrain the transformation of matter-energy within their boundaries. Systems are regularities within less regular environments.

Astrophysicist Eric Chaisson (2001) has developed an interesting way to measure what he calls the 'complexity' of systems. He measures the energy per unit volume that the system transforms over a constant time period and calls it an "energy rate density," or "power density." Essentially what he found was that systems consistently process more energy than their environments. For example, one average gram of solar system processes much more energy than one gram of galaxy, one gram of sun processes more energy than one gram of the solar system and one gram of brain processes more energy than one gram of sun. Chaisson's findings support the idea that systems are specially maintained concentrations of energy constraints within environments of less dense energy relationships. In the upcoming "Toward a Universal Theory of Directionality" chapter we will take a deeper look at Chaisson's power density metric.

However, it does seem that calling energy rate density a measure of complexity is problematic because although the constraints (rules or regularities) that make up the system cause more transformation (i.e. process more energy), the constraints themselves are actually simplifications. They represent regularities, or 'information,' which is a *reduction* of uncertainty -

[70] See Appendix 1 for more on the scale-free universe.

and thus a *reduction* of complexity. For example, of all the possible things one gram of elementary particles in your brain could be doing right now, what they are actually doing is a miniscule, highly improbable sample from a vast space of possibilities. And they are doing it regularly, again and again. They are in a vastly simplified state of transformation.[71]

This semantic bind presents opportunities for further investigation. Specifically we now need two words for complexity. One needs to refer to the comparison of systems to each other (i.e. how many different regularities are included in one system as compared to other systems). Another definition of complexity needs to refer to the comparison of the system to its environment (i.e. how many, or how 'dense,' are the regularities in the relationships that differentiate the system from its environment). In this case we are talking about the latter definition. The difference between systems and their environments is the way, and degree to which, matter-energy is constrained within the system as compared to its environment. System matter-energy is subject to extra levels of constraint. These constraints are the rules that define the system both to itself and from its environment. An atom is distinct from its environment not because it contains electrons, quarks and gluons, so does its environment. An atom is distinct because it represents a fixed and limited set of electrons, quarks and gluons that relate to each other in a fixed and limited set of ways and together they represent a fixed and limited set of ways they can interact as a whole with their environment. This fixed and limited set of constraints (i.e. information) is the rules that define that atom. For this reason it is problematic to call systems complex. They are actually made of simplicities. Outside, all different particles are doing all kinds of things. Inside a regular number of particles are doing a regular number of activities, regularly.

Organisms are highly organized localities within their environments. Even though they are built from the same 'stuff' as their environments, that stuff is organized into a distinct set of energy constraints (Kauffman, 2000, pp. 4-5). This makes internal interaction conservative in the sense that it conserves this limited set of transformations. External interaction (i.e. environmental interaction) represents a far greater set of transformations that includes not

[71] See Appendix 1 for more on complexity.

only all other nearby organic organizations, but the complete set of abiotic interactions in that area (e.g. quantum, chemical and astronomic). All kinds of things happen outside that don´t happen inside. External environments are noisy.

A basic distinction between systems and their environments is the noise level. Noise here will refer to unstructured or irregular interaction, something like the common sense definition of chaos (e.g. work was chaotic today). Environments contain a lot of noise compared to their systems.[72] In fact, systems could be defined as the low noise, or high signal, parts of their environments. Systems are information in the sense that they reduce the uncertainty of what their environment can do within their boundaries. This distinction between systems and environment will become pertinent when the question of how organism´s navigate their environments is examined further on.

Organisms are their environments. They are localities of highly constrained matter-energy transformation. The difference between an organism and its environment is not substance, but relationship. Both are composed of the same elements organized to different degrees and by different rules. In general, internal rules are regular, conservative and historic while external rules are more chaotic and creative.

## Communication, Cooperation and Reproduction

*"Nothing in Biology Makes Sense except in the Light of Evolution."*
*Theodosius Dobzhansky (1973)*

Life evolved from combinations of chemical interactions unique to the Earth´s cooling surface. Like the pathways that quantum, chemical and astronomic symmetries followed to emerge from the cooling universe, the exact historical pathways these structures and their rule-sets followed to attain persistence remains unknown. Knowing them would require teasing out googols of broken symmetries. However, at some point quantum, chemical

[72] To some degree environments are noisy because environments are by nature not well defined. They are usually defined as whatever surrounds the system. They are not chosen by their continuities, like systems are, but by their localities.

and astronomic forces did combine to enable tiny matter-energy aggregates that not only maintained themselves, but also reproduced according to the rule-set created by the novelty of their own component and environmental relationships. This new rule-set is called biology.

In biology, matter-energy aggregates that transform according to the rules of biological symmetry are not called things, systems, or aggregates they are called organisms. Likewise, interaction between organisms is often called communication. Communication, like interaction, can be defined as the process by which the function of one organism alters the functions of other organisms.[73]

Communication is at the bottom of all biological symmetry, just as interaction is at the bottom of the other symmetry domains. It is the constantly fluctuating states of ubiquitous and varied interaction that maintain the spectrum of interaction from which all cooperative systems are selected. The chart of intermediary metabolism is the rule-set that constrains all biochemical communication.[74] For organisms it is the communication substrate. It is to organisms what the periodic table of elements is to molecular arrangements. Biology is based on molecular communication in the same sense that chemistry (molecular communication) is based on atomic communication.

In biological systems, constant, ubiquitous and varied communication creates the conditions for the selection of cooperative systems. Cooperation is ordered activity over time between two or more biological systems, often between different organisms. In the inanimate domains, cooperation is referred to as synchrony, and like synchrony it is measured as the duration of the ordered communication (i.e. order over time). Cooperation is synchrony. Cooperation is persistent, regular communication over time, which also means it can be seen as canalized communication. Canalized simply means that selection has maintained only regular structural transformation, and as such, canalization is a biological term for positive symmetry.

Furthermore, constant, ubiquitous and varied cooperation creates the conditions for the selection of reproductive systems. Reproduction is the organismic canalization of communication and cooperation pathways that

[73] For more see chapter "Interaction."

[74] See Morowitz (2002).

enable organisms to re-create the component relationships that characterize their own unified biological structure. This canalization is the chemical, developmental and behavioral processes that are selectively preserved (death to those who don´t reproduce). Reproduction is the most persistent process. It is the kind of biological transformation that preserves structure the longest.

Together, communication, cooperation and reproduction represent a nested hierarchy of biological transformation that reflects the universal hierarchy: interaction, synchrony and propagating symmetry. First of all, each process is a means of informational symmetry with the amount of preservation of structure (i.e. information) increasing up the hierarchy. Communication can be brief, simple and uncertain. But if the communication is reinforced by positive selective feedback then the duration of the regular interaction makes it cooperation. Cooperation is the result of canalized communication. As such, it is characterized by more extended, more complex and more regular communication routines.

Reproduction is the most highly conserved symmetry process of all. It is defined by information conservation. It is highly conserved cooperation. Reproduction is extended, complex[75] and regular. The communication within a reproductive system is highly canalized to ensure the exact communication and cooperation of those components that re-produce the biological system. Reproduction is communication and cooperation to an impressive degree of regularity. It is communication and cooperation to a persistent degree.

It is worthwhile to note that the amount of information (i.e. the reduction of uncertainty) goes up at every level. Communication maintains trivial amounts of information - limited regularities. Communication describes interaction that does not significantly reduce the uncertainty of the adjacent possible. It doesn't constrain future action. Cooperation maintains more information than communication -regularities within lifecycles. Cooperation describes interaction that reduces the uncertainty of the adjacent possible. It constrains future action. Reproduction maintains the most information

[75] In this context 'complex' refers to the common definition of 'more and different components.'

- regularities over generations. Reproduction describes interaction that persistently reduces the uncertainty of the adjacent possible. It permanently constrains future action by limiting interaction to what is re-creatable.

As we shall see, this process of communication selecting for cooperation selecting for reproduction occurs at all levels of biology: from organic chemistry to autocatalytic cycles to prokaryotic cells; from individual prokaryotic cells to symbiotic prokaryotic cells to eukaryotic cells; from eukaryotic cells to cooperative eukaryotic cells to multicellular metazoa; and from metazoa to cooperative metazoa to eusocial metazoa. Novel biological structure is created through the selective preservation of communicative relationships to make cooperative relationships and the selective preservation of cooperative relationships to make reproductive relationships. This is how biological structures found and maintained the major emergent transitions of life.

Specifically, every major transition named by John Maynard Smith and Eörs Szathmáry (1995) is either a result of or a side effect of selection for increased cooperation.[76] It is exceedingly difficult to think of any biological structures, physical or informational, that were not formed by the process of selection for increasing degrees of preservation of relationships (i.e. increasing degrees of symmetry). As Dobzhansky says in the quote at the beginning of this chapter, "Nothing in biology makes sense except in the light of evolution." If you think of evolution as the relentless selection of communicative relationships for more persistent cooperative relationships, and the relentless selection of cooperative relationships for more persistent reproductive relationships, then Dobzhansky was more right than he knew. He could have said 'the universe' in place of 'biology' in the quote above.

Communication *is* interaction, cooperation *is* synchrony and reproduction *is* propagating symmetry. Interaction is when one 'thing' alters the functions of other 'things.' Biologists call this communication when those things are organisms, but there is no substantive difference. Biologists often use the term 'communication' to mean a regular pattern of organisms altering the function of other organisms, but this is better described as cooperation. When interaction is regular it is synchronous. When communication is regular it is cooperation. When cooperation

[76] The following chapter will explore Smith and Szathmáry's findings in more detail.

is selected for even more regularity, it becomes reproduction. When synchrony is selected for persistence, it becomes propagating symmetry. The names for these processes can change in different disciplines, but the processes remain fixed in their relationships. Communication, cooperation and reproduction represent not only the same continuum of increasing regularity[77] that interaction, synchrony and propagating synchrony does, but also the same process for increasing that regularity - selection for persistence.

Again, communication is interaction. Cooperation is regular interaction over time. Cooperation is canalized, behavior-generated synchrony that benefits system maintenance (i.e. maintains positive feedback). When the system is biological, this maintenance takes the form of metabolism and reproduction. By these definitions, reproduction is just persistent cooperation and cooperation is just persistent communication. Another way to look at it is that communication is the processual substrate upon which cooperation works to maintain organic structure and cooperation is the processual substrate upon which reproduction works to maintain organismic structure.[78]

On all biological levels, communication selects for cooperation selects for reproduction. Figure 14 is the same cycle of emergence presented in the previous section, but using biological terms instead of abiotic terms.

[77] Regularity means fewer degrees of freedom or increasingly constrained interaction. It is closely related to Shannon information in that it also represents a reduction of uncertainty or, conversely, an increase in certainty. "Increasing regularity" = decreasing freedom = increasing constraint = increasing information.

[78] This could also be interpreted as interaction being the processual substrate upon which synchrony works to maintain systems and synchrony is the processual substrate upon which propagating symmetry works to maintain the system domains (i.e. quantum, chemical and astronomic 'species').

FIGURE 14: CYCLE OF BIOLOGICAL EMERGENCE

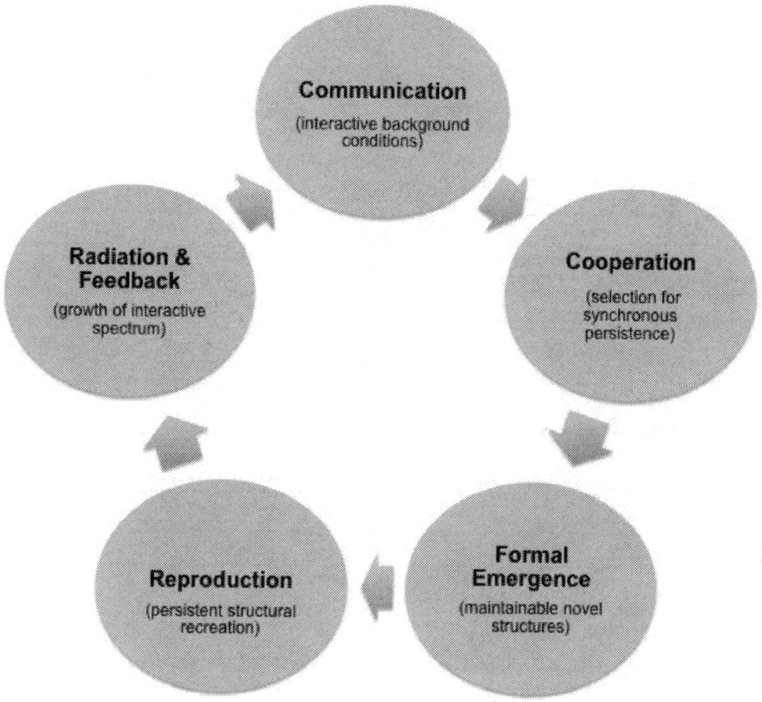


Source: author illustration

## "The Major Transitions in Evolution"

In 1995 John Maynard Smith and Eörs Szathmáry published a landmark book for understanding evolutionary history. This book outlines the major changes in living structures since the dawn of life on Earth. It is a concise description of the emergence of true novelty in the history of life.

Every day, every organism is transforming. We are in flux all the time. However, most of these changes are functionally trivial. They are minor alterations in our cycles of maintenance and reproduction. On a much larger scale, however, scientists are beginning to see how a very few select relationships between individuals have completely transformed the biosphere every few billion years or so.

Table 1 is a summary of Smith and Szathmáry's findings. What is fascinating about these results is that, with one possible exception, each saltation is a move from individuals to groups.[79]

**TABLE 1: MAJOR TRANSITIONS IN BIOLOGICAL HISTORY**

| Transition from | Transition to | Notes |
|---|---|---|
| Replicating molecules | Populations of molecules in compartments | Cannot observe |
| Independent – replicators (probably RNA) | Chromosomes | RNA world hypothesis |
| RNA as both genes and enzymes | DNA as genes; proteins as enzymes | |
| Prokaryotes | Eukaryotes | Cannot observe |
| Asexual clones | Sexual populations | Evolution of sex |
| Protists | Multicellular organisms – animals, plants, fungi | Evolution of multicellularity |
| Solitary individuals | Colonies with non-reproductive castes | Evolution of eusociality |
| Primate societies | Human societies with language enabling memes | Sociocultural evolution |

Source: Smith & Szathmáry (1995)

These saltations are also characterized by increasing information in the connectivity between individuals. Each saltation moved from individuals mildly constraining each other, to individuals constraining each other, to individuals highly constraining each other. Each saltation was from individuals to social groups to social groups that reproduce as a whole.

Replicating molecules became populations of replicating molecules in compartments. They did this by maintaining a novel catalytic language, or rule-set, that kept reproducing only those individuals that functioned together to make unified populations (e.g. DNA or RNA). Eventually this new chemical communication system canalized to the point where cooperative strings of autocatalytic communication could be consistently maintained through generations. These chromosomes were a new reproductive system,

[79] The obvious exception to this is the saltation from RNA to DNA. I would argue that this was a side effect or a result of the 'social' chemical environment of protocells and thus is an adaptation to sociality, if not a specific move to sociality. The move was from independent RNA molecules to DNA groups arranged in the twisted helix structure - chromosomes.

but only because they canalized molecular communication and cooperation to such a high degree.

Downstream we see the emergence of prokaryotes, asexual clones, protists and individual metazoa (including the solitary precursors to social insects and social primates). In each case the transition was from individuals to social groups. New levels of organisms emerge from a background of communicative interaction from which cooperative systems are selected until a novel reproductive language can canalize an emergent cooperative group as a functional unit of selection in the biosphere. Each cooperative group is preserved from what was interacting before. Each of them is the result of communication selecting for cooperation selecting for reproduction. This is selection for cooperative persistence, and it is the symmetry of life. It is also a fractal subset of universal symmetry (i.e. symmetric persistence).

Table 2 shows an example of the human descent line's major transitions as informed by Symmetry Theory:

**TABLE 2: MAJOR TRANSITIONS IN HUMAN DESCENT LINE**

| Individuals | Emergent Group Process | Emergent Group Structure |
|---|---|---|
| Molecules | Binary fission | Prokaryotic cells (groups of molecules) |
| Prokaryotic cells | Mitosis | Eukaryotic cells (groups of prokaryotes) |
| Eukaryotic cells | Meiosis | Multicellular organisms (groups of eukaryotes) |
| Primates | Symbolic language | Human society (groups of primates) |

There are two important features of this chart that differentiate it from Smith's and Szathmáry's. The first is the focus on the move from individuals to groups clearly showing that selection for groups is characteristic of major transitions. Although Smith and Szathmáry (1995) acknowledged that reproductive groups were the result, they spent a lot of time searching for how this could be possible. They believed that, "transitions must be explained in terms of immediate selective advantage to individual replicators" (p. 8).

Apparently, they were reluctant to accept the fact that individual replicators are themselves reproductive groups and that long before biology started, quantum, chemical and astronomical structures evolved group structures that maintained and reproduced themselves.

The second important feature is the focus on reproductive processes. Emergent social structures seem to need emergent social processes to maintain themselves. Without the emergent process to reproduce these emergent structures, they would not exist. Smith and Szathmáry's comparison considers both emergent reproductive processes (like sexual meiosis) and emergent organisms (like fungi, plants and animals) as separate major transitions. For them some major transitions are new reproductive processes and others are new kinds of organisms. This indicates that they were either hesitant or had no clear idea of how to link the two. This investigation considers the emergent structure as the canalized result of a process of individual cooperation that was differentially selected for increasing synchrony until persistent synchrony became embodied in the structure. Things are regular relationships. Embodiment is persistent synchrony. Things emerge by the relationships between their components being selected for more and more regularity over time. Reproduction is the result of continuous selection for persistence. The canalization of the cooperative process *is* the emergent reproductive process that maintains the emergent structure. Thus, the specific emergent structure is both a result and a cause of the emergent process. Structure is process, and vice versa.

Table 1 above is highly simplified. In truth, each of these transitions was highly complex and thus is underdetermined in this chart. For example, the first two transitions are almost completely unknown. The presumed transition from RNA to DNA or from individual DNA/RNA units to the looped double helix structure are both unrepresented, even though both (if they did happen) are indicative of group selection and positive symmetry. Also, the last transition from primates to human society is, at this point, hypothetical. The middle transitions from prokaryotes to animals are all well established. They establish the trend that allows us to hypothesize about what to look for in the origin of cells three and a half billion years ago and in the origin of humanity a mere hundred thousand years ago. Alone, this trend of group selection in the history of life is at best intriguing. It points to heretofore unexplored terra nova in those recalcitrant problems of the origin of life and the origin of humanity. Combined, however, with the evidence from the evolution of the universe, this trend becomes an almost fourteen billion year process of consistent group selection.

Table 3 below shows the cosmological major transitions. Again, this chart is highly simplified. As with the "major transitions in biological history" chart, the "Individuals" column only mentions the main cooperative players. There are also a considerable number of unknowns in the early transitions. One surprising result this kind of symmetry group selection process suggests is the emergence of gravity. Cosmologists generally treat gravity as always present, not emergent.

**TABLE 3: MAJOR TRANSITIONS IN COSMOLOGICAL DESCENT LINE**

| Individuals | Emergent Group Process | Emergent Group Structure |
|---|---|---|
| Elementary particles | Strong nuclear force | Protons and neutrons (groups of elementary particles) |
| Protons and electrons | Electromagnetism | Atoms and molecules (groups of protons and electrons) |
| Atoms and molecules | Gravity | Stars, solar systems, galaxies (groups of atoms and molecules) |

This investigation knows of no good evidence that gravity as we know it exists in a dispersed universe with nothing bigger than beryllium atoms in it. Quantum gravity is currently a dream. Even if gravity ends up not fitting this model it is already treated as unique in so many ways that this case would be just another case of its unique status.[80] Most important, however, is that the history of the universe looks like the history of life. Emergent systems are composed of pre-existent structures entering into new positive symmetry and reproductive relationships.

[80] Gravity is considered 'unique' in inflationary theory where it temporarily changes during inflation. Gravity is also considered 'unique' in those quantum theories that deal with gravity (e.g. most grand unified and quantum gravity type theories).

Ironically, perhaps the most important implication of Smith's and Szathmáry's survey is exactly that which they denied and struggled so hard to explain; that all major transitions in the history of life are the result of *some kind* of group selection process. Group selection is not a popular concept in biology (and was far less so in 1995 when they published their book). However, those who disbelieve in group selection face the unsavory task of explaining why all major biological emergences reflect a transformation of individuals into unified groups.

The best answer to this puzzle is that group selection is very rare, most selection is not group selection, but when it happens, it can produce radical results. These radical results are the opening of vast new phase spaces in the biosphere. The fact is that selection for each of the major domains of abiotic structures also appears to be the result of group selection. Solar systems are cooperatively constrained molecular groups, molecules are cooperatively constrained atomic groups and atoms are cooperatively constrained quantum groups. Each emergent domain also has its own reproductive group process or rule-set or 'language of maintenance and reproduction' as per gravity, electromagnetism and the strong nuclear force.

## Toward a Universal Theory of Directionality

Symmetry Theory hypothesizes directionality. It claims that selection for symmetric persistence from expanding spectrums of interaction results in the growth of order. Order is the emergence of new regular relationships embodied in the systems that have evolved since the Big Bang. The claim is that the evolution of the universe can be measured by a single metric, symmetry. The growth of symmetry ultimately relates all different types of systems. Such a metric faces serious challenges if it is to be operationalized. How does one measure and test something like symmetry?

How do we measure and test Darwin's theory of evolution? This is a similar question. The classic answer is to examine all species and test the evidence to see if any of them did not have predecessors that varied and passed on this variation to select descendants. So far, all of the evidence points to the validity of evolution. But what exactly do scientists test when they compare the variation of amoeba over time to the variation of rattle snakes over time? Variables like size or complexity change in different ways with different organisms. How big is a bee? Do you have to consider the hive? How complex is the lancet liver fluke parasite (*Dicrocoelium dendriticum*) when you consider its multiple nefarious interactions in the lifecycles and behaviors of its various host species?[81] Even a straight count of the amount of DNA in different species becomes problematic when some frogs have more DNA than humans and recursive reading of the same chromosomal base pairs can change function in different somatic cells during ontogenetic development. There is no single universal variable that measures the origin of species through natural selection. What we do measure and test are many different variables that to different degrees, in different types of organisms, all point to the *process* of evolution.

Despite the obvious difficulty of operationalizing a process as complex as evolution, researchers still like to search for discrete variables that could be used to formalize complex dynamical transformations like those found in evolution. Recently Daniel W. McShea and Carl Simpson (2011) posed such a question in their article *The Miscellaneous Transitions in Evolution* asking, "What is it that is increasing over the history of life?" (p. 20). They conduct a survey of multiple theories on biological directionality and are particularly interested in operationalizing these theories so that a variable can be identified that can be measured and tested. The result of their survey is that, despite numerous attempts, no one viable candidate variable has been identified that captures biological directionality. McShea's and Simpson's ultimate concern is similar to the one presented here, 'What exactly is it that is increasing through these various major transitions?'

This investigation will not operationalize directionality into a single variable. This seems to be very much an ideal from a different type of paradigm than the one proposed here. Essentially, the argument made here

[81] Lancet liver fluke parasite larva are first picked up by snails that eat ruminant (often cow or sheep) feces. The snails react by encasing the juvenile larvae in cysts which they then shed in the grass. Ants then ingest the cysts when they use snail mucous trails as sources of moisture. One of the many juveniles then migrates to, and infects, the ant brain (i.e. ganglion) making the ant climb to the top of grass blades nightly until it is eaten by a passing ruminant.

dodges this requirement by showing that although the variables change, the process by which the variables invariably change, remains consistent. Thus, the variables may be expressed in terms like degrees of constraint, complexity (in whatever form), energy, information, autonomy from environment, cooperation or so on, but the result is incontrovertible. Every major transition demonstrably shows the addition of an emergent form of constrained release of energy that both maintains and reproduces an emergent structure. The operational variables involved in these emergent steps differ, but the results do not. The transition from protons to atoms involves the addition of electromagnetic constraints. The transition from bacteria to protist involves the addition of mitotic constraints. Clearly, each transition includes new types of systems in new types of relationships.

The rest of this chapter will be dedicated to discussing some of the potential variables in Symmetry Theory in the hopes that they can eventually be woven into a coherent whole. Although it is unlikely that we can reduce all of them to a single operational algorithm somewhat in the spirit of Newton's famous operationalization of force and Einstein's famous operationalization of space-time, it may be possible to sharpen our understanding of the underlying process in the spirit of Darwin's famous definition of natural selection.

The first thing to realize is that propagating symmetry occurs in a continuous spectrum of interaction. This means that all kinds of things happen and we should expect the natural world to present a wide array of interaction spanning from extinction (e.g. entropy) to emergence (e.g. propagating symmetry). Most transformation is somewhere between these two extremes. It would be as useless to look for a variable that increases between fish and reptiles as it would be to compare planets and moons. Both fish and reptiles are multicellular animals that reproduce through meiosis and within this category, or domain, there is a lot of transformation that is not major emergence. The same goes for planets and moons. There are billions of unique varieties, but all are large spheroid masses reproduced through gravity (solar and galactic cycles). We must be clear about what we are comparing. We are not looking to compare systems that share the same mode of maintenance and reproduction. So whatever variable(s) we are looking for should differentiate garden-variety transformation from major transformations.

We should also reiterate that novel interactions are the norm. Growing spectrums of interaction should produce a whole range of almost emergent systems before they discover an emergent system. For example, we should expect to find many forms of multicellularity (e.g. archean, bacterial and edicaran) that in many ways resemble the transition that did eventually result in metazoan multicellularity. Likewise, there should be many forms of prokaryotic symbiosis that both parallel and presage eukaryotic symbiogenesis. Likewise, there should be many forms of metazoan sociality that closely resemble the eventual result of human symbolic sociality (e.g. ants and naked mole rats). Questions like, 'Why didn't ants evolve minds?' or 'Why didn't this or that type of autocatalytic cycle evolve into the first cell?' are false comparisons. Many different (and ultimately less successful) experiments in similar processes should accompany any major transition. Nature had to try many different more or less successful combinations of similar variables before she discovered major emergent systems.

So we should start by recognizing that any candidate variable for what increases in major emergences is not likely to be exclusive to those emergences. Candidate variables can be expected to be present to differing degrees, in many different closely related forms. We should expect many possible variables to be evident to different degrees in systems that are not part of major transitions. So we find ourselves in the unenviable position of trying to identify common variables that identify uncommon results. As such, these variables may be something like the ingredients milk, eggs and flour. They are present in most of the pastries we eat, but it is the way they are combined, the process - or recipe - that makes their eventual result so interesting to us that the emergent pastry is immortalized in cookbooks.

The second thing to realize is that this investigation is not presenting a finished answer. This is just the beginning of, hopefully, a larger discussion. The ultimate goal would be to answer McShea's and Simpson's question of what variable is increasing in the major transitions and a measurable way to test the increase of this variable. However, this investigation proposes that it is a process, not a variable that we are looking to elucidate. As mentioned, the Theory of Evolution does not describe specific variables that increase as it proceeds (e.g. endothermy, feathers and flight). Rather, it is a rule-set that describes how different variables came to exist in general. The way that directionality in the symmetry process will be presented here is a rather hopeful suggestion that all of the variables below can eventually be united. The following is a short list of candidate variables to be discussed:

1) Size
2) Range of action
3) Reproductive process
4) Cycle times
5) Variation
6) Energy rate density
7) Information
8) Symmetry

Take size for example. Within the major domains size varies. Planets, suns and galaxies all vary in size. What does not vary is the increase in size in each of the major transitions. All astronomical systems are larger than the atomic-molecular systems that preceded them. This is necessarily so because astronomical systems are composed of atoms. They are atomic systems with an added level of gravitational relationships. The same goes for biological systems. Major transitions invariably increase the size of the biotic systems that constitute an emergent domain (e.g. autocatalytic, prokaryotic, eukaryotic, multicellular, symbolic). There are two possible exceptions, but both are invalid disproofs for completely different reasons.

The first false exception is the origin of life itself. Cells are not bigger than the solar system that preceded them. This exception to the rule is a false exception for the simple reason that cells did not emerge directly from the growth of the spectrum of astronomic systems. They emerged from the feedback of the growth of the astronomical domain into the chemical domain. Cells emerged from the growth of the Earth-bound chemical domain. Although it is true that the preexisting requirement for the growth of the chemical domain was the growth of the astronomic domain (the formation of many galaxies with enough solar systems to eventually find the stable push-pull, goldilocks chemical conditions that a planet like Earth presents), the actual domain from which life emerged was chemical. Chemical systems, like molecules, are smaller than even the most basic biotic autocatalytic systems. What this shows us is that while all major transitions evidence an increase in the size of the emergent system, the actual emergence event can happen within any of the pre-established domains. This is very reminiscent of fractal structure emergence.

The second false exception is the emergence of the symbolic system. Arguably, the symbolic system emerged in our brains and humans are no bigger than any of the other large mammals. Symmetry Theory however, suggests a different point of view. The point of view this investigation takes is that the symbolic system emerged in groups of social humans and only secondarily was it canalized in individual brains.[82] Essentially, our ancestors used symbols to organize group activities long before they themselves could think symbolically. Thinking symbolically was a result of feedback from group behavior into individual behavior. In a way, this is not unlike the feedback from the domain of astronomy into the domain of chemistry to enable the emergence of life. The evolutionary constraints on individuals within symbolically organized groups created the growth and selection conditions for the neural maintenance and reproduction of symbolic behavior.

In a similar fashion, the evolutionary constraints on molecules within Earth-bound chemical groups created the growth and selection conditions for the autocatalytic maintenance and reproduction of genetic behavior. This argument will be developed later. For now, comparing individual humans to other individual animals is a false comparison. The transition of individual meiotic animals to multi-individual symbolic groups should be compared to the transition from individual mitotic cells to multicellular meiotic animals

[82] In fact, the proposal here is that the unification of individual humans into consolidated social units is still incomplete. This is one of nature's many "sloppy," "quirky design" exaptation processes that Steven Jay Gould (1993) made famous. It is also the inspiration for my next book, co-authored with Charles John Castillo.

(e.g. not from individual mitotic cells to different mitotic cells). So the valid comparison here is not from animals to humans, but from animals to symbolically unified human groups.

Now let us look at range of action. Range of action is obviously related to size. By definition the limit to the range of action (i.e. the behavioral halo) of any unit of selection is the upper limit to the distance at which that unit of selection can maintain any regular relationship. If major transitions are selection for emergent symmetric relationships then the range of action of any potential component systems limits the size of the emergent unit of selection. For example, the strong nuclear force is operationally limited to extremely short distances, thus the systems that emerge as a result of it (e.g. protons, neutrons and nuclei) are limited to the sizes circumscribed by that range of action.

Like size, range of action increases with each major transition. Atoms can act at distances greater than quarks and protists act at distances greater than bacteria. The only exceptions are the previously mentioned cases of the emergence of life and the emergence of humanity, both of which were disqualified for reasons previously explained. This topic gets more complicated when one considers the process of unification and the blurred boundaries within the transitions from individual components to unified wholes. In brief, range of action appears to increase in all major transitions.

This brings us to reproductive process. From the Symmetry Theory point of view, the terms reproduction, re-creation and maintenance all refer to the same process of transformation-with-preservation-of-structure and they will be shortened here to just 'reproductive.' As mentioned, a reproductive process is the canalization of component cooperative forces into a cycle of maintenance and reproduction that ensures the persistence of a unified system. Reproduction is the embodiment of cooperation. Reproduction is symmetry embodied.

Symmetry Theory implies that the appropriate scale of comparison, and thus measurement, is between system domains defined by reproductive process. All systems emerged from the interactive environments of previous systems. The difference between normal transitions and major transitions (i.e. the transition that differentiates system domains) is the mode of reproduction. Mode of reproduction can be thought of as the way the symmetry structures of that level are maintained and reproduced, their cycle of persistence. More specifically, major transitions are characterized by distinct *additional* modes of reproduction. All systems maintained and reproduced by the strong nuclear force systems make up the quantum domain. All electromagnetic force systems make up the chemical domain (see Figure 8). They depend on strong nuclear systems like protons, so the electromagnetic reproductive process is an addition to the strong nuclear force reproductive process.

From this point of view the appropriate methodology for the classification of which systems belong to which domains is inclusion within the reproductive or maintenance cycle. The reproductive cycle in turn is defined by system size and range of action. All matter-energy that is included in a system's reproduction and/or maintenance cycle is part of that system. What can be included is constrained by the system's size and range of action. The process by which that matter-energy is regularly included is the reproductive process and all systems that share the same reproductive process constitute a domain. The "emergent reproductive processes" listed in Tables 2 and 3 in the previous chapter thus define the major transition domains.

This methodology places systems, within systems, within systems, and it is only the added reproductive process that differentiates emergent domains. For this reason emergent domains are named after the added reproductive force even though they include previous reproductive forces. Obviously this kind of taxonomy constructs overlapping domains in nested hierarchies where system definitions can change depending on the scale level you define them from. One gram of weight is as translatable into quantum terms as altruistic behavior is into chemical catalysis terms. This scale problem makes system classification relative to how that system is nested and from which level one starts the comparison. The resulting relativity of this methodology will be further discussed toward the end of this chapter. For now, it is important to note that increased size and range of action result in larger systems which include more matter-energy whose organization into cycles of reproduction and maintenance takes more time. So cycle times slow down. On average smaller systems from older domains preserve less structure so they transform faster.

There is also the possibility of adding variation to the list of variables that increase in major transitions. There are only two types of quark systems, neutrons and protons. There are about a hundred different types of atom in the periodic chart of elements. There are billions of types of molecules and even more types of astronomic objects. The biological descent line also evidences an increase in variation with bacteria being highly physiologically constrained and metazoa being highly physiologically labile. This trend should be related to larger systems with more components and longer ranges creating higher dimensional behavioral halos that in turn support extended phase spaces from which emergent domains can more easily be selected. This is basically the argument for the statistical increase in the spectrum of interaction from which novel persistent symmetries can be discovered, maintained and reproduced. At this point the operationalization of this viewpoint is too complex for a practical discussion. Some of these complications arise from the same problems we will presently find in operationalizing measurements of available energy and information. However, variation clearly has great potential as a measurable and testable variable that consistently increases in all major transitions.

Domains then are defined by groups of systems that share the same reproductive process. As a result, domains could be measured by the averages of their system sizes, ranges of action and cycle times. Each of these variables can be measured and compared in terms of averages between extant domains or as averages between the specific component systems and the specific emergent system in any one major transition. What is even more intriguing is the apparent link between the three measurements (e.g. size, range and cycle time) and the eventual possibility of unifying the three measurements into one.

Operationalizing a methodology that can measure and test the nested, relative and scale free nature of systems is problematic. Eric Chaisson's energy rate density metric is an interesting case of a variable that seems to be clearly physically defined. As a physicist, Chaisson seems to have gone to pains to operationalize this variable as the average amount of available energy the system processes per second per gram.

**TABLE 4: AVERAGE ENERGY RATE DENSITIES TABLE**

Average Energy Rate Densities

| System | Age (Gya) | $\Phi_m$ [erg/s/g] |
|---|---|---|
| Human society | 0 | 500,000 |
| Animals, generally | 0.5 | 40,000 |
| Plants, generally | 3 | 900 |
| Earth's geosphere | 4 | 75 |
| Sun | 5 | 2 |
| Milky Way | 12 | 0.5 |

Source: Chaisson (2010, p. 28)

Table 4 shows the average energy rate density of different kinds of systems as they step back in evolutionary time. What is interesting is that Chaisson identifies a variable that is both measurable and testable and seems to imply growth or directionality through time. By his metric, the older the system, the smaller average energy rate density it has. Read that in reverse and you get the generalization that newer systems show an increase in energy rate density. This increase is the basis for Chaisson's argument for the growth of complexity. Figure 15 is a Chaisson graphic that clearly reflects the directionality of energy rate density growth as measured from the ancient Milky Way Galaxy to modern human society.

**FIGURE 15: ENERGY RATE DENSITY GRAPH:**

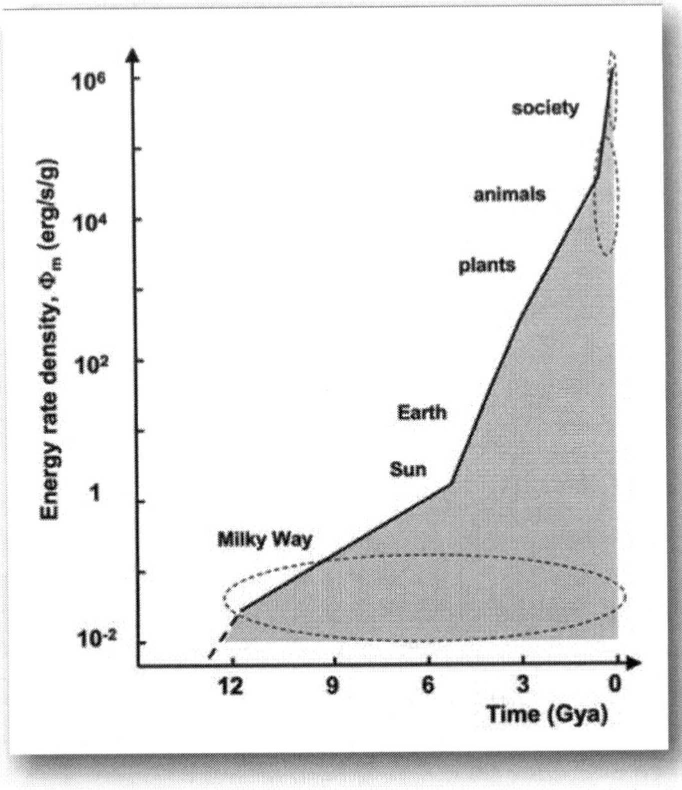


Source: Chaisson (2010, p. 28)

There are a few problems with Chaisson's metric though. The first is the rather unclear relationship of the metric to emergent systems. How exactly are increasing energy rate densities formed and then maintained? He does make an appeal to statistical emergence in the form of Layzer's comparison of potential entropy to actual entropy to conclude that, "the very expansion of the Universe, then, provides the environmental conditions needed to drive order from chaos; the process of cosmic evolution itself generates information" (Chaisson, 2001, p. 131). This formulation is very close to the argument made here: first that "environmental conditions" are central to the process, second that the "expansion of the universe" is central to the process, and third, the implication that there is a need for a spectrum of interaction between chaos and order. In Chaisson's and Layzer's case this spectrum is usually expressed in terms of the growth of order, information and/or complexity from available energy. However, the connection between the generation of energy available for work and how that energy was captured in the complex forms that did emerge seems to be insufficiently explained. Ultimately, Chaisson's argument may lack the philosophical grounding necessary to link the correlation of energy rate density growth to physical causality. Is energy rate density a correlation, or cause of emergent order?

The second problem with the energy rate density metric is the previously mentioned problem of defining it in terms of complexity, as it may or may not be definable at different scale levels. Basically, this is related to the problem of what exactly is internal and what exactly is external (see chapter "Environment and Organism: external and internal"). Since this topic was previously discussed, let us take a different tack here. How do we determine where the Milky Way stops and the rest of the universe begins? The Milky Way is constantly radiating matter-energy into, and receiving matter-energy from, the rest of the universe. Where and when do we draw the boundary between outside and inside? In any one case, a shrug and a practical 'summing over' of the issue is good enough to back pocket this problem of measurement. Unfortunately, when the whole set of transformations is presented together (as in Table 4 above), the compounding of various summing-overs can produce complications. One example is Chaisson's dubious calculation of the geosphere as just including the surface of the Earth and not the volume of the Earth itself. Although his method of drawing boundaries around what needs to be included in his measurements is, on the whole, agreeable; it lacks a consistent methodology and may be too arbitrary for standard scientific application.

Another related aspect of the complexity metric is the problem of scale. The gram unit seems both useful and sensible for classical physical systems, but it runs into problems at the quantum and astronomical scale levels. Molecules, atoms, protons and neutrons all contain far more energy (be it electromagnetic or strong nuclear force) per unit volume than any subsequent systems that evolved. Witness the nuclear bomb. It is only if one takes the relatively large measure of a gram, which would necessitate a rather large amount of empty space between the quanta packed into that gram, that the energy rate density can be seen to consistently increase. If you shrink the space inside the molecule down to the atoms as defined by their outer electron halo and their straight line force links to other atoms, then the molecule contains far more energy per *relative* unit volume than any realizable gram sized arrangement of molecules. This scale problem is intimately related to how one chooses to draw system boundaries. Presumably, one would run into similar problems trying to operationalize the measurement of galaxy clusters and their relationships.

These two problems just touch on the difficulties of making Chaisson's energy rate density metric a standard measure of the growth of order, or as he likes to call it, growth of complexity. The term 'complexity' itself is questionable for the previously stated reason that it can also mean simplicity if one chooses to measure the constraints on the degrees of freedom that components experience as a result of becoming parts of a unified system. The word complexity itself only sensibly applies to the comparison of pre-evolved systems to each other and not to the comparison of themselves to the environments that maintain, reproduce and will eventually become part of them in new emergent structures.

Unquestionably, Chaisson's energy rate density metric points to a consistent growth of energy use per unit volume over evolutionary time. The questions that remain are; is this metric causal or correlational to emergence, how do we identify the appropriate units to be measured and compared, and can this metric be unambiguously called a measurement of complexity?

Here is an example of how to combine Chaisson's energy rate density metric with Symmetry Theory. The appropriate scale of measurement for the directionality of universal evolution should be the energy rate density as calculated by averaging over the various structures that are maintained and reproduced within a reproductive domain. The comparison should not be between systems within the same domain, but between the different domain averages. By this metric the major transitions should be evidenced by a jump in the average size of domain systems and thus an average *loss* of energy use per unit volume. This loss is because the components of the emergent system are only as dense as the previous domain average, but now they are spread out over an increased range of action, a much larger unified whole. This new larger unit of selection necessarily contains a lot of unenergetic empty space within the extended range of action now included within its new boundary limits. So, as systems jump to bigger and slower domain levels, with bigger and slower reproductive processes, they should get less energetically dense. Initially at least, emergent systems in major transitions get less energetically dense

However, in the subsequent reproductive radiation of that domain there may be good reason to expect a within-domain *increase* in energy density. In Symmetry Theory, emergent structures are discovered stochastically. The chance that the initial structure will be optimally organized to maximize energy use is highly unlikely. It is significantly more likely that such initial structures will be stochastically discovered in their most simple form, which is probabilistically far removed in phase space from the complex forms that would optimize energy use for their own reproductive benefit. Thus, major structural transitions should typically be discovered near a Gouldian left wall of minimal energy density.[83] Although these emergent structures will be larger, wider ranging and on average less energetically dense than their component predecessors, they will also be constrained to evolve in the direction of more energetically efficient. New structures within this domain will eventually become more energetically dense on average as competition for resources within the domain will favor energetic efficiency. Examples are the emergence and radiation of the heavy elements, and the emergence and radiation of oxygen burning cells.

So, from a Symmetry Theory perspective reproductive domains are the appropriate units of comparison. This may or may not ultimately make Chaisson's energy rate metric easier to operationalize, but it should clarify the units of measurement problem and provide better predictive power. Instead of a an absolute increase in energy rate density over time the metric should be adjusted to measure the relative loss of energy rate density between emergent domains and to measure the absolute gain of energy rate density

[83] This refers to Gould's (1996) canonical "left wall of minimal complexity."

between subsequent systems within the domain. This break down would potentially allow us to include the abiotic systems that emerged previous to astronomical systems like the Milky Way as well as the fractal branching of the chemical domain into living systems here on Earth all within the same clearly defined metric. It may also give Chaisson's adjusted metric a more parsimonious theoretical foundation.

The topic of energy begs the discussion of information. This comparison is a bag of snakes. What we would like to know is can we essentially equate the growth of Chaisson's energy rate density over time to the growth of information? And if so, can we refer to the directionality of universal evolution in terms like the 'growth of complexity' as Chaisson and others do (Christian 2011; Spier 2010)? The short answer is no - not in its present form.

The first difference between energy and information is the degree of abstraction. Energy exists in all physical, real world systems. Information is our interpretation of how to count what we believe exists in real world systems.[84] Energy is the phenomena we count. Information is the method we use to count it. Information is like mathematics. It is a formula that we have invented that helps us describe any natural structure. However, real, natural energetic structures are inclusive. They include all levels of energetic difference. I am composed of my simultaneously interacting gravitational, electromagnetic and strong nuclear force differences. Conversely, information structures are exclusive. They exclude all levels of difference that are not explicitly counted as bits. To measure my information you would need to decide, to some arbitrary degree of fine graining, which bits made which differences on which changescale. Energetic structures can be thought of as the pattern itself and information as the formula we use to describe the pattern. As a result, matter-energy can always be frozen, separated and mathematically represented as information, but information can never fully represent matter-energy.

Additionally, information can represent a lack of energy. Previously we thought space was void. This void contained no measurable energy, yet it did contain information in the form of distances between planets or galaxies. This information was measurable, testable and actionable in the form

[84] This is a gross simplification, but a lot more explanation will not clarify what I hope is intuitively conveyed in this simple comparison.

of red shifts and gravitational action. Information can represent zero (0). Information can represent absence, energy cannot.

Energy and information also differ as per the conservation of energy and the non-conservation of information. The first law of thermodynamics is unequivocal, matter-energy cannot be destroyed or created; it can only transform.[85] Information, on the other hand is known to be destroyed all the time, as per chaos and entropy. The creation of information, in the form of selection for symmetric persistence from expanding spectrums of interaction, is the topic of this investigation. Energy is conserved. Information comes and goes.

This difference between energy and information is notable in the often referenced term 'available energy.' Available energy is the amount of energy that is displaced from equilibrium so that its inexorable entropic return to equilibrium (i.e. entropy) can be used to generate work. Basically, some energy just sits around and other energy works. The only way to create the distinction between the two is by playing with boundaries. You need an internal and an external and an energetic difference that makes this distinction meaningful in order to get available energy, or work energy.[86] It is this work energy that comes and goes in the form of information.

In fact, Shannon's original definition of information is mathematically based on Boltzmann's original formulation of entropy and the displacement of available energy. At a fundamental mathematical level, the calculation of information and work energy are both part of the same human artifice. By our definition, matter-energy states in thermodynamic equilibrium contain no differences. However, in this universe it is impossible to find an equilibrium state that does not have both micro internal and macro external differences. They all do. When we decide what is inside and outside we generalize nature's high dimensional spectrum of interactions to a low dimensional scale of bits and voids. Energy is nature's way of making differences that make a difference. Information is the back of the envelope way we have of representing those differences.

---

[85] Technically this law applies to closed systems, but it is rarely, if ever, considered inapplicable to the universe as a whole by working physicists.

[86] See Appendix 1 for more.

This whole bag of snakes is a scale problem. Where do real boundaries exist and how can we operationalize the measure of available energy in which situations and at what scale levels and to what level of fine graining, in order to be able to equate energy with information? Right now we cannot. Right now information is energy's poor cousin. Information can only be loosely correlated with available energy.

Chaisson seems to be on the right track with his energy-information-complexity metric. The universe began with no systems in it and ended up with billions and billions of them. In the words of Layzer (1975), "why has it [the universe] not settled into equilibrium?" (p. 68). If entropy were the only rule then we should all be in an equilibrium gas state right now. In a very real sense, new forms of ordered matter-energy have emerged. All of these emergent forms of order need work energy and information to maintain and reproduce themselves, and the growth of this order could be called complexity. But if information and energy cannot be equated, then the connection between Chaisson's energy metrics and terms like complexity, which canonically depend on measurements of information, become dubious. Appendix 1 goes into even more detail on this subject, but for now let us say that terms like 'growth of information' and 'growth of complexity' are problematic. At this time, such terms are inadequate to assert the directionality of universal evolution.

So, can such terms be useful in the eventual identification and formulation of 'What exactly is it that is increasing through these various major transitions?' This investigation argues that they can, if adjustments are made.

Symmetry Theory argues that energy, information and complexity must all be thought of in relative terms. Much like Einstein 'relativized' the world previously described in terms of absolute Newtonian mechanics, Symmetry Theory argues for a relativization of the world now described in terms of absolute measurements of energy, information and complexity. The key to this new metric of relativity is the scale free approach described in Appendix 1. The energy, information and complexity of systems can only be measured in the relationship systems have to their environments - in the relationship the determinate internalized systems of the past have to their statistical externalized emergent future. This investigation is a struggle to clarify the absolute scientific terms that describe the structures of the past and to transform them into a science of emergence whose terms can describe emergent processes of future transformation.

To operationalize a new metric that could capture the directionality of universal evolution we need to think in the parallel, relative, processual way that the universe itself is unfolding. Symmetry theory proposes to start the discussion of how to operationalize such a process with the following candidate guidelines. The growth of symmetry can be defined as emergence. Emergence is those rare major transitions of component systems into novel unified systems characterized by:

1) an increase in average system size that
2) increases average system range and
3) increases average system cycle times and
4) requires a new reproductive force.

Less concretely - depending on how one defines energy, information and system categories - there may be an overall increase in absolute energy rate density. However, Symmetry Theory predicts that if domains are defined by reproductive process, then testing major transitions will produce evidence of a decrease in *relative* energy rate density followed by a smaller yet distinct increase in *total* average energy rate density. As systems within a domain evolve, they get more energy efficient. Obviously information and variation both seem to increase in major emergent structures, but we need a new way to understand and operationalize the measurement of information and variation if we are to operationalize terms like complexity.

Symmetry Theory is a new way of metaphysically grounding what it calls the growth of order. Order is what David Layzer (1975) calls "macroscopic order" when he says that the, "evolution of the universe represents a growth of macroscopic order." Order is simply those systems that have emerged since the Big Bang. In the same way that Darwin did not propose to explain the actual origin of species - just the process by which species transform - Symmetry Theory does not propose to explain the origin of order itself, just the process by which systems transform to generate this order. Although the four characteristics listed above will be referred to throughout this investigation, they will not be given definitive formulation. Right now they are just guidelines for further discussion. As this investigation moves forward each characteristic will be referred to in a variety of different ways in the hopes that such epistemic flailing about can provide variation for the future selection of a robust (and hopefully expanded) theory of universal directionality.

## Symmetry Domains and Reproduction

> *"Working together, cells become colonies and colonies become individuals at ever higher levels of organization."*
>
> *Lynn Margulis (1998, p. 98)*

Quantum, chemical and astronomic symmetry domains all have signature characteristics. Each domain represents a slowing of transformation rates (i.e. cycle times) with chemical interactions occurring many orders of magnitude slower than quantum interactions and astronomic interactions occurring orders of magnitude slower than chemical transformations. Each also has signature rule-sets and product structures. Rule-sets, like dense plasma thermodynamics, and products, like black holes, are unique to astronomic symmetry - neither have analogues in either quantum or chemical symmetry domains.[87]

Terrestrial life is, to a great extent, a subset of the chemical symmetry domain. Like fractal designs, the closer we zoom in on the tiny subset of the spectrum of chemical symmetry called life, the more we find that it is ever more finely patterned by its own historical symmetry domains with their own signature characteristics. As with the universe, each life domain is characterized by unique rules of transformation, rates of transformation and product structures.

[87] Do they have different rule-sets or are they just affected by existing rules to different degrees? For example: either space-time isn´t warped (e.g. gravity) by tiny quanta or it is warped, but so slightly that it cannot measurably disrupt interaction at that level. This is the old quantity or quality problem that is only complicated by big data and measurement issues.

The first life domain to emerge was prokaryotic. The rules by which prokaryotic cells transform and preserve structure are the basic rules of the chart of intermediary metabolism and binary fission (i.e. prokaryotic reproduction). Binary fission occurs fast by biological standards and the products are constrained to the relatively small, un-nucleated, single celled organisms commonly referred to as bacteria.

The second domain to emerge was eukaryotic. The rules by which eukaryotic cells transform and preserve structure are the expanded rules of the chart of intermediary metabolism and mitosis. Mitosis includes the genetic principles of binary fission, but it is far more complicated and thus takes longer. The single cell products of mitosis are called eukaryotic protists and they are orders of magnitude larger and more complex than prokaryotes.

The third domain of life is multicellular. The rules by which multicellular organisms transform and preserve structure are the rules of the chart of intermediary metabolism and meiosis (sexual reproduction).[88] Meiosis is a version of mitosis that is highly complicated by the cooperative inclusion of two separate sets of DNA from two separate organisms. Meiosis happens very slowly compared to binary fission and mitosis, especially if one considers the duration of development from embryo to adult. The multicellular products of meiosis are the immensely complex plants, fungi and animals that give macro-structure to the Earth´s surface today. Like the emergence of all universal structures, each level of interaction is maintained by the differential preservation of added levels of constrained energy release. Each process of reproduction is constructed by adding levels of control and constraint to previous maintenance and reproductive processes.

As with universal symmetry domains, life began with small, fast, simple systems whose interactive products created larger, slower, more complex systems with novel rules of interaction.[89] Like the early expansion of the universe, processes and structures unfamiliar to us characterized the first two billion years of life-on-Earth. If we could travel back it would feel like an alien world operating on alien logic. The next billion to

88 Not all multicellular organisms reproduce through meiosis, but most do.

89 'Complex' here means comparing systems to other systems, not systems to their environments. See chapter "Environment and Organism: external and internal" and Appendix 1 for more on complexity.

billion and a half years would still feel alien, but familiar macrostructures that require extensive growth and development would just start to be appearing at the end of it. Only since the last billion to half billion years would we begin to see the characteristic structural and processual features that constitute the phenomenological and sensual environment we are currently adapted to.

## Inclusion: how symmetry propagates

> *"When we calculate the present entropy of the universe, we find it is staggeringly low; that is, we can conceive of distributing the forms of energy in the universe in ways that are far more disordered."*
>
> *John Barrow (1994, p. 27)*

> *"I find that at least one further phenomenon seems to be identifiable and either universal or remarkably common, namely, ordering (regularity or periodicity) in the time dimension"*
>
> *Nobel Laureate Phillip Anderson (1972)*

Selection for positive-symmetry-relationships creates systems whose rule-sets have boundary conditions. Under these conditions, foreign structures (e.g. free, unsynced systems) often enter these boundaries and become included in the rule-set interactions. Matter that enters a gravitational field is converted into characteristic celestial systems. Calcium that enters a metazoan (i.e. animal) is converted into bones and carapaces. Slowly, the symmetric interactions of persistent systems can collect unrelated structures from their ambient and synchronize them - that is, include them within their own system of symmetry. Another way to say this is that the reactive face of emergent systems presents new opportunities for regularities to form between it and other systems in its environment. This is inclusion.

Inclusion is the flip side to entropy, and like entropy, inclusion happens in all symmetry domains. Protons attract and collect neutrons to form nuclei. Nuclei capture electrons to create atoms. Atoms couple to other atoms in bonded molecules that further collect other free atoms and so on up the line. Inclusion is really just a more accessible word for positive symmetry. In biology, inclusion first took the character of metabolism. The first cells needed to eat to avoid death. They ate the chemicals they encountered in their environments. Most often the molecular structure of the cell wall would bond appropriate molecules and then envaginate them. Once inside, the molecules were systematically processed, incorporating the useful bits and jettisoning the waste. What was free floating chemical arrangements interacting noisily became included in the highly organized metabolic structure of organisms that grew and reproduced, incorporating more free chemicals along the way. Basically, external noise was captured, internalized and synchronized.

Waste is also produced in this process. Entropy is ever-present. However, for living systems to exist they must grow and reproduce more than they die. Extinct organisms do not exist. Growth and reproduction are almost by definition the inclusion of more matter-energy within living cycles. This is why positive symmetry is called 'positive' symmetry. Most of the universe is dead. The biosphere and its living components are in the very rare, but statistically necessary business of including more matter-energy within their cycles of maintenance and reproduction. Clearly, there are limits to growth. Life as we know it cannot exist in Earth's molten interior, nor can it exist in space (yet). Life has limits, but then all systems have boundaries. Most of the universe does not support solar systems or galaxies. There is not enough matter-energy at most locations. Most of the universe is empty space.[90] Organisms produce waste, but they must produce more order to persist.

As organismic boundaries expanded to include more and more differentiated components with specialized functions, this process of inclusion becomes more and more complex. Metazoa in particular have highly complex motility and sense organs that allow them to extend their

[90] We now know that space is not empty, it has unknown dark energy and unknown dark matter. However, we do not know what is in empty space. Whatever it is, it is empty of the kind of matter-energy we know the rest of the physical universe is made of.

abilities to find and capture ambient information and material. Every organism has a *sense halo* within which valuable environmental regularities can be recognized.[91] A sense halo is the combined limit of what an organism can sense. Organisms can also be said to have a *behavioral halo* which would be the extended limit of what is actionable for them, what they can do to capture (i.e. internally synchronize) or couple to (i.e. externally synchronize) material and information. These two concepts are intimately related in simple organisms, but become increasingly distinct in complex organisms who have highly differentiated and specialized modes of both sensing and capturing/coupling important saliencies. For most bacteria the two are combined in its cell wall that both recognizes and chemically binds valuable molecules. Protists can extend their sense halo with cilia and specialized chemical gradient receptors. The metazoan sense halo is the most extensive and variable, extending through numerous specialized senses to recognize cryptic environmental saliencies over vast distances.

In a more comprehensive sense, inclusion occurs on other levels of interaction too. Other than eating, organisms have a long history of developing social networks. These networks actually organize themselves to include more individuals within the behavioral logics of the group through learning and reproduction. In this case inclusion occurs over generations as information, in the form of successful behaviors, is coupled to and differentially preserved by internal mechanisms. This internalization of successful behaviors is variously referred to in biology as the Baldwin Effect, genetic assimilation and/or stereotypy (Dennett, 1991; Wilson E. O., 1975). Each of these terms refers to the differential survival of those phenotypes whose internal arrangements happen to better predispose it to perform certain adaptive behaviors. Over generations these adaptive behaviors, or "good tricks" as Daniel Dennett (1991) calls them, become physiologically fixed in the descent line as their phenotypes are selected for predisposition for the good trick. The good trick becomes internally conserved in the phenotype (i.e. hardwired). Ultimately, information

[91] See chapter "Adaptive Response and Moving into the Adjacent Possible" for a working definition of sense halo.

in the form of adaptive behavior coupled to environmental and/or metabolic triggers becomes included in the physiological regularities of the organism itself.

Social networks, be they organized horizontally through space or vertically through time (e.g. generations), naturally grow, differentiate and specialize like all organic systems do. They embody novel reactive faces that sample their environments. As mentioned, organisms are social networks, both of molecules and cells. By modifying each other's environment and thus their behaviors, organisms can include more and more 'others' within their social system. In a sense, social networks organize interaction through time and space on a metalevel. As Smith and Szathmáry's major transitions imply, it is from this metalevel (i.e. from this 'background') of social interaction that all major and truly novel life forms have been selected.

One thing to note is that over long periods of transformation environments themselves can become more ordered and more varied as the systems within them include more and more of what was external noise in their internal synchronies. One example is the Earth's surface. For four billion years the Earth's surface has been including extrinsic information and material both from space, in the form of meteors and sun energy, and from its core in the form of volcanic minerals and heat. For four billion years the Earth's surface has been including these external sources of information and material to propagate the symmetry of life.

Inclusion is a universal process, but it takes on a special significance in biology. Universal inclusion is to meet and bring within the system regularities - to create information or to increase symmetry. Biological inclusion is a special class of universal inclusion. It begins the same way, but the feedback process of natural selection differentiates and specializes the ways organisms can meet and include external saliencies at an extraordinary rate. Ultimately, these external saliencies can be material and/or information in the form of food, each other and/or descendants. On many levels of interaction, and thus at many time scales, inclusion propagates symmetry throughout the biosphere.

## Cooperative Boundaries: inner space

> *"The most important feature of metazoan cells for complex multicellular development may have been their ability to seal off the outside world and generate extracellular space internal to the organism."*
>
> *John Gerhart and Marc Kirschner (1997, p. 18)*

> *As soon as boundaries are defined sharply, elements must be attributed either to the system or to the environment. Yet relations between system and environment can exist. Thus a boundary separates elements, but not necessarily relations. It separates events, but lets causal effects pass through. ... Boundaries can be differentiated as specific mechanisms with the specific purpose of separating yet connecting.*
>
> *Niklas Luhmann (1995, p. 29)*

Another signature feature of the emergence of life-domains (e.g. prokaryotic, eukaryotic and multicellular domains) is the consistent extension of internal boundaries to include external regularities. This is related to the phenomena of inclusion and to the inversion of external environments to find and maintain internal self-referential cycles or systems (e.g. organisms) mentioned in the "Maintenance, Feedback and Reproduction" chapter.

In the larger scheme of things, boundaries can be thought of as those constraints that structure the exchange of energy (i.e. the transformation of matter-energy). What this means is that both the biosphere and the organisms within it are structured by the continual preservation of additional levels of constrained release of energy, of boundary building. What constrains the energy? New boundaries constrain them. In this sense boundaries are just novel rule-sets. They are the difference in how and what structural relationships are preserved between two systems. Boundaries denote the state differentials necessary for energy transformation. In a cooperative context, boundaries embody the transformations between rule-sets that both constrain and maintain the energy flow between two synchronous systems. The regularity of the interaction over boundary couplings is the language of cooperation within synchronized yet distinct systems.

Niklas Luhmann (1995) defines boundaries as those, "mechanisms with the specific purpose of separating yet connecting." In this context, rule-sets connect distinct systems (Luhmann's "elements"). A connection is a regular interaction, a relationship. By definition, the rule-set of regular interactions within one system are separate from the rule-set of regular interactions within another system. The difference between the two rule-sets forms the classical idea of a boundary, a separation. However, under the right conditions the two systems can also connect (i.e. form a regular interaction) over this boundary. In this case the boundary itself is the mediation of the cooperative interaction between the two separate system's rule-sets. It constrains the connectivity that is selected for persistence between the two systems and thereby defines the nature of the relationship, the emergent regularity the two as a coupled whole now represent to its environment. In this context, boundaries separate rule-sets, but they can also mediate the emergence of novel rule-sets. They 'separate' individuals 'yet connect' those individuals to make larger scale unified systems. They separate yet connect the internal to the external.

The original biological boundary is the cell wall that enclosed the chemical autocatalytic cycles. This cell wall mediated the cooperative coupling of those autocatalytic cycles to their chemical environment. Together the wall and the cycles evolved into prokaryotes.[92] The next emergence was the inclusion of external cells within prokaryotes to make symbiotic eukaryotes. After that came the enclosure of whole cell groups within an extended boundary, an epithelia. In each case the foreign system that is captured represents what Dennett (2003) calls a, "horizontal transmission of design" (p. 145). He continues with, "symbiotic visitors got a huge gift of *competence designed*

[92] Saltations always include *additions* of cycles of maintenance and reproduction. In each of the major transitions the emergent reproductive structure was larger than its components, resulting in an expanded 'inner space.' However, the steps taken to achieve this expanded inner space need not be additions in and of themselves. Prokaryotes had to *lose* their cell wall to gain cooperative connectivity. The reactive face can be expanded by physical additions or subtractions as long as the resulting process increases cooperative dimensionality.

*elsewhere*." In a sense, the resulting symbiotic system's power of interaction (read: reactive face) can be multiplied through cooperative union.

Organisms need energy input to maintain their metabolic steady state, that is, to maintain their displacement from death (i.e. thermodynamic equilibrium). From the very first proto-cell, cells that captured and internalized the structures in their ambient that could be chemically synchronized to their own metabolic systems survived longer. Everything needs to know what and how to eat. A variation of this happened in Margulis' Serial Endosymbiosis Theory when one prokaryote ate another and the first proto-eukaryotic cells were born. Another variation happened when a group of eukaryotic cells coupled to chemically synchronize their internal systems within a common epithelial boundary. Obviously, one of the common ways major complexifications of organismic structure happens is through cooperative ventures that synchronize separate systems within an expanded boundary.

One of the characteristics of these novel cooperative boundaries is their increased ability to differentiate and specialize the interactions between the external and internal ambient. Prokaryotic cell walls are notoriously limited in their range of reactivity to external stimuli. They have a limited behavioral halo. Protist boundaries have a far more varied range of reactivity to what they can recognize and how they can couple to external stimuli. They have a more extensive behavioral halo. Multicellular organisms have by far the most varied range of reactivity with highly specialized sense organs and reactive routines. They have the most extensive behavioral halos. Corresponding to the increasing external reactivity potential is the increasing potentialities of the internal metabolic systems to couple to and synchronize an expanding variety of incoming information and material. This topic will be further developed in the "Internal Systems" chapters.

New boundaries accompany emergent systems and emergent languages of interaction in all three of the domains of life. These new boundaries consistently extended the range of cooperation between the organism's internal metabolic structures and their external environment. The metazoan cycles of maintenance and reproduction include much more of their environment than protists do. They are connected to their environments in novel and expanded ways. Protists are connected to much more of their environments than their bacterial ancestors are. Emergent communal organisms present novel reactive faces. They embody expanded languages of synchrony. They embody expanded phase spaces of cooperation.

## Multicellular Differentiation and Specialization

The evolution of multicellular metazoa represents a basic change in biological transformation. Up until two billion years ago slowly evolving single cell generalists, bacteria, populated the world.[93] Up until seven hundred million years ago these bacteria shared the biosphere with slightly faster evolving single celled eukaryotes - protists. However, starting more or less with ediacaran type fauna seven hundred million years ago, wave after wave of adaptively radiating, rapidly evolving and rapidly going extinct multicellular specialists have permanently altered the pace of biological transformation.[94]

This pace has been facilitated by the growing spectrum of interaction between organisms. As such, one of the features of this new tempo of biological transformation is the vastly augmented roll that processes of differentiation and specialization play in the formation of inter- and intra-organismic relationships.[95]

Metazoa are aggregations of synchronized eukaryotic cells and eukaryotic cells are aggregations of synchronized prokaryotic cells. However, each instance of cooperative union involves the attendant emergence of novel component features that support the synchronization of the conjoined external and internal systems. Eukaryotes evolved plasma membranes fixed to microtubule cytoskeletons to compartmentalize their internal milieu. The separation of plasma compartments inside them allowed different metabolic processes to occur under different catalytic

[93] Protists (free living eukaryotic cells) are more complex than prokaryotes, but less than metazoa so a better characterization of slowly transforming prokaryotic generalists to quickly transforming multicellular metazoa would have an additional step in between.

[94] Metazoa have altered the rules by adding multicellular organisms to the biosphere. Bacterial and protist levels of organization still exist, but now within an expanded interactive background which includes highly plastic multicellular organisms.

[95] For a fascinating view of this expanded spectrum of inter-/intra-organism interaction see McMenamin & McMenamin (1994).

conditions at the same time. This differentiation of the internal milieu allowed eukaryotes to sample a greater range of specialized interactions for selection. Additionally, cilia and other specialized tubulin structures allow the eukaryotic cell to become far more complex in structuring its external relationships than prokaryotic cells. The eukaryotic cell presents a much more plastic reactive face to its environment than the prokaryotic cell. The plasticity of the eukaryotic reactive face translates into more variety being offered to the environment for selection. It translates into more and different kinds of interactions in the biosphere. Ultimately, it translates into eukaryotes evolving faster.[96]

Multicellular organisms are full of eukaryotic cells. The rate at which these cells can transform each other to make new phenotypes is revolutionary. Multicellular cells, phenotypes and species are differentiating every day at every level in every population and eco-niche in the world. Multicellular plasticity is a biological rubicon that, once passed in the Cambrian Explosion, changed how the business of biological symmetry is done in this world. In an analogous example, symbolic language will be examined as it expanded the phase space of human interaction with their environment. Symbolically synchronized aggregations of humans may have upped the rate of differentiation and specialization in the biosphere in much the same way that multicellularity did. However, this topic belongs to the "Human" and "Mind" sections.

These examples of increased rates of differentiation and specialization are in no way unique to biological interaction. When the sixty to sixty one elementary particles differentiated after the Big Bang, very few of them formed the specialized relationships that make atoms. Those atoms then differentiated into the elements of the periodic table and specialized in the molecular relationships that form classical matter arrangements. A very few of those molecular relationships then formed the specialized relationships that make organisms (e.g. carbon and a few of the other most common elements). However, in each case the emergence of

[96] Gerhart and Kirschner (1997, p. 9) refer to the speeding up of evolutionary transformation from prokaryotes to eukaryotes. Christopher Wills (1994) also refers to Alan Wilson's study of increasing brain size as it correlates to increased rates of differentiation and specialization in speciation. For more see chapter "Human Speciation" in Section 4.

new communal systems marked the opening of new phase spaces of adjacent possibles accompanied by increased rates of differentiation and specialization.

The process of differentiation and specialization is really the false dichotomy of structure and process mentioned in the "Processes and Structures" chapter. Ludwig Von Bertalanffy (1968) says, "the contrast between structure and process breaks down in the atom as well as in the living organism whose structure is at the same time the expression and the bearer of a continuous flow of matter and energy" (p. 248). Differentiation is a static measure of form. It is a snapshot of relative arrangement. Specialization, on the other hand, is a dynamic measure of transformation. It is the rules of interaction by which different structures change each other over time. Once again it is the chicken or the egg. Structure is a snapshot of process, state symmetry is a snapshot of transformational symmetry, and differentiation is a snapshot of specialization.

Multicellular metazoa are fantastic examples of how static structural change equals dynamic functional change. Internally, metazoa have taken the basic eukaryotic cell structure and differentiated it into highly specialized cooperative networks like the brain, the endocrine system and the immune system. All of these networks are dynamic processes, not static structures. Externally, social metazoa have evolved systems of differentiating the physical and behavioral structures of other group members to produce highly specialized social phenotypes like the queens, workers and soldiers of ant colonies. These dynamic social networks are the very definition of specialized yet they are composed of differentiated component structures.

Differentiation and specialization is a feature of the universal pattern of symmetry. It is part of the growing spectrum of interaction. Biologically, differentiation and specialization is a major feature of how and why Metazoa transformed the biosphere from slowly evolving simple microstructures to the ever changing, complex macrostructures that surround us. As life-on-Earth unfolds, differentiation and specialization seems to be expanding and speeding up.

## The Direction of Biological Novelty: genotypes and phenotypes

> *The important point to keep in mind is that such phenomena as the hormonal mediation of behavior, the ontogenetic development of behavior, and motivation, although sometimes treated in virtual isolation ... are really only sets of adaptations keyed to environmental changes of different durations. ... These organs have themselves evolved to serve the requirements of special multiple tracking systems possessed by particular species.*
>
> *E. O. Wilson (1975, p. 145)*

> *"A physical system that exhibits constraint is in that configuration due to extrinsic influences."*
>
> *Terrence Deacon (2007b, p. 3)*

Novel interactions can maintain novel structures. Novel structures persist if they feedback positively into the conditions that maintain the novel interactions that maintain them (i.e. reproduce them). This means that statistically, most persistent novel organic features are initiated by phenotypic interaction, not genotypic interaction.[97]

Biological interaction happens in two very different contexts, internal and external. The internal context is characterized by historically bounded processes of conservation. It is the cycles of maintenance and reproduction that bound the organism as a unit of selection. The metabolic system functions to conserve the organism in a steady state of energy transfer, maintaining itself for as long as possible despite constant entropic deterioration. The reproductive system functions to conserve the organizational state of the organismic type after entropic death. In the short term, the organism itself is conserved and in the long term,

[97] Ruth Garrett Milliken (1984) also argues this point in her theory for the evolution of language. Environmental (external) interaction is key to producing emergent structure not only in the quantum, chemical and astronomic symmetry domains, but also in the biological and linguistic symmetry domains.

the organismic type is conserved. This is accomplished by limited and highly predictable cycles built from the chart of intermediary metabolism.[98] Internal interaction is canalized to conserve those regular relationships and arrangements that successfully functioned to maintain that organism´s structure in the past.

The external context is largely unbounded and thus is characterized by highly varied interactions. The external boundary of the organism functions both, to protect the conserved internal systems and, to differentiate useful patterns from the chaotic external soup of patterns.[99] The useful patterns are those that can be metabolically or reproductively synchronized to help maintain that organism, its group or its progeny. Once differentiated (i.e. recognized as distinct from other external patterns), the useful patterns can be synced in one of two ways: captured (i.e. internalized) or coupled to (i.e. cooperated with). In either case, external order, usually in the form of information or material, is synchronized by internal processes.

Take humans for example. We are all surrounded by an ambient, be it Manhattan or the Gobi desert. Within these ambients we recognize important saliencies like money or vultures circling. We use our senses to distinguish and then try to bring these saliencies into sync with ourselves. In the case of money, we extend our sense of probable futures to constrain our actions in such a way that leads to the accumulation of it. We then barter this money for houses, washing machines and healthy families. In this case 'sync' is the ordering of features of our now and future ambients to contribute to our greatest metabolic and reproductive advantage. Ultimately, the goal is long life and a healthy family. Naturally, degrees of failure to synchronize

[98] The term 'highly predictable' here refers to the kind of deterministic chaos that biological cycles usually follow. These cycles are unpredictable in detail, but very predictable in general. For example breathing and water drops from a tap will both occur with statistically predictable frequency yet each individual breath, like each drop, is deterministically unpredictable due to the amplification of tiny differences in previous breaths (i.e. chaos).

[99] See the upcoming chapter "Phenotypic Behavior" for more on the ectodermal specializations of brain and epithelia that constitute the 'external boundary.'

external patterns routinely result in early death and unprosperous offspring, whether these are genetic or memetic[100] offspring.[101]

What this means is that there is directionality to the emergence of biological structure - from external to internal. The rather chaotic external soup of patterns that comprise any organism's environment is reduced to ever more conserved internal patterns by selective levels of recognition, capture and coupling. As organisms became more differentiated and specialized, so did their levels of capture and synchronic conversion. In general, complex organisms can recognize more external patterns that can be synchronized in more ways to produce more specialized functions than simple organisms. In fact, by definition they *are* additional levels of specialization. They are additional levels of constrained release of energy.

The genotype is a good example of this. The genotype is the fundamental code by which the structural patterns of organisms are arranged. It is the biochemical substrate upon which different levels of external patterns trigger cascades of stereotypic chemical interaction that ultimately instigates the development of new organisms and new parts of organisms. For example, on the prokaryotic level the genotype is constrained at one cellular metabolic level. On the eukaryotic level the genotype is constrained at two levels: the intra-nuclear level (e.g. transcription) and the extra-nuclear level (e.g. translation). On the multicellular level the genotype is constrained at additionally external levels as per both somatic differentiation-and-specialization and germatic haploidy/diploidy where foreign DNA is introduced to reconstitute the entire genotype. As a substrate, the genotype has evolved to depend on increasingly extended and external levels of interaction to activate what, in binary fission was a simple, serial, line

[100] The meme is a concept coined by Richard Dawkins (1980) to identify a unit of coded information passed from mind to mind in much the same way as genes are passed from generation to generation. The term 'memetic offspring' refers to our desire to reproduce our name/ideas by creating socially memorable lives, careers, books, theories or works of art. In a sense genetic children are physical memories of our individual selves. They 'embody' us into the future. In another sense writing *War and Peace* is a social memory of Leo Tolstoy. He is reproduced in our collective consciousness. Leo Tolstoy lives on through both his genetic and his memetic offspring.

[101] This is not to say that internal interactions like cancer are not selective. This is only to say that *assuming internal maintenance;* it is external interactions that statistically end up selecting for survival.

transcription and in meiosis has become a long and dizzyingly complex, simultaneous parallel multilevel process of recursive transcription, translation and transduction.

As organisms complexified, they added levels of constraints for triggering and organizing these external signals - syncing them in new ways. As noted in the example above, what were *intra*-cellular triggering processes in prokaryotes became *inter*-membrane, *intra*-cellular processes in eukaryotes; these became *inter*-cellular, *inter*-membrane, *intra*-cellular triggering processes in protists; that became *inter*-organism, *inter*-cellular, *inter*-membrane, *intra*-cellular triggering processes in social animals. In each case the basic mechanisms of the ancestral arrangement are conserved and complexified by higher levels of organizational control (i.e. by higher levels of constrained release of energy). As we have seen, these more complex levels of organization are the canalized results of what were at one time external negotiations. They are the result of the process of communication canalizing into cooperation. Ultimately, these new levels of communication and cooperation end up being conserved by exactly the new process of reproduction they embody.

The genotype is the substrate upon which level after level of added external cooperation (i.e. synchronization) work. DNA is constrained at the cellular level. Cells are constrained at the hormonal or nervous system level (i.e. epigenetically). The endocrine and nervous systems are constrained at the behavioral level. The behavioral level is constrained by the features of the environment (e.g. social conspecifics, predators, disease vectors or other saliencies). Of all these levels of constrained release of energy, the genome is typically the last (i.e. oldest), slowest and most highly conserved *biological* level. It is the innermost conserved core of what has worked for that organism in the past.

The phenotype, on the other hand, can be thought of as the external expression of the organism. Where the genotype conserves the organism's past internally, the phenotype works with the organism's present and future externally. The phenotype is the incarnation of the timing and broken symmetries that triggered the genotype through multiple layers of organization to become the specific organism that lives, dies and reproduces according to conditions right now. It is the phenotype that novelly interacts with its ambient. It is the phenotype that either dies or reproduces, thereby leaving a record of its successful interactions in the genotypes of its progeny.

Another way to see the external to internal direction of biological emergence is through spectrums of interaction. The spectrum of interaction inside the organism is highly constrained by what has worked to maintain that organism and its ancestors. The spectrum of interaction outside the organism is massively expanded. Selection for emergent systems depends on sampling an environment of ubiquitous and *varied* interaction. External interaction (i.e. phenotypic interaction) occurs within an expanded (i.e. more varied thus noisier) spectrum of interaction than internal interaction. The result is that exploring the external environment for emergent synchronous relationships should be more fruitful. Truly novel cooperative relationships are more likely to be sampled from phenotypic interaction than from genotypic interaction.

Canonically, organismic novelty begins with genotypic mutation. However, is it more likely to discover novel synchronies within a small range of interaction or within a large range of interaction? Internal accidents in the form of mutations are constant because they need to provide a sustainable variety of interactions that can be selected to track changing external conditions. In this sense mutation is a necessary, but far from sufficient answer. In the same sense that a background of constant and varied interaction is necessary for selection for persistence to occur, genetic variation (i.e. mutation) is necessary for the selection of novel persistent organisms. This mutationally interactive background does not, however, create the novel relationships themselves. It merely presents the samples from which the successful versions can be selected. Genes record successful external interactions; they do not make them.

The organizational information necessary to sustain the energy flow that maintains and reproduces life is captured and synchronized from external relationships, and preserved in internal relationships. These internal levels of organization (i.e. relationships) are characterized by ever more historic and highly conserved behavioral and epigenetic processes, until the ultimate substrate, the genome is reached.

The genome is one of the oldest and most highly conserved of metabolic systems. As long as entropy and chaos exist, accidents will happen, but the genome is built to minimalize them. Most true novelty in biological structure comes from outside, from novel communication that leads to novel cooperation that results in novel maintenance and reproduction systems. It is a process of maximizing mutual advantage through differentiation and specialization that is preserved by cooperative feedback loops that can be canalized into highly conservative reproductive systems. The genome is at the end of this process, not at the beginning. It preserves what has worked. It isn´t creating what will work.

In the late eighties neuroscientist Antonio Damasio (1989) proposed a theory of cognition that sowed the seeds of change for both neuroscience and artificial intelligence. His convergence divergence zones (CDZ) framework proposed that cognition is not simply feed forward neural signaling, but is actually a complex process of simultaneous, globally coordinated neural patterns that are triggered by external homeostatic and environmental feedback. The last thirty years of neuroscientific work have largely supported Damasio's theory.[102] Jeff Hawkins, apparently unwittingly, popularized Damasio's idea under the rubric of the 'memory-prediction framework.' He calls the brain a pattern recognition machine that, "learns the structure of the world by basically learning what patterns flow through time together" (Hawkins, 2008, min. 34:20). They both seem to be arguing that by memorizing experience we embody that experience in our brains. But what about organisms that have no brains? Don't they also embody the successful experiences of their ancestors? Bacteria too preserve the functional autocatalytic interactions of their predecessors. Bodies are systems that preserve functional interaction with their environments. The brain is a very notable part of this process of internalization, but it is not the only one. Novel metabolic systems are conserved because they either maintain functional relationships with external saliencies (e.g. neutralizing invaders or replenishing necessary molecules) or reproduce those functional relationships in other organisms (e.g. their children or social partners).

[102] For more on Damasio's CDZ framework see chapter "Neural Digital Simplification: the CDZ framework."

Ultimately, the direction of biological novelty is from the external to the internal, from the ambient to the genome. The genome itself is just a highly organized part of its ambient. It is maintained only by being nested within successive boundaries of denser energy relationships. Like the tip a moving wave, the genome is defined by, and made of, the rest of the wave below it. The genome changes through space and time only as a result of its environments moving beneath it. It records the wave's movement; it doesn't cause it. In fact, at some point in the history of the genome its component molecules (i.e. adenine, cytosine, guanine and thymine) must have constituted each other's ambients. All structure begins with (i.e. is selected from) external interaction. The genome preserves what has worked. The phenotype preserves what is working and what will work.

Schrödinger (1944) asked, "How can the events in space and time which take place within the spatial boundary of a living organism be accounted for by physics and chemistry" (p. 1)? He believed we could use the physics and chemistry, "within the spatial boundary of a living organism," to explain "What is Life?" Fortunately, he was wrong. Environments are also crucial to explaining what life is.

Presumably the idea that emergent novelty comes from outside will be contentious. Many careers are predicated on the idea that what is inside a thing is exclusively what the thing is. Take brains for example. Neuroscientists are correct to emphasize that the brain is built by the expression of neuronal DNA as cells divide, differentiate and specialize. This DNA helps build our neurons and these neurons spontaneously begin to fire, generating highly conserved patterns of oscillation deep within the sub-cortical (i.e. old) parts of the brain. These oscillatory rhythms come standard in the 'set up' package for infants. They *seem* to be internally predetermined. However, there are two senses in which they are not exclusively internal. Neurons and brains do not come to exist in vacuums. Each neuron needs both its parent cell to create it, as well as its neighbors to help determine its predetermined rhythm of firing. The emergent information measured by those conserved neural oscillations is actually highly constrained by external cells (e.g. parental and/or neighboring cells) that are in turn dependent on further external constraints.

Children acquiring language are much the same. Many theorists correctly assert that language is to some degree internally predetermined (e.g. Chomsky, 1976; Pinker, 1995). For the entire history of humanity as we know it, language ability has been one of the most important abilities that determines how many of our progeny survive. Language ability should have preferentially survived. After thousands of generations, children will be born with some degree of predisposition for language. Brains have deep, highly conserved oscillating rhythms. The question is where do these patterns come from and how do they function? As we shall see, language comes from social behavior, from historically captured external relationships. There is no internal, preexisting pattern or élan vital. Emergent pattern is the capture and canalization of previously independent patterns into a reproductive cycle. All internal patterns are descended from canalized external patterns. Systems evolved from environments, not the other way around.

# Internal Systems: the guts

## Organisms as Systems

> *"Nature is nothing if not rhythmic, and its rhythms are many and varied."*
>
> *Ian Stewart (1995, p. 93)*

External interaction is crucial in considering the evolution of emergent phenomenon, like the human mind. In biology this external interaction is typically characterized as behavior. However, before the behavior of metazoa is addressed it is important to yet again emphasize that metazoa themselves are constituted of groups of cooperating organisms, cells. These cells mediate their collective interaction with external features through a boundary. This boundary allows their internal interactions to proceed without disruption and delineates the internal environment (i.e. boundary conditions or rule-sets) within which those cooperative arrangements are maintained and reproduced.

The internal milieu of animals is a dizzying array of interlocking cycles of interaction. When an animal functions normally, there is no interaction within it that is not part of oscillating, synchronized systems. Even sequences that happen only once in the animal´s lifecycle (e.g. haploidy/diploidy) must be repeated by their offspring at generation level cycles. Ultimately, to trace the patterns of biological development that lead up to human emergence it is important to understand some of these internal systematics that create, bind and organize the society of cells that constitutes us. These are the communication processes that selected for the preservation of those systems that came before us. They are also the internal processes and regularities that constitute us.

Like all systems, animals must persist to be anything. Collections of cells have two options for persistence. They can maintain themselves individually and go it alone or they can maintain themselves collectively and pool their resources. The problem with going it alone is that outside is often noisy, chaotic and disruptive. As a result, collectives of free living individuals organized by external forces are often amorphous and simple, like fungi or algal mats. Constant hap-hazardous change disrupts selection for and maintenance of complex relationships. Under these conditions communication cannot be easily canalized into cooperation.[103]

The other option is that collections of cooperative cells can come to persist by constraining external disruption. In this case, the collective is selected for within some kind of boundary that allows the regularities in cell interaction to increase and differentiate internal connectivity. The cells do this by positively feeding back into their own interactions (e.g. connections that improve maintenance and reproduction of the whole are differentially preserved). This feedback results in internal cycles of interaction. These cycles persist until, inevitably, entropy results in enough loss of order within the system for it to die. If the collection is to survive it must maintain itself despite constant entropy - despite constant statistical steps towards death. Eventually, only those groups of cells whose internal cycles include the re-creation of the cellular relationships that maintain global function will survive. Eventually, an organism's maintenance cycles must include a reproductive cycle if it is to persist as a unit of selection.

Maintenance and reproductive cycles are the first and primary functions of any organismic whole. Animal cells are even identified this way as either soma (i.e. body) or germa (i.e. reproductive) cells. When the first multicellular animals evolved they were basically cooperative groups of cells that shared canalized systems of maintenance and reproduction. As connectivity increased amongst the cells, size constraints were increased. Increased size and connectivity maintained a substrate of more components, interacting

[103] For this reason self-organized criticality, as described by Per Bak, is characterized by environments or conditions within which interesting relationships occur. Gasses are too interactive while solids aren´t interactive enough for life to have emerged. Liquids are the phase in which push-pull interaction can sample a range of interactive relationships from which persistent synchronies can be selected. Constant haphazardous change, as in gasses, is as unlikely a place to find emergent cooperative novelties as environments of low change, as in solids.

more variably, to select from. This 'growth of the interactive spectrum' process ultimately resulted in an expanded sampling of phase space and differentiation of cell types. With differentiation came specialized functions. These functionally specialized groups of internal cells are the regulatory systems of animals. They are all interconnected through the process of syncing external regularities to produce the work necessary to maintain and reproduce their own unified system of synced oscillations.

In animals like us, common internal systems are the cardiovascular system, the autoimmune system, the digestive system, the nervous system and the endocrine system. By any metric one could chose, each system oscillates in the sense that it varies repetitively about a central value. Also, each system´s oscillation is regularly cued and constrained by the others.

In a typical animal, all of these systems are coordinated to accomplish specialized functions.[104] The cardiovascular system captures air, isolates the useful elements and distributes them to cells in the body. The immune system identifies harmful intruders and eliminates them. The digestive system processes and distributes food energy. The nervous system coordinates the immediate muscular/mechanical capture and coupling routines necessary for behavioral as well as internal interaction (i.e. homeostasis). And finally, the endocrine system hormonally mediates growth and development and other longer cycle internal interactions. It also potentiates external interactions (behavior) through moods and emotions.

Again, each of these systems evolved by selectively conserving certain forms of exchange with environmental regularities. In fact, it was the interaction with these external regularities that determined which metabolic structures were represented in further generations. The problem of getting nutrients and oxygen to internal cells constrained the evolving structures of the various vascular and digestive systems in animals. All of these systems mediate external interactions. Even the endocrine system, which is traditionally associated with internal regulation, feeds back into external relationships. As Antonio Damasio (1999) is famous for pointing out, emotions potentiate behavioral routines like decision making. Or more to the point,

104 There are a few exceptions (e.g. vestiges like the human appendix), but they will not be considered here.

behavioral routines that are consistently successful are potentiated and canalized by emotional cycles.

Like all emergent social systems, multicellular animals started out simple.[105] Their environment contained the behaviorally simpler and smaller protists and bacteria, but nothing more complicated than themselves (e.g. nothing had a bigger phase space for complex communication/cooperation than them). As they radiated, differentiated, and specialized, their behaviors became more complex and distinct. When they began to constitute their own environment, their own behaviors became a major factor in the complexification of their environment. This continual increase in environmental complexity has driven the conservation of systems that mediate the complex interactions necessary to survive these environments.[106] Nervous systems developed as they did largely to mediate these complex external interactions. The endocrine system probably began as a simple inter-cellular chemical transduction system that was later exapted to potentiate global reactive routines. It is highly likely that the complexifying of the environmental background has driven the development of all the regulatory systems of animals. Had interaction remained stable amongst the first animals, they would have become more complex in size and/or behavior much more slowly.[107] Deep sea thermal vents often exhibit these kinds of 'arrested evolution' micro-environments.

What this all boils down to is that animals are constructed of interlocking internal systems whose principle functions are to maintain and reproduce themselves. Those animals better able to discern and make use of the various environmental features necessary to maintain and reproduce themselves are those that reproduce most often. Since animals themselves are often the most

---

[105] Lynn Margulis (1986) notes that, "all five kingdoms - bacteria, protists, fungi, plants and animals have multicellular members. But in the first four kingdoms - even among plants - the multicellular organisms have only minimal strands of communication between their body cells" (p. 169). The reference to 'animals' here is to Metazoa.

[106] For a good example see Parker (2003). By his account the Cambrian explosion of multicellular animals was fueled by the evolution of a new sense - sight.

[107] Organisms that start out at minimum size have no option statistically but to get larger as they diversify. There is no other direction for them to go in. However, this situation is distinct from one where organisms have reproductive incentives to get bigger and more complex. In the latter case, complexification is an active process, not passive like the former case. For more see Stephen J. Gould's (1996) beautiful book, *Full House*.

complex features in their own environments, they themselves are a principle feature driving their own complexity. Thus, ultimately, many of the most complicated features of animal internal systems (e.g. nervous and endocrine systems) function largely to mediate animals´ complex relationships with each other. In this way internal metabolism is intimately synced with the external behavior.

As we shall see in the next few sections, the human trick of internally representing the behavior of others through symbolic representation is an emergent way of preserving a more prolific sync between the internal and external. In this sense, language becomes just another emergent tool in a long history of evolutionary tools preserved to sync the internal to the external - the conserved past to the adaptive future. Language becomes just another cooperative medium in a long list of cooperative mediums that propagate symmetry.

## Internal Processes

> *With each level of organization (molecular, cellular, structural, functional) in a complex multicellular body there is an exponential increase in the amount of information necessary to control its construction and operation. In addition, the range of possible adaptive reactions also grows exponentially with the complexity of the organism, including the complexity of its genome.*
>
> *Terrence Deacon (1997, p. 107)*

Human interaction is animal interaction. To understand how we interact, animal interaction must be seen in light of what it is. It is the synchronizing of the external to the internal, and vice versa. To a great extent, animals are made of the relationships their ancestors captured and coupled from their environments. Many of those relationships were internalized or otherwise canalized through positive selective feedback. What was autocatalytic behavior four billion years ago is now cellular metabolism. What was cellular behavior a half billion years ago, is now animal metabolism. What is animal behavior now is candidate for novel emergent social organisms of the future.

However, before moving on to animal behavior, it is necessary to understand a little more about some of the processes that characterize these internal relationships. If internal patterns are constrained by, and composed of, ancient external patterns, a better understanding of the one should help illuminate the other, especially as we keep moving outwards to explain behavioral, social and human patterns.

Internal cycles are conservative because metabolic and reproductive maintenance is a highly complex process. Disruption of these processes is often fatal. As such, external and internal systems must evolve to mediate disruption so as to preserve function. For this reason internal maintenance and reproductive systems only select the forms animals take in an immediate and trivial sense. Statistically over time (i.e. historically) they preserve transformation, they do not initiate it. Maintenance systems keep the animal alive and reproducing only as long as the animal interacts successfully with its environment. It is adaptive interaction with the environment that selects which combinations of internal relationships are preserved.

Adaptive interaction is mediated at two levels, internal interaction (i.e. metabolism) and external interaction (i.e. behavior). In animals, internal systems are isolated from external systems (i.e. environment) by the physical boundary of the epithelia. Within the epithelia, systems can grow their own relational logics with minimal random external interference. In this way the internal is its own environment with its own spectrum of interaction. Outside the epithelia are the multitudes of environmental systems whose uneven distribution of slings, arrows and opportunities they must navigate. Like a well-designed house, the structure of the epithelia must let limited fresh air and friendlies in, while keeping rain, earthquakes and unfriendlies out. The history of which epithelia successfully mediated which slings and arrows, built the second level of interaction, behavior.

Behavior is the result of internal systems preferentially preserving successful boundary interactions. All animal senses were created this way. Feeling is a way for an animal to recognize good or bad interactions with external saliencies.[108] In animals like us, seeing, hearing and smelling are

[108] "Good" are those interactions that metabolically or reproductively maintain the organism. "Bad" are those that stop or interrupt organismic maintenance or reproduction.

specialized extensions of what was originally feeling in our protist ancestors. Essentially, these extensions serve to expand the boundary of successful interaction beyond our epithelia. They discern the external distinctions we need to sync or react adaptively to. Together these physical senses form a sense halo that surrounds the animal to the degree that those senses overlap to create a boundary limit within which salient features can be distinguished and reacted adaptively to. In fact, the sense halo is an extension of the epithelia. Ontogenetically both the brain and the epithelia develop from the same ectodermal cells and together they mediate the sense halo. Perhaps the brain's sense halo is the most important part of the epithelia. The relationship of the brain and epithelia to the sense halo will be explored in depth in the upcoming chapter "Phenotypic Behavior."

This idea of a 'sense halo' can be favorably compared to the concept of "niche" presented by Maturana & Varela (1980, p. 8). Maturana and Varela are also interested in showing that organisms are cycles of interactions and that these cycles of interactions are not just internal; these cycles also include external interactions. For them,

> Living systems are units of interactions; they exist in an ambience. From a purely biological point of view they cannot be understood independently of that part of the ambience with which they interact: the niche; nor can the niche be defined independently of the living system that specifies it (p. 9).

For Maturana and Varela, organisms *include* that part of the environment they need to maintain and reproduce themselves - what they need for autopoiesis. Maturana and Varela go on to say, "The nervous system *expands* the cognitive domain of the living system by making possible interactions with 'pure relations'" (p. 13, italics added). For animals like us the nervous system is our brain and it is the brain that "expands the cognitive domain ... by making possible interactions with 'pure relations'" (p. 13). Brains expand "pure relations" in the form of the senses. Animal sense halos are essentially extensions of their niches, of those external interactions that define, maintain and reproduce them. Their example of a cognitive pure relation is the sensory relationship between a cat's eye and a distant bird. The two are not in a physical relationship, but visually the cat can expand his interactive domain to include the bird in a cognitive, or "pure," relationship.

Sense halos are also comparable to Dawkins' (1999) 'extended phenotype.' In essence, Dawkins extends the influence of his all-important genome to include relationships beyond organism's bodies. He argues that the relationships that animals maintain with external elements in their environments are also part of what an animal is. Disregarding his arbitrary choice of genes as prime biological movers, his message is similar to Maturana's and Varela's; organisms are defined in part by their extended, external relationships. One set of key modalities that maintain these extended relationships are the senses. Senses extend what an animal can do in its environment. They expand the dimensionality of an animal's cycles of maintenance and reproduction. Senses expand what an animal *can be* in its environment. The extension of the human niche, or sense halo, through the development of an emergent symbolic sense will be explored in the upcoming "Human" and "Mind" sections.

Adaptive reaction to saliencies within the sense halo (i.e. behavior) is a prime constraint for internal differentiation and specialization. That is, of course internal systems have innate historical physio-chemical and mechanical constraints, but ultimately, the organization of how these systems constrain the flow of energy through the organism was preserved by successfully canalizing external interactions. The internal cycles of maintenance depend on external cycles of interaction not only immediately but also historically. Ultimately, the species phenotype itself emerged through canalized external relationships. Internal systems like the nervous system are largely constrained by their maintenance of adaptive external reactions.

Typically these adaptive reactions are initiated by localized nerve centers. In animals like us these centers are called brains. The brain is tuned to receive environmental information transduced into electro-chemical patterns. The senses are the brain's information channels that register differences in external patterns (e.g. dark/light, acid/alkaline, etc.). These distinctions are a tiny selection of the total distinctions that pattern the animal's environment. They are just 'niches' in the overall environment. This tiny 'sensorium' was selected historically, by the survival of those ancestors whose sensitivity to those specific distinctions helped them react adaptively to them. Part of the sense halo then, is the set of the brain´s external distinctions that contribute to the animal´s maintenance and reproduction. The brain, through its sense halo, is a key part of the animal´s immediate interface with its environment.

Which environmental features are recognized as important is first historically constrained by the kinds of information the senses are tuned to. Visual features are as unimportant to a shrew as echolocative features are to us.[109] The simplification of massive amounts of environmental information into sense channels is the first reductive step in the process of producing behavior. The second reductive step is the sorting of the channels into hierarchies of connectivity. Those senses that contribute greatly to adaptive reaction are highly represented in brain connectivity. Those that don´t, are not. Both of these processes serve to reduce the complexity of external information patterns. Both are also more or less hardwired in the brain.[110]

How the brain further reduces this information to produce behavior is complicated and unknown. That the information is further reduced is deduced by the limited computational power of the brain and by the unity of reactive routines. Even with the hardwired reduction of external information by the 'distinctions that can be sensed' and the sorting of the 'senses that are more important,' the brain must be potentially presented with a massive amount of information from the senses. It is more likely that this information goes through a filtering process that reduces it to a manageable number of distinctions (e.g. bits) than that they are simultaneously considered en masse. An important part of this filtering process may be limiting attention to one sense or another.[111] Evidence for this can be found in the orienting

109 Some species of shrew have rudimentary echolocative senses that help them navigate the dark tunnels they live in.

110 To different degrees, neural systems can be genetically determined, predisposed or potentiated (i.e. trained) to certain arrangements. In general, the older the part of the brain is (phylogenetically) the more genetically and epigenetically predetermined it is. "Hardwired" implies that either the neural system is more genetically-epigenetically predetermined than trained or it is trained very early by invariant saliencies (like gravity).

111 For more on the reduction of information in sensory pathway to the brain see the chapter "Neural Digital Simplification: the CDZ framework."

reaction animals apply when they stop and focus on one sense to the exclusion of the others (e.g. 'Shh! I hear something!').

Even though the details of brain processing are still unknown it is relevant here to mention a few features of how the brain hardware works. The first is that brains are neural networks. They are massively interconnected neuron networks within networks within networks. Some of these networks are physically defined (e.g. left, right hemispheres) and some are functionally defined (e.g. Wernicke´s area). Still other networks are informationally defined (e.g. patterns of electromagnetic firing like theta rhythms). One physical region can host various functional or informational events as well as one functional or informational event can be found in various physical regions. All areas are interconnected by chemo-electric signal pathways that host rhythmic patterns of neural firing. Brains generically generate rhythms of oscillation (i.e. communication patterns). This is like the automatic 'music' of the brain. As the neurons interact with each other, the body and their environment, external signals disrupt this music to sample and canalize more and more adaptive behaviors by introducing rhythmic constraints.

Neuroscientist Georgy Buzsaki (2006) says this about how the brain is calibrated through the interaction of internal and external rhythms:

> Adjusting internal connectivity and computations to the spatial and temporal metrics of the external world ... can be referred to as 'calibration.' ... 'Representation' of external reality is therefore a continual adjustment of the brain's self-generated patterns by outside influences, a process called 'experience' by psychologists. From the above perspective, therefore, the engineering term 'calibration' is synonymous with 'experience' (p. 11).

The brain generates "experience" by calibrating its historic generic patterns to its immediate external patterns. Over time, which signals constrain which patterns in which adaptive ways end up surviving preferentially. Through interactive experience, neurons learn to play the songs that constitute successful maintenance and reproduction in the body and environment they survive in.

Basically, brains function by inciting cascades of chemo-electric signals where single cells can incite up to tens of thousands of other cells in complex patterns of distribution. Typically, many cascades organized on many levels in both space and time, are coordinated to produce most brain functions.[112] These functions include helping to regulate (i.e. synchronize) internal systems and coordinating the capture or coupling of external saliencies, eventually bringing them into synchrony with metabolic and reproductive systems. Somehow all of these functions together incite adaptive behavior.

## Capture and Coupling

Capture and coupling is the behavioral analogue of inclusion or positive symmetry mentioned previously.[113] Matter-energy systems maintain certain patterns of relationships by definition. These systems present an ordered reactive face to whatever they interact with. This reactive face constrains the kinds of interactions possible. Positive symmetry is when matter-energy systems bring extraneous matter-energy systems into sync with their own organizational schemes and some percentage of the resulting interactions will be synchronic to some degree. It seems that areas of positive symmetry interactions often tend to feedback into further positive

[112] Interestingly, the space aspect of neural connectivity has long been recognized as important (in the sense of identifying the synaptic map of neural connectivity) whereas the time aspect has not. Only recently have neuroscientists like Georgy Buzsaki (2006) made some headway in establishing the rhythms of brainwave activity as being *constitutive* of neural activity. Symmetry Theory proposes that processual and informational relationships are the foundation of physical structure and thus it predicts that the rhythmic timing of neural relationships will become increasingly recognized as fundamental to thought in the future. The glacial recognition of the fundamental nature of timing or synchrony relationships was recently reflected in the belated realization that junk DNA is not 'junk.' Crucially, junk DNA orders ontological development in the time dimension. It is a snapshot of rhythm, of informational symmetry.

[113] The comparison between a planet collecting asteroids and an organism seeking food is described here as "analogous" which is being used in the evolutionary sense of 'performing the same function, but not sharing the same descent line.' Planets belong to a different domain than organisms. In the astronomical domain, gravity is necessary to maintain solar systems or galaxies or nucleosynthetic function. In an analogous fashion, seeking food is necessary to maintain cellular or organismic function. Both are the acts, or the behavior, of collecting from environments the matter-energy necessary to maintain themselves.

symmetry interactions bringing exogenous material into sync with their own endogenous organizational schemes. Planets collect asteroids. Molecules collect free atoms. Boney organisms collect free calcium. In a sense, persistence presumes this process of inclusion because entropy is inevitable. Only systems that constantly remake order can exist because disorder is pervasive. Another way to say this is that a prerequisite for existence is maintenance up to and including reproduction. Animals, however, make order at an unusually complex level compared to other universal systems. Animals actively seek what they need to collect in order to maintain themselves. To do this they evolve senses to recognize what they need and appendages to collect what they need. They also evolve internal regulatory systems to transform what is collected into usable energy arrangements. This is what capture is. It is the inclusion of external matter-energy into the organism's own system of constrained release of energy. Coupling is slightly more complicated.

Coupling is a less direct form of cooperation than capture. Instead of taking from the environment what is needed and changing it to internal specifications, coupling is the use of pattern produced outside the organism to regulate internal patterns. Where capture is the disruption of the external pattern to remake it as an internal pattern, coupling is the preservation and linking of both sources of pattern that causes novel regularities in one or both sources. Typically these regularities affect the metabolic or reproductive systems, but do not directly sustain them like captured oxygen and nutrients do. An example of coupling is seasonal adaptations. Both the reproductive and the metabolic systems of many animals fluctuate according to seasonal conditions in their home ranges. In this case, there is nothing physical to capture from the environment so the animal must couple its internal cycles to external conditions.[114]

Another interesting instance is evasion. Evasion is incited by the recognition of external regularities that will disrupt organismic maintenance. It is an internal reaction to a recognizable external danger. A rabbit sees a fox and freezes or flees. Contrary to common sense, this too is a form of cooperation

[114] It is fascinating to note that ecosystems themselves may contribute to stabilizing climates. The coupling can go both ways even with immense, abiotic systems. Critical, push-pull conditions seem to be a crucial criteria for existence.

in the sense that it is canalized communication. Sight incites freezing or flight every time.

It is important to realize too that even though capture and coupling are relationships between internal and external systems this doesn´t mean capture and coupling does not occur within animals. Animals themselves are made of around three hundred different cell types each of which can be seen as distinct systems (Gerhart & Kirschner, 1997, p. 588). Furthermore, there are the physically separate and distinct organal systems and the functionally distinct regulatory systems. Internally, animals are a multitude of different and coupled systems. Changes in any one of them are to some degree external to the others. Novel couplings and capture (e.g. absorption of non-functional cells) are common internal events. To a great extent internal differentiation and specialization occurs in just this way, novel communication becoming canalized into novel cooperative relationships through positive feedback.

## Regulatory Linkage

> *"Organisms that could enrich their extracellular signals could further enrich their evolutionary options."*
>
> *John Gerhart and Marc Kirschner (1997, p. 595)*

This quotation from John Gerhart and Marc Kirschner´s textbook *Cells, Embryos and Evolution* reintroduces one of the most important features of internal relationships - communication. Most of what follows is drawn from Gerhart's and Kirschner´s excellent and comprehensive textbook. Extracellular signals are the body´s communicative network. They are the substrate for the coordination of all the internal homeostatic systems. In fact, they themselves are built on the even more ancient substrate of intra-cellular coordination. The coordination of intra-cellular events linked to extracellular events and extracellular events linked to further regulatory systems is called by Gerhart and Kirschner 'regulatory linkage.' Regulatory linkage is just what it sounds like it is, the communication pathways that coordinate the internal systems.

Regulatory linkage exists in different forms at all levels of internal systems. At the bottom there are intra-cellular processes like transcription and translation that transform DNA into proteins or the production of ATP by the mitochondria. At the top there are intersystem processes like the endocrine induced chemical mediation of neural processes (e.g. hormonal mediation of the nervous system). The mechanisms that facilitate these transformations are immensely varied, but most of them function by signal transduction. Signal transduction is an overarching term for, "the conversion of a signal from one form into another so that it can produce a response" (Gerhart & Kirschner, 1997, p. 90). Yet again symmetry, transformation with preservation of structure, rears its beautiful head.

Essentially, signal transduction is the language of internal cooperation. As such it occurs between distinct systems, this is just another way to say that transduction happens between different rule-set environments over distinct boundaries. Within animal bodies, these boundaries are often cell membranes and variations of epithelial boundaries. Loosely, what happens is that these boundaries function to both separate and selectively connect their internal cycles with others. These connections are between distinct systems that have their own organizational rhythm, code and/or language. For one system to coordinate with another, signals from one system need to produce a stereotypical reaction in the other. Coordination happens by converting the signals of one system into signals for the different system. This conversion not only typically occurs at a boundary, but often the boundary structure itself mediates this transformation (Gerhart & Kirschner, 1997, pp. 102-103). Once again structure is process.

One of the key features of how boundaries mediate signal transduction is by reducing the complexity of external information (i.e. from the sent signal) for the internal system (receiving system). This is common sense. Boundaries evolve to separate signal from noise in their environments. Boundaries don´t let everything in. They react selectively to specific external elements. They stop most and let a select few pass through. The signal that arrives internally is distinguished first by what it is not (everything else outside). This process deserves further discussion, but for now let it be noted that boundaries usually help produce the internal signal by reducing external noise. Once the signal has been transduced and internally received, it is often re-elaborated.[115] For example:

> In the end, the ion channels and their associated second messenger controls filter the complicated signals coming from the environment and produce a simple calcium signal that results in a stereotyped output, such as release of neurotransmitter in a nerve or contraction in a muscle (Gerhart & Kirschner, 1997, p. 99).

Such neurotransmissions and muscle contractions initiate highly coordinated and elaborate interactions on the other side of the communicative boundary.

Regulatory linkages are communication pathways internally mediated as signal transduction. Some of the key elements that characterize signal transduction as a language of cooperation are: distinct systems that communicate over boundaries, the reduction of information at those boundaries and the incitation of synced (i.e. cooperative) reactivity over this boundary. All of these characteristics will also be shown to characterize the evolution of human language.

## Exploratory Processes

The word 'constraint' is often used in a completely different context. Instead of being real structures that reduce the potential next reaction steps of energy/matter transformations to make control pathways, in a completely different sense constraints can be seen as the obstacles to evolving certain functions. This kind of constraint is not creative and it does not exist in real structures. It is simply a measure of difficulty for organisms to evolve certain functional capabilities (e.g. growing two hundred feet tall, breathing air not water, thinking symbolically, etc.). Richard Lewontin referred to this kind of constraint when he said, "evolution is best viewed as a history of organisms

[115] That interaction can go from simple to complex or complex to simple is not the point. The point is boundaries typically ensure minimal trespass. Once the boundary has been passed the signal can be elaborated.

finding devious routes around constraints" (Lewontin as cited in Gerhart & Kirschner, 1997, p. 595). What he implies here is that a main feature of evolution is finding creative solutions to very difficult problems.

It is normal for species to adjust the limits of their phase spaces by varying their phenotypes through reproduction. Often this process has been referred to as exploring their fitness landscapes.[116] New phenotypes have unique limits to their phase spaces and through these phenotypes the species, or population, searches for new ways to synchronize itself to new conditions (or synchronize new conditions to itself). These new conditions are by definition outside their traditional phase space. Traditionally, this search has been thought to be incremental. Each generation is selected for slightly different features (e.g. bigger beaks, more photosensitivity, more resistance to antibiotics etc.).

While this is certainly true, we have learned that the old slow and steady version of organismic change is not the whole story. Niles Eldridge and Stephen Jay Gould recently popularized the notion that in times of flux, species can, and often do, change very rapidly. Their idea is called 'punctuated equilibrium' and it highlights what paleontologists have been noticing since before Darwin, that fossil species seem to appear suddenly and then remain phenotypically stable for remarkably long periods of time until they abruptly disappear again to be suddenly replaced by others (Gould, 2002). We will return to punctuated equilibrium later, but for now this view emphasizes two points: that mechanisms that can help species change their structures rapidly will be preferentially selected and that it is the environment that initiates the selection for novel biological structure.

These points have been amply and eloquently shown by John Gerhart´s and Marc Kirschner´s descriptions of exploratory mechanisms. They describe the development of seeming impossibly complex vascular systems where the blood vessels seem to know where, when and how much to branch off to supply the correct amount of blood to every cell of any type of animal by implementing a surprisingly simple developmental process. In Gerhart & Kirschner (1997) words,

116 For more see chapter "Adaptive Response and Moving into the Adjacent Possible."

> [This process] is to generate without strict bias a large number of possible states and select the most appropriate one… In many biological systems, several, and often a large number, of alternative responses to external stimuli are in fact produced, and one is selected (p. 147).

In the case of vascular systems, blood vessels grow largely randomly[117] into tissue and it is the structure of cell division that develops some vessels and leaves others undeveloped. Gerhart and Kirschner call these processes, "exploratory processes that employ continuous functional feedback," and identify them as ubiquitous in biological systems.

These exploratory processes have developed using completely independent substrates. Cells use filaments and microtubules, immune systems use antigens, nervous systems use neurons, brains use brain-body nervous system networks, bees use waggle dances and ants use hormonal trails. In each case, environmental feedback trains the system to become functional at a level of detail far exceeding that which it could differentiate and specialize itself (Gerhart & Kirschner, 1997).

What these exploratory processes do is allow organisms to be born with a simple form of plasticity that can be programmed or trained by their environments. In the case of vascular systems, the plasticity is built into a simple rule, something like, 'keep growing vascular connectivity wherever the blood flows the most.' And then the blood flow itself, as constrained by cell division, creates the complexity we see in vascular systems. Now the phenotype is free to vary because any of billions of body types can be equally well serviced by one simple developmental program.

Exploratory processes save the animal having to pre-prescribe the necessary information genetically. At the same time it allows them to achieve complex, flexible reactivity to their environments. In a sense, exploratory processes help both the organism and its species to explore their phase spaces. These exploratory processes expand organisms' adjacent possibles by allowing them to in-corporate (i.e. embody) cryptic regularities in their

[117] There is also a chemical bias called 'chemotaxis' which draws the capillaries into the tissue. For more detail see Gerhart & Kirschner (1997, p. 190).

environments. These regularities can be informational as well as physical. Organisms can develop functional reactions to physical arrangements (e.g. antigens, hormone trails, etc.). They can also develop functional reactions to informational arrangements (the flow of blood, the rhythm of synaptic firing, the vector of the sun relative to the hive, etc.).

This kind of expansion of the adjacent possible, or growth of the interactive spectrum, is what drives creative evolution. Evolvability is selected for. What Eldridge and Gould suggested was that we should be willing to see biological change (e.g. speciation in this case) as capable of very rapid change. What Gerhart and Kirschner (1997) suggest is that many internal and external biological systems have evolved to facilitate rapid changes by preserving certain mechanisms that allow them to explore their environment. These mechanisms are designed to change the animal as its environment changes. Like they say, "the capacity of an organism to respond to new conditions is ultimately limited by the diversity of its responses and the mechanisms of contingency it employs" (p. 146). Their survival depends on it.

If truth be told, exploratory processes are just natural selection internalized. Natural selection is also characterized by producing adaptive complexity by producing many organisms and then letting those that do not adapt die off. Exploratory processes will become especially important when the human brain and language start to be considered. As we shall see, exploratory processes are potentiated by external patterns that differentiate and specialize organisms' capacities to coordinate cooperative adaptive reactions to their environments to impressive degrees of novel complexity. Perhaps the greatest exploratory process is behavior itself.

# External Systems: the glory

## Universal Behavior

In the larger picture behavior is a system's reactive face. It is how that system interacts as a whole with its environment. In temporal order of emergence, systems in each domain behave (i.e. interact) at greater distances than their component systems. Chemical interactions behave at greater distances than quantum interactions and astronomic interactions behave at greater distances than chemical interactions. This consistent extension of symmetric interaction seems to be a design feature of, or constraint on, emerging domains. If all emerging domains are part of the same fractal process of symmetry unfolding, there is good reason to expect that over time emerging biological domains would also extend their fields of behavior.

In fact, this has already been mentioned. Eukaryotic cells have vastly greater sense and behavioral halos than prokaryotic cells. They can recognize and capture/couple a greater variety of salient features at greater distances than prokaryotes. Each major transition in life has further increased the interactive potential of this behavioral halo. The latest major expansion of the behavioral halo in the history of major transitions is human behavior. Humans behave uniquely and very extensively. We can imagine the Big Bang, over thirteen billion years ago, and we can use this information to cooperate to a degree that allows us to plan voyages to Mars decades in advance. Despite the fact that physiologically we are remarkably like other animals, in the biological world there is nothing out there that behaves to our extent.

What appears obvious is that the major transitions in organisms, and their expanded behavioral halos, are all part of the same process - communication selected for cooperation selected for reproduction. Even though the behavioral phase spaces opened up by these emergent communal organisms are unpredictable, the process that produces them is always the same. That is to say, even though individual behavior is always unique, the way to produce truly novel successful behavior has always been the same - increase cooperation. Successful emergent systems include new levels of matter-energy relations in their cycles of maintenance and reproduction.

Canonically, only biotic systems behave. However, abiotic systems can be seen to behave as well. The moon regularly 'pulls' our oceans. Neutrons 'hold on to' protons. From this point of view behavior becomes a scalar phenomenon with emergent behavior being contextualized by previous behavior. For this reason it is critical to take 'social' chemical behavior into account when considering cellular behavior. Likewise, social animal behavior must be taken into account when considering human behavior (think ethology, sociobiology and evolutionary psychology). The trajectory of this investigation is towards human behavior and human experience. To make sense of human behavior one must make sense of its context and its context extends back over thirteen billion years to the emergence of our body's most basic systems - quantum systems. In this sense behavior is a nested universal hierarchy. It is regular reactivity selected for expanded cycles of regular reactivity selected for ever more expanded cycles of regular reactivity. This nested universal hierarchy of behavior is the background for animal interaction, which is the background for *Homo sapiens* interaction, for human behavior.

## Phenotypic Behavior

> *The important point to keep in mind is that such phenomena as the hormonal mediation of behavior, the ontogenetic development of behavior, and motivation, although sometimes treated in virtual isolation ... are really only sets of adaptations keyed to environmental changes of different durations.*
>
> *E. O. Wilson (1975, p. 145)*

Metazoa almost invariably use meiosis to reproduce. They also invariably depend on the preexisting chart of intermediary metabolism (i.e. protist biochemistry) as the biochemical substrate for their more complex metabolisms.

Despite this simple biochemical base however, metazoans are highly differentiated and specialized organisms with many additional levels of controlled constraint of energy expression. The most external level of expression is behavior. As our ancestors evolved from chordates to vertebrates to mammals to anthropoids to hominids to hominins to us, they evidence a consistent increase in behavioral functionality. They consistently expanded their sensory and behavioral phase spaces. They consistently expanded what their cycles of maintenance and reproduction included. Many other phylogenies did not develop so - ours did.

Eukaryotes do more than prokaryotes and metazoa do more than eukaryotes. This four billion year process of behavioral extension is the substrate for human language and thought. It is the historical context in which language and mind evolution should make sense. However, to trace this transformation there must be some understanding of what behavior, in all its technicolour glory, is.

Biologist E. O. Wilson (1975) illustrated the levels of adaptive reactions that animals use to track their environments. The following chart (Figure 16) and quote are from his book *Sociobiology.*

**FIGURE 16: WILSON'S HIERARCHY OF BIOLOGICAL RESPONSES**

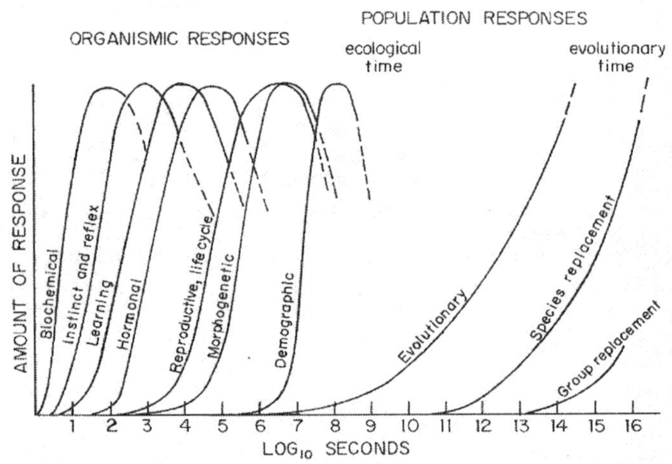


Source: Wilson E. O. (1975, p. 145)

> The full hierarchy of biological responses. Organismic responses are evoked by change in the environment detectable within a life span, population responses to long term trends. The hierarchy ascends with an increase in the response time; that is, any given response tends to alter the pattern of the faster responses. Beyond evolutionary responses are replacements of one species by another or even entire groups of related species by other such groups. The particular response curves shown here are imaginary (p. 145).

Wilson's "biological responses" are the various methods different organic units of selection use to capture and couple to environmental saliencies. One can divide this list in a few ways. The obvious division is between responses on the intra-organismic, the organismic and the extra-organismic levels. The intra-organismic level includes biochemical, hormonal, reproductive-life cycle, and morphogenetic behaviors. The organismic includes "instinct and reflex" and "learning" behaviors. The extra-organismic level includes behaviors at the population or species level like demographic, evolutionary, species replacement and group replacement responses. The intra- and extra- organismic levels of transformation are important, but will not be focused on in this chapter. In this chapter 'behavior' will be treated as adaptive reaction as it exists at the organismic, or phenotypic, level.

Wilson organizes his list of responses by their cycle times. Biochemical adaptive reactions can happen in less than one second while instinct and reflex responses generally take longer, and morphogenetic responses occur over the whole life cycle, from infancy to senescence. This is a good way to organize the various levels of adaptive reaction, but it avoids any kind of useful definition of what we commonly refer to as behavior. Is physiological maturation brought on by hormone expression really 'behavior'? It certainly plays a part in how we behave. Sexually mature adults behave differently than adolescents and infants. However, there seems to be something about behavior that is different from what occurs within our bodies.

Behavior in the *common* sense is a phenotypic expression. It is how the whole organism relates to its environment. It is not how parts of the organism relate to each other even though these parts working together help maintain behavior. Phenotypic behavior describes the external relationships the organism has to its environment, not its internal metabolic relationships. Wilson probably chose the phrase "biological responses" because behavior is usually limited to phenotype-environment relationships and he wanted to include behavior within a wider spectrum starting with biochemical relationships going up to social and species relationships. In this investigation behavior refers to the unit-of-selection's reactive face. In this specific case we are talking about the organism's reactive face. However, it must be acknowledged that this level of behavior exists as just one level within a range of "biological responses" extending from biochemical behavior up to group and species behavior. The phenotype is not the only unit of selection nor does it exist at the only level of selection within this natural range. For metazoa like us, our cells can exist as units of selection at a metabolic level while our social groups or even our species can exist as units of selection on greater changescales.

Although phenotypic behavior is just one level of biological response, it is the level that we know the most about. As such, it is worthwhile to take a look at how it is organized to get an idea of how intricate the interplay between different levels can be.

The first aspect of phenotypic behavior is internal and historic. It is the metabolic and reproductive level. This level of behavior is simply those internal systems that have helped the organism´s descent line respond to different environments successfully. Behavior itself is expressed in external relationships, but these relationships are supported by an internal and historical substrate of previously selected metabolic, ontogenetic and reproductive relationships. Each descent line canalizes those internal systems that predispose its members to successful external adaptive reactions. E. O. Wilson (1975) refers to this process of internalizing good tricks as "stereotypy - 'instinct' formation" (p. 156). The common use of the term 'instinctual behavior' hints at this milieu of conserved internal systems, but it mistakenly implies a unified system, an instinct system. In reality, the phenotype conserves these helpful mechanisms wherever, however and whenever it can, making them a gemisch of cobbled together molecular, cellular, organal and social interlocking systems that sample communicative interaction to select for cooperative couplings that can be further sampled to canalize successful morphogenetic and behavioral adaptations. In reality, the term 'instinctual' is a black box term, which indicates that we don't really understand how the genetic and epigenetic filters that constrain behavior work. They are the result of billions of generations of preferential selection for successful behaviors in billions of distinct (and now extinct) historical niches.

The complexity of these internal relationships is just starting to be recognized by scientists like Gerhart and Kirschner (1997), Deacon (1997) and Goodwin (1994). Ultimately, those internal relationships that are selected because they encourage adaptive behavior are preferentially conserved over generations. In a sense, the internal milieu is a history of captured and conserved predispositions through which external stimuli are filtered to prime future adaptive reactions.

Together, mutation and reproduction provide an ample substrate of variation. In fact, the history of major transitions shows that the emergence of each new cooperative unit was supported by the attendant emergence of its novel mode of reproduction. Novel modes of reproduction like mitosis and meiosis each introduced mechanisms for increased variation. Mitosis introduced a longer chain of transcription and translation which could incorporate more mutation as well as recursivity at the chromosomal level which allowed the same stretch of DNA to be reused for different developmental tasks. On top of this, meiosis introduced haploidy/diploidy. This process splits and recombines two completely different DNA codes. Selection for emergent mechanisms of variation has resulted in the increased pace of differentiation and specialization we see in each biological domain. This history of complexification has resulted in the extremely plastic metazoan group and its most behaviorally plastic of species, humans. Increased variation leads to different phenotypes. Those different phenotypes that allow for improved reactions to salient features feedback strongly by leaving more descendants. This process is largely responsible for the growth of the second aspect of behavior - the external aspect.

The second aspect of behavior is external and immediate. It is the various capture and coupling mechanisms that recognize those ambient patterns that can be co-opted to contribute to internal metabolic or reproductive ends in the here-and-now. This is the level that most people associate with behavior. Even though this behavior is seen as external, it is also intimately related to the history of how behavior has been genetically and epigenetically internalized throughout its evolutionary history. The internal aspect of behavior is historical and conserved in the phenotype's various specialized structures – in its physiology. The external aspect of behavior is more immediate and embodied in the phenotype's various external relationships.

The external level of behavior can be divided physically and functionally. Physically, an animal's reactive face ends at its epithelia. The epithelia is the physical boundary of the animal and thus it has evolved to negotiate proximate boundary interactions. Hair, eyes and skin all evolved to embody adaptive negotiations with the temperature, light and chemical features of our environments. Physical boundary specializations like hair can help maintain adaptive behaviors to temperature differences. Eyes maintain adaptive behaviors to light differences. Feet maintain adaptive behaviors to ground differences and tooth enamel maintains adaptive behaviors to food differences.

Behavior is not, however, limited to interactions that occur at the epithelial boundary. Functionally, behaviors can evolve that mediate interactions beyond the epithelia. The nervous system and its senses are good examples of the functional extension of adaptive boundary interactions. Originally, all the senses evolved from what we would probably characterize as feeling in our bacterial ancestors. Bacteria behave according to what touches their cell walls. Over the last four billion years, the bacterial descent line that resulted in humans has consistently extended its potential for recognizing valuable information at greater distances. This 'interacting at a distance' has resulted in the canalization of behaviors that have led to the inclusion of more and more extended information and material in the organism's cycles of maintenance and reproduction. Thus behavior not only happens at the physical boundary, but also at the extended functional boundary, the boundary of our sense and behavioral halos.

Although it is well known, it is rarely noted that both the physical and functional aspects of behavior stem from the same embryological source. In complex animal ontology, the fertilized egg (i.e. zygote) divides creating a multicellular blastula, which rather quickly (first week) differentiates into a gastrula. The gastrula is characterized by the differentiation of three unique cell groups, the endoderm, the mesoderm and the ectoderm. To grossly simplify a massively complex and largely mysterious process, the endoderm cells differentiate into the gastrointestinal and internal organs. The mesoderm cells differentiate into the blood, muscles and bones. The ectoderm cells differentiate into the skin, eyes and nervous system. Obviously, the endoderm and the mesoderm should be closely associated with internal differentiation and specialization. We would expect these internal levels to mediate most of the essential and historical aspects of behavior constraints. Specialized metabolic structures that mediate internalized ancient external negotiations like lung, vascular, immunological, gastrointestinal, genetic and endocrine systems would all be expected to be mediated at these internal, conserved core levels. These internal specializations develop to keep the organism alive like their ancestors did in previous environments.

The ectoderm, however, is involved in developing the specializations that explicitly mediate the organism's day-to-day external negotiations. Interestingly, in animals like us the ectoderm itself differentiates into two different developmental cell groups. The 'surface ectoderm' cells develop into the skin, hair, tooth enamel, eyes, etc. Obviously, these specializations correspond strongly to the proximate physical boundary of the organism. The 'neuroectoderm' cells develop into the brain and nervous system. Obviously, these specializations correspond strongly to the extended sensory, or functional, boundary of the organism, its extended sense halo.[118] What is interesting is that our brains and skin are both part of the same developmental system that specializes in negotiating external interactions. These external specializations learn to keep the organism alive in the current and future environment.

As mentioned in the "Cooperative Boundaries" chapter, it is the boundaries that mediate the cooperation between external and internal systems to maintain the organism. In animals like us the boundary that maintains this internal - external cooperation is mostly embodied by the two outermost specializations of our ectodermal cells, our various epithelial specializations and our brains. These two levels of external interaction are what most of us think of when we think of behavior. We either behave because of what our body surface sees, hears, feels, tastes and smells or we behave because of what our brains 'tell us' to do in response to what we see, hear, feel, taste and smell. The brain and nervous system may appear to be very different than our skin, eyes, and lips, but both developmentally and functionally these systems are different parts of the same external behavioral system. Both systems sync us behaviorally to our now and future, extended external environments.

The goal of this ectodermal system is to capture and couple the environmental saliencies necessary for survival. Its goal is to establish and maintain those external relationships necessary for survival. The ectoderm system itself has differentiated and specialized over the last half billion years to extend the sensitivity of its increasingly communal emergent organisms to their environments. This extension of the senses, and the brain that coordinates them, is the history of behavior. It is the history of added levels of inclusion in ever-extending cycles of maintenance and reproduction.

For the entire history of life-on-Earth, the single most successful strategy for persistence in an unknowable future is social behavior. Where many animals have evolved adaptations that select for inclusion of abiotic features like snow or moonlight, the most successful strategies

[118] This is not to say that the nervous system exclusively mediates the senses, just that behavior that is constrained by the senses is mostly mediated by the nervous system.

have behaviorally included other organisms. Where single descent lines seek to capture and internalize beneficial ambient patterns through differentiation and specialization of their phenotypic components vertically through time, social organisms can further functionally specialize by differentiating and specializing each social member horizontally through space. Social groups can contain specialist hunters, workers or reproducers that provide an extra level of adaptive reaction to the capture of beneficial ambient patterns. Through social cooperation, these groups can extend the reactive faces of many individual members to maintain a communal reactive face to the group's environment. This extended level of social behavior serves to enable complex relationships above and beyond those that individual organisms and their descent lines are capable of. Each of the major domains of life (e.g. prokaryotes, eukaryotes, and multicellular organisms) are descended from these kinds of socialized ancestors.

Phenotypic behavior is the relationship of the whole phenotype to its ambient. It can be broken down into two basic aspects. The first aspect is the internal structures that preserve the constraint constructs that maintain what we refer to as instinctual behavior. Simple animals often behave very little and what little they do behave can be characterized as historically constrained 'automatic' surface reactivity. The second aspect of behavior is the specialized external structures that negotiate the animal's daily interactions with its environment. This level is characterized by ectodermal specializations in the form of the brain and/or the epithelia. These specializations are especially prevalent in complex animals that have evolved not only complex sensory surfaces, but also complex brains that can interpret the long distance information these surfaces provide and maintain complex suites of behavioral contingencies.

Phenotypic behavior links the phenotype to its environment by synchronizing internal historical systems to external future systems, or vice versa. The evolution of behavior in general is characterized by the extension (both in space and time) of adaptive reaction capacities over the entire history of life. In the following chapters the word 'behavior' will refer to the common idea of phenotypic behavior unless otherwise indicated.

## Synchronizing Behavior

Behavior is an adaptive reaction. Behavior is adaptive because ultimately it feeds back either into the maintenance of the organism or the reproduction of the organismic type. Behavior that doesn't feedback into either metabolic or reproductive benefit dies out (or needs highly persistent and unlikely sets of conditions to maintain it). Behavior is a reaction because it is initiated by changes in immediate external or internal patterns.[119] Externally, behavior can be triggered by immediate sensory input that identifies environmental saliencies like drops in barometric pressure or nearby predators. Internally, behavior can be triggered by historically established metabolic, ontogenetic or endocrine patterns resulting in states of hunger, sexual maturation or motivational states (e.g. moods or affective states).

Defining behavior as an adaptive reaction divides behavior into two contexts, before and after. The *before* context of behavior is the unit of selection's adaptive history as embodied by those internal metabolic and reproductive systems that have been selected for survival over the last three point eight billion years. The *after* context is the unit of selection's immediate future as embodied in its reaction to whatever surrounds it at that moment. Thus, behavior is a term that applies to the external reactions of whatever unit of selection is interacting as constrained by its adaptive history. Since every unit of selection is itself composed of component systems, the behavioral reaction of the unit of selection in question is always at least partially supported by the historical behavior of its components. The relativity of what is inside and what is outside is what makes behavior hard to define in absolute terms.

For most animals their external environment is a chaotic flux of patterns.[120] In contrast, their internal environment is a highly regularized pattern. The boundary that separates these two environments serves to mediate the capture/coupling of those external patterns that can be synchronized (i.e. those patterns that are in the animal's physical, physiological or behavioral

119 In a universe in which change is the norm, a lack of change (i.e. stasis) in internal or external patterns can also be a change.

120 Parasites and symbionts are often exceptions to this rule.

adjacent possible). In animals this boundary is their ectoderm with its differentiated and specialized sense organs to recognize salient features (e.g. brains and senses) and their evolved specialized epithelial organs to capture/couple to those features (e.g. horns, eyes, appendages, mouths, noses, sex organs, etc.). Behavior is the external expression of the selective interaction of the external and internal environments through this highly specialized boundary.

In animals, behavior is highly varied. The sea squirt simply attaches itself and waits for food to come to it. It is limited to synchronizing the parts of its environment that happen to pass within its epithelia. The beaver, on the other hand, spends a lifetime building an elaborate environment for itself and its children. Through this behavior, the beaver essentially extends its boundary of controlled interaction. It takes a complex external pattern like a creek and regularizes it into a dam pattern that it and its descendants can couple to. It essentially synchronizes the ambient beyond its epithelia. It inverts an external, chaotically reactive pattern, into an internal, synergistic pattern. Instinctually, beavers create an extended internal environment by including an extended part of its environment within its cycles of maintenance and reproduction. Richard Dawkins (1999) called this the 'extended phenotype.'

Eagles can see for miles. Naked mole rats are virtually blind, but communicate through extensive social nets to let each other know about distant features. Both have evolved behaviors that allow them to capture and couple to more distant and more cryptic features of their environments. Eagles use the visible properties of distant saliencies to include these saliencies in their cycles of maintenance and reproduction. Naked mole rats use the social properties of distant relatives to include these relatives in their cycles of maintenance and reproduction.

Later we will examine the difference between the kind of vocal/aural adaptations that bats and whales evolved to survive in new environments and the human vocal/aural adaptation that evolved to help humans survive their own new environment. As we shall see, both bats and whales learned to sense distant saliencies by reading their audible properties, much like eagles do - visually. Humans learned to sense distant relatives by interpreting their audible properties, much like naked mole rats do - socially.

All of these evolved behaviors allow organisms to extend their sensory halos. All animals evolve to synchronize external patterns to their own metabolic and reproductive functions. One way or another they behave to bring external features into internal sync through adaptive reactions - through behavior.

## Adaptive Reaction and Moving into the Adjacent Possible

Behavior can be seen historically as an adaptive reaction that is maintained by generations of differential selection. It is like DNA in the sense that the transformational structures that both preserve are the result of selection through histories of interaction. However, day-to-day behavior is at the beginning of the cascade of conserving adaptive reaction. If behavior samples a new adaptive interaction, that interaction is first preserved in learned behavioral routines and only eventually, over generations, can it move into instinctually and epigenetically conserved cycles. Only in the end, if the interaction remains adaptive for a very long time, can the genome find a way to preferentially preserve a predisposition for that behavior in its DNA.

So behavior exists at the beginning of the conservative internalization process by sampling and maintaining those here-and-now interactions that maintain a persistent future. In the species case, behavior is evolving at generational cycle times and its effects are seen in population genetics and speciation/extinction events. In the individual case, behavior is evolving at biochemical up to reproductive cycle times and its effects are seen within the life cycle of the phenotype. Another way to see adaptive reaction is in relation to past and future - memory and prediction. Working from Damasio's CDZ model, Hawkins (2008) argues that brains are mechanisms that learn to produce adaptive reactions by, "learning what patterns flow through time together" (Hawkins, 2008, min. 34:20). ). By both Hawkins' and Damasio's accounts, brains conserve historically functional (i.e. synced) responses to extend future prediction capabilities. Brains conserve memories that inform prediction. As far as behavior is concerned, the convergence divergence zones (CDZ) framework could equally be called the memory-prediction framework or the adaptive reaction framework.

To understand the evolutionary history of the brain as it functions to produce adaptive reactions (i.e. predict the future) one has to understand that many of the same proteins that human neural synapses use also exist in brainless single celled protists. Even more importantly, these proteins perform largely the same function in unicellular eukaryotes that they do in human brains. They mediate behavior.

Individual yeast cells have no nervous system, no brain. However, roughly twenty five percent of the proteins that constitute vertebrate synaptic receptors have been found in these independent cells. Astoundingly, these 'brain' proteins also appear to mediate single celled behavior, or how yeast cells react to environmental saliencies.[121] The conclusion that molecular biologist Seth Grant draws is obvious, the biochemical mechanisms that produce adaptive reactions are far more ancient than brains or animals. Behavior is older than brains. These molecular mechanisms started generating behavior up to a billion years before multicellular animals evolved (Grant, 2008, min. 20:00). Since before multicellular animals existed, the same molecular systems have been working to extend memory induced prediction through sensory boundaries. This process can be called 'adaptive reaction' or 'behavior.' Essentially the history of the senses, memory and brains is far older than previously thought. In fact, the entire history of life is characterized by the extension of adaptive organic response through both time and space. Life is a record of increasing positive symmetry interactions.

To understand another aspect of how behavior works we must look at how individual organisms move into their adjacent possibles. A fitness landscape is a mathematical abstraction of a range of reproductive potentials. It is usually presented as a graphic plane where successful lifestyles (i.e. high reproductive success) are represented as peaks and unsuccessful lifestyles as valleys. As an abstraction, it can be applied to

[121] Seth Grant traces the chemical origins of behavior back to its single celled origins in this interview (Grant, 2008, min. 20:00).

different scales: to individuals within a range, to different populations within species or even to different species entirely. What is interesting about fitness landscapes is less their somewhat questionable ability to capture complex dynamics and more their ability to capture abstract relationships in general. Fitness landscapes are a great introductory conceptual tool for how to imagine behavior.

As an abstraction, fitness landscapes are to biological metrics what phase space is to mathematics and conjugational space is to physics. First, each abstraction attempts to bound all the possible states of some system. Second, they all attempt to capture a measure of potential change within that boundary. An organism is represented on a fitness landscape not to show how well it is reproducing, but how it can change its location to improve its reproductive potential. If the goal is to understand how organism´s move into the future, these kinds of conceptual tools are helpful.

To describe how organic systems move into their futures, it is useful to imagine them as points on this graphic landscape. From the point they occupy these organisms are surrounded by possibilities for change, but are limited in moving to these adjacent possible states by reaction steps. These reaction steps begin with the minimum actions organisms are capable of making and can scale up to extinction events depending on the coarse graining of the graph.

A reaction step is an abstract measurement of change along a natural continuum that realizes two requirements. First, state change must occur. Second, the new state must be distinct enough that it interacts differently than the original state. Thus, reaction steps are relative to the coarse graining of the environment the landscape itself is representing. This means that at some point a threshold is passed and a novel reaction becomes possible. An example is the canonical subliminal image embedded in a video. It can pass so quickly that the subject neither consciously nor subliminally recognizes it. No reaction is possible. No step has been made. It can pass slow enough that only the subliminal systems recognize it and the subject reacts to the image but doesn´t consciously know why. An unconscious step has been made. In the last case, the subject recognizes the image both subliminally and consciously and reacts to it on multiple levels. Both conscious and unconscious steps are made.

How organisms step into their adjacent possible is not only constrained by the conserved rules and thresholds of their physical structure, but also by the novel stimuli presented by their environments. The business of stepping into the adjacent possible is the business of using internally conserved past information to step into the present future adaptively. Animals do this by identifying and capturing or coupling salient features of their external environments. Animals like us move into their adjacent possibles by reaction steps. These reaction steps are constrained by what environmental features are present, what the animal can sense and how it can respond to what is sensed.

Environmental features are of two kinds, recognizable and unrecognizable. Those features that are recognizable are those features that the animal's ancestors evolved to recognize as salient for survival (e.g. for us these are objects that we can see, feel, hear, taste or smell - and those that someone told us about).[122] Those that are unrecognizable are present, but have no significant history of affecting the animal's survival.[123] They are invisible, or irrelevant, to it. The recognizable features make up the set of distinctions that animal can make about its environment. They are what the animal can sense. Obviously, all senses have limits. Put all the senses together at their maximum limits and the combined set of all distinctions an animal can make at a certain location and at any one moment is that animal's sense halo.[124] Environmental information from the sense halo enters the brain in as many different ways as there are senses.

---

[122] The comparison of symbolic communication to the other senses will be addressed in the next section.

[123] Features that are too large, too small or too distant to sense like ice ages, harmful bacteria and distant volcanoes do have a history of affecting our survival, but they are only 'recognizable' in an actionable sense by their telltale signs. We *feel* ice ages coming in increasingly severe winters. We vomit when we *taste, smell* or *see* something rotten and we move away from regions that we *see* and *feel* are darker, colder or otherwise ash damaged.

[124] Obviously, a further division of 'sense halo' into all those environmental features the animal is able to sense and those it does actually sense at any one minute would also be a useful distinction. Unless otherwise stated the phrase 'sense halo' will refer to the former case. It is a sensory phase space of all possible sensory distinctions an organism can make.

The brain has evolved to learn to sort the different signals coming in on the different channels and to produce a behavioral response that is adaptive. In this way, what is behavable is intimately related to what is sensable.

Animals move into their own futures by behaving, by taking all the incoming information patterns (e.g. sensory and internal) and choosing one outgoing pattern of phenotypic response (see the "Behavior Bottleneck" chapter for more on this). Like any physical system animals can only occupy one state at any one time and thus they can only step into one adjacent possible at a time. This serial behavior is adaptive if it contributes to the animal's survival (i.e. to its maintenance and reproduction). Animals that behave so as to maintain and reproduce themselves create a positive feedback cycle that allows them to explore their adjacent possible while preserving themselves and ultimately their species. Like all systems, animals are in the business of preserving structure while they transform.

Behavior is how the organism predicts its future by using its past. In animals like us, our brains mediate most behavior. Our brains have a genetically, epigenetically and environmentally conserved history of creating adaptive reactions by creating brains capable of being trained by the salient patterns in our here-and-now ambients. Thus, our adjacent possible is significantly constrained by what we recognize as salient in our environments and from this adjacent possible our brains select adaptive behaviors. This is how organisms like us step out of their adaptive pasts and into their adaptive futures.

## The Behavior Bottleneck

*"To behave is to choose one pattern among many."*

*Jose Delgado (as cited in Montagu, 1980, pp. 11-12)*

*"A broken symmetry is when one of many possible results becomes actual."*

*Camelo Castillo (see chapter "One Hot Mess")*

Behavior starts with external stimulus. Genes are external to base pairs. Cells are external to genes. Organs are external to cells. The ectoderm is external to the mesoderm. The sense halo is external to the epithelia, the environment is external to the organism, and etcetera. Each extended level represents an expanded space of regular reactivity composed of ever smaller, more historical and more constrained components. When we speak of phenotypic behavior, or the reactive face, we are considering the level of organismic reactivity within its environment. At this level there are the more conservative metabolically embodied internal 'instinctual' levels as well as the more creative 'brain and sensory' external levels. Where the conservative internal and the dynamic external meet is where we would expect selection for emergent synchronicities to be particularly rich. This is the area of the interactive spectrum where conservative pull meets dynamic push in that balance of criticality necessary for self-organization. Like the biosphere sandwiched in between a swirling galaxy and a molten heavy-metal core, the organism is sandwiched between a dense but conservative internal nature and a far less dense and more chaotic environment. Behavior starts with external stimulus because all systems emerged from external interaction, but the external is constrained by the internal at the reactive face. This chapter focuses on how the historic and internal behavior systems meet and sync chaotic external systems.

The internal constraints on behavior in animals have both ultimate and proximate sources. The proximate source of most behavior is usually recognized as the nervous system that triggers and coordinates specific adaptive reactions. However, all adaptive reactions are composed of a regress of proximate causes ending in the coding of their genome that preserves the deep history of their ancestor´s successful responses. Adaptive behavior is preferentially preserved by nervous system controls, but upstream to the nervous system is the endocrine system controls. As cycle times get longer and history counts back through learning and ontogenetic development, the process ends in the morphogenetic fields[125] that influenced how the organism's genes were expressed. The genes are not the oldest, but they

[125] The reference to "morphogenetic fields" is a direct reference to Goodwin (1994) and should not be interpreted in any other sense, e.g. in terms of Sheldrake (1981).

are the most obvious historical repository of what has functioned for that organism's ancestors.[126] Organisms like us with extensive nervous systems combine the stimuli entering our bodies from numerous sensory channels with an array of pre-evolved internal biochemical and biomechanical systems we have either learned or ontogenetically developed to constrain our behavior adaptively.

The main place where all this information is organized is the brain. To produce behavior, the brain takes the multiple external and internal information inputs and reduces them rather mysteriously, to a single output. Like a biological version of the Pauli exclusion principle, animals can only occupy one state at a time. This means that they are functionally limited to serial action. Although they are surrounded by legions of adjacent possible states they can occupy only one of those states at a time as they transform into the future.

Philosopher Daniel Dennett calls the reduction of multiple, simultaneous inputs to one serially coordinated output a "Von Neumann bottleneck." Dennett (1991) describes the Von Neumann bottleneck as, "the place where all activity of the system has to pass single-file through a narrow gap" (p. 214). He uses this computer metaphor to describe the serial nature of thought behavior, but it could just as easily be applied to all behavior. All phenotypic, that is 'whole body,' behavior is a series of single unified expressions of coordinated action. The neural process that supports this "unity of expression" is, as of now, poorly understood. However, it does seem clear that not only are whole organisms constrained to serial actions, but also, any attempt to engage in two or more behavioral routines simultaneously would cause the organism to experience conflicts between multiple bodily systems essentially freezing or muddling them at crucial evolutionary moments. Thus, animals are either hunting or defecating at any one time. They are walking or sitting at any one time. Parts of them are doing different

[126] There is a good argument to be made that the quantum and chemical symmetry patterns that exist as the substrate for DNA are even older ways of maintaining order that DNA depends on. Goodwin (1994) calls these deeper patterns "morphogenetic fields." McFadden (2000) is another fascinating exploration of behavior's deep patterns.

things all the time, but the whole, the phenotype itself, engages in a single series of states.[127]

Interestingly, the behavioral bottleneck can be seen as the boundary between two system contexts; the internal and the external context - the organismic and the environmental context - the past and the now context. As previously noted, boundaries typically constrain the information transferred between two complex systems. Boundaries adapt to link complex state changes in one system to complex state changes in the other system. In this case the information (i.e. reduction of uncertainty) the behavior bottleneck represents is similar to the information the epithelial and sense halo boundaries represent. The brain and the epithelia are two parts of the same outermost ectodermal boundary of an animal. In fact, the brain seems to be something like the headquarters of an extended virtual epithelial boundary that reduces incoming sensory information only to re-elaborate it on the other side - the inside.

Basically, the brain coordinates the collapse of a multiplicity of external and internal information channels into a single unified channel of behavior. These behavioral routines may serve multiple functions, but they remain a single series of coordinated states of being. We will return to this point when we discuss the mysterious 'unity of self' experience in upcoming chapters.[128] For now it is sufficient to note that all behavior is constrained to serial unified action.

## Social Behavior and the Delta Metaphor

All behavior can be seen as 'social' behavior in a general sense because, as Leibniz (1962) said, everything is acting on everything else all the time. True individual behavior would require no interactive environment and, thus, outside of considering the universe itself as an individual, individual behavior does not exist.

[127] Animals can lick their lips and walk and hunt at the same time, but these are all partial descriptions of what the entire body is doing at any one time. At any given moment the entire body can only be in one state at a time even though parts of it can be doing different things (e.g. lips and legs and 'intentional' neurons) simultaneously.

[128] See chapters "The Origin of Self" and "Language Simplification."

As the universe evolved, certain areas of propagating symmetry formed emergent social systems. Quanta formed social groups called atoms and atoms formed social groups called molecules. These social groups formed a nested hierarchy of social behavior where social behavior in the astronomic domain (e.g. gravitational behavior) contained within it social behavior in the chemical domain (e.g. electromagnetic behavior) that contained within it social behavior in the quantum domain (e.g. strong nuclear force behavior).

'Social' is traditionally defined as being applied to groups of humans or animals who exhibit some degree of relationship. This investigation denies that humans have a privileged status in this universe. 'Social' here is equated with synchronic interaction and cooperation. Any group, human or not, that shares some degree of regular relationship can be seen as social. Social is 'the process by which the function of one thing regularly alters the function of other things.' Social is synchrony. Social is cooperation.

In this universe there is a nested hierarchy of social systems that is organized through both space and time. In the *time* dimension the smallest, simplest and fastest transforming systems appeared first and the larger, more complex, slower systems appeared farther downstream (i.e. down the ever expanding evolutionary time stream). This same nested hierarchy of social systems can be organized through *space* as the smallest, oldest, simplest and fastest systems organize transformation at very short distances and the bigger, younger, more complex and slower systems organize transformation over ever expanding distances.

A nested hierarchy is often depicted by a pyramid metaphor with a base of many simple structures constructing ever-higher levels of fewer, more complicated structures.

**FIGURE 17: PYRAMID METAPHOR**

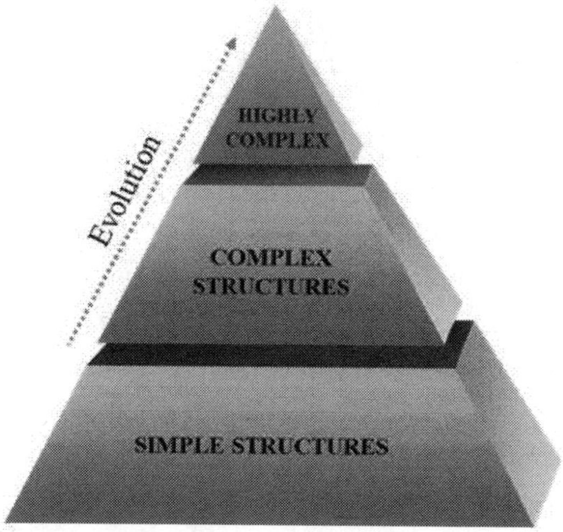


Source: author illustration

The problem with this way of picturing social systems is that it is too static. It fails to represent the symmetry process, the transformation through time that finds and maintains emergent social systems. It simply shows where each system is in relation to the others at this moment.

To change this static pyramid metaphor to a more dynamic metaphor, a new way of picturing this transformation is desirable. A more isometric metaphor for social systems is the flow of water through a river delta. Upstream is a large source of simple structures, in this case atoms. As these atoms interact over time they begin to divide (i.e. be constrained) into various chemical structures that further divide into astronomic and biological structures. As long as there is a cycle of atomic re-production these atoms will continue to flow (interact) into molecular, astronomic and eventually planetary and biological structures creating bifurcating patterns of emergent novelty downstream.

The important features of this new metaphor are threefold. First it does not imply 'construction' like pyramid levels do. Patterns of emergent structure are selected for persistence from a constant stream of interaction, they are not constructed or created by something. The pattern of propagating symmetry is downhill to universal interaction. Second, the delta metaphor represents increasing variety and pattern downstream so that the same amount of matter-energy gets more and more finely represented in distinct structural patterns as the flow is repeatedly constrained. The nested hierarchy pyramid implies construction to a peak while the delta metaphor implies ever-branching diversity.[129] The last important feature of the delta metaphor is its dynamic nature. In the delta metaphor the universe is maintaining the structure it has, while constantly producing finer structure downstream. The universe is not patterned, it is patterning.

The frame shift down from the fractal pattern of universal flow to the fractal pattern of biological flow is like moving from a satellite photo of the Yukon River delta to a close up photo of one section of that same Yukon River water branching out to make its way over a tiny section of beach. Both represent the same pattern (i.e. they look like they are structured by the same rules), but each is transforming at different time scales. The overall macro structure of the Yukon River delta branches over years and decades while the microstructure of its beach transforms over hours and days. If time is transformation, then each scale level is transforming at different transformation scales.

[129] In fact, Kauffman's (2000) candidate law of open thermodynamic systems: "on average, biospheres persistently increase the diversity of what can happen next. ...biospheres may maximize the average sustained growth of their own 'dimensionality'" (p. 4) compliments Gould's 'branching bush' metaphor for evolution nicely.

**FIGURE 18: YUKON RIVER DELTA FRACTAL**

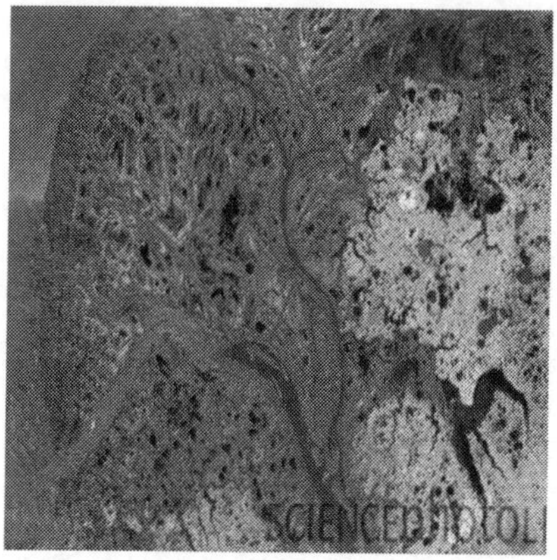

Source: Science Photo Library (n.d.)

**FIGURE 19: WATER FRACTAL ON A BEACH**

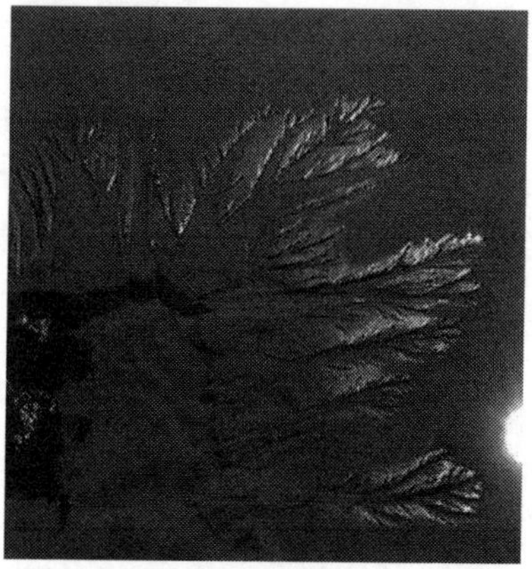

Source: Fitzgerald (2011)

Biological social behavior is at the end of this evolutionary time stream because it depends on all of the rest of the universal symmetry domains preexisting within certain boundary conditions. It is at the end because it is maintained by the regular interactions of all of its upstream component systems. Social behavior in the biological domain is maintained by the rest of the delta pattern, galaxies and solar systems included.

Upstream in the biological domain the rules for social behavior are chemical. Chemical social behavior is represented by autocatalysis and the chart of intermediary metabolism. Downstream in the biological domain the rules of social behavior keep bifurcating and complexifying. Mitosis, mitochondrial ATP production, signal transduction, senses, behavior and symbolic language are all downstream from the original chemical behavior that constituted ancestral bacteria. Further social behavior becomes even more highly bifurcated and complex as it flows downstream.

Revisiting Wilson's behavioral chart "The full hierarchy of biological responses" (see Figure 16) it becomes evident that all biological responses can be equated with social behavior. All responses are social in the sense that responses are synchronic interactions. 'Responses' imply coordinated interaction between external stimuli and whatever unit-of-selection is responding. Coordinated interaction between two or more systems is social.

What is interesting about Wilson's "hierarchy" is that it extends from biochemical responses at the individual level up to group replacement responses at the population level. All of these levels are interconnected. Social biochemical behavior occurs over very short timescales and social group replacement behavior occurs over very long timescales. Group replacement social behavior is like looking at the satellite photo of the Yukon River delta and biochemical social behavior is like looking at the beach photo. When the whole process is seen in a static way, as in the hierarchy metaphor, no one part is seen to be primary, or 'upstream.' There is no flow, or 'fractal symmetry.'

Wilson's biochemical responses are group replacement responses in the sense that groups of autocatalytic chemicals were selected for persistence (i.e. they *replaced* other *groups* that were not selected) to create those biochemical responses. From a dynamic point of view, the delta itself has evolved to have those behavioral response patterns. The biological domain is a branching of the universal domains and it too has a history of development. The development of social behavior in the biological domain is a result of a localized cellular source from which various other structures branched. Social-population and social-individual behaviors always exist contemporaneously, but the product structures that result from the feedback between the two levels of selection change over time. Social behavior has changed and its changes are embodied in the many levels of emergent social structures.

In a more conventional context, social behavior is usually thought of in terms of animal relationships. Most of Wilson's book is dedicated to explaining social behavior within the context of modern animal species. Brilliantly he implies a story of human sociality from what is alive and around us now. Bees communicate by waggle dances to coordinate the activities of conspecifics so that the hive itself can cooperate to the level of reproducing itself. In this context social behavior is the horizontal extension of cooperative communication to preserve the reproductive whole.

As Wilson convincingly demonstrates, social behavior is a pervasive feature of the biological domain. The media that each species uses to coordinate cooperative behavior varies, but the result is the same. By canalizing cooperative communication these species can capture and couple environmental features that are otherwise outside the adjacent possible of individual capture or coupling. Social behavior is used to extend the sense and behavioral halo of individuals to access ever more distant and cryptic environmental regularities. It is used to include ever more distant and cryptic saliencies within the social unit's cycles of maintenance and reproduction. Social behavior is one of many ways organisms can use to find, "devious routes around constraints" (Gerhart & Kirschner, 1997, p. 595). Yet again, cooperation paves the road to success.

In its most general sense, behavior is social because it syncs the phenotype to external features. The best way to represent the continual growth, differentiation and specialization of social behavior in the history of life is not as a nested hierarchy because social behavior is continually transforming and a nested hierarchy is a static metaphor. A better metaphor is the delta metaphor that castes the development of social behavior as self-determining, ever diversifying and dynamic. Unfortunately, our understanding of dynamic, simultaneous and ubiquitous transformation as represented by the Delta metaphor is constrained by the serial, digital form of understanding humans have, by the language we evolved. For this reason the term 'nested hierarchy' will still be used in this investigation as a static, but comprehensible, stand in for the delta metaphor. This only reinforces the need for philosophers and scientists to start developing a language capable of describing complex dynamic transformation.

Traditionally, social behavior is restricted to the study of social animal species, usually humans. Seen from another point of view, social behavior is just one more way for phenotypes to differentiate and specialize the ways they can capture and couple to their environments. Caste in the light of the four billion year history of major transitions in evolution, social behavior is the only class of interaction that has generated truly novel domains of life.

## Conclusions: Life

The biological domains mirror the universal domains because life is a fractal expression of universal pattern. From the biochemical firmament of primordial Earth was born (i.e. persisted) the prokaryotic domain of life. This domain was maintained by Earth conditions, the metabolic rules of the chart of intermediary metabolism and by the rules of binary fission. The structure it maintained we call 'cells' and as they interacted and reproduced they included more and more of the matter-energy of the last three spheres of the Earth within their own systems. Eventually they constituted a major part of their own environment. At that point they regularly interacted with each other creating a new interactive substrate from which new cooperative unions could be selected.

From the firmament of prokaryotic cell interaction was born (i.e. persisted) the eukaryotic domain of life. From the firmament of eukaryotic cell interaction was born (i.e. persisted) the multicellular domain of life. From the firmament of multicellular animal interaction humanity was born. This is the historical context of human existence. The truly significant changes in our descent line came from the interaction of new life forms with each other. It came from communication from which cooperative systems were selected for their capacity to maintain the whole organized structure through novel reproductive processes (e.g. binary fission, mitosis, meiosis). These reproductive processes were really just the canalized communication processes that maintained those emergent cooperative systems. This is the language of life: communication, cooperation, reproduction.

Life is the history of maintenance through cooperative reproduction on Earth. What organisms want to do is maintain and reproduce themselves. But how do they do this?

Internally, organisms are networks of chemical, electrical and mechanical relationships that oscillate through informational processes. That is, the patterns that trigger and constrain the flow of energy through the system are information in the sense that these cyclic patterns of transformation represent a predictable and recursive set of interactions. In terms of classical information theory this means that these metabolic cycles 'reduce uncertainty.' They embody constrained degrees of freedom. Of all the ways the components of your body could interact, only a very limited range of these possibilities is maintained. This conserved set of interactions was selected for persistence from the interactive background of the three universal symmetry domains constrained by the historical conditions here-on-Earth. As such, organisms are universal systems. Internally, they are the definition of synchrony, order through time. From our new perspective on time, this is order through transformation. More to the point, internally, organisms are systems of transformation that preserve structure. Organisms are complex symmetries.

Multicellular organisms like humans have highly differentiated internal networks that operate together on multiple levels of interaction. As Terrance Deacon (1997) notes, "with each level of organization (molecular, cellular, structural, functional) in a complex multicellular body there is an *exponential increase* in the amount of information necessary to control its construction and operation" (p. 107). All of these information relationships cycling together constitute a functional unity, a society of cells that we call 'animal.' However, the information that constrains this exponential increase in the organism is not exclusively internal. At each level of organization different environments contribute information necessary to maintain that level of "construction and operation." This makes the organism a hierarchy of environments that define and constrain ever denser, smaller and quicker levels of component organization. Externally, this same organism is partially constructed and operated by its own relationships to its own environment.

As a result the unity of the organism is real, but it is not complete. The organismic point of view is necessary, but not sufficient to explain how life transforms. Organisms, like all systems, are constructed by components that were historically separate. Historically, each component was part of the environment of other components. In becoming a unified cooperative entity they had to synchronize with one another. This synchronizing of environmental regularities is perhaps the most basic transformation of living systems. All organisms synchronize matter-energy to maintain and reproduce themselves. All organisms eat, and in eating additionally constrain the food´s self-relationships within the organism´s self-relationships.

All organisms are also products of the synchronization of ancestral organisms.[130] All organisms came from symbiogenesis. Ultimately, the first organisms came from chemical symbiogenesis. Because interaction, in the form of the synchronization of environmental regularities, is fundamental to organismic transformation, organisms cannot be defined solely by their internal relationships. Deacon (1997) finishes the above quote by saying, "in addition, the range of possible adaptive reactions [to external features] also grows exponentially with the complexity of the organism" (p. 107). He is saying that the addition of levels of complex internal interaction, "molecular, cellular, structural, functional" is accompanied by the addition of levels of complex external interaction, "the range of possible adaptive reactions." Internal interactions define the organism in terms of itself, but external relationships also define the organism in terms of its environment.

Externally, organisms are engaged in identifying and capturing or coupling environmentally salient features.[131] This process is as essential to any definition of organism as the canalized and conserved internal cycles are because they are two sides of the same coin. If salient environmental features are not synchronized, there is no energy flow through the organism and it dies - there is no organism. Furthermore, without synchronization, organismic transformation loses its only known road to major diversification, cooperative union. Even internal differentiation and specialization begins with novel cooperation between component systems (e.g. molecular, cellular, organal, etc.). Thus organic life, like all the interesting parts of the universe, is systems within systems within systems. What life is becomes comprehensible not only by considering what is internal, but how the internal interacts with the external.

[130] Arguably this is not true of the first cells, but only because of the convention of calling the first cells 'alive' and their chemical precursors 'dead.'

[131] Salient environmental features are features outside the organisms that contribute to the maintenance or reproduction of that organism or its offspring/species (i.e. those features of the environment that contribute to maintaining or reproducing that 'type' of biological organization).

The relationship between internal and external interactions in organisms is a complex one. Rather than spending endless time detailing this relationship it is expedient to sum them up. Internal interaction is conservative. All metabolic and reproductive systems are preserved because they maintain specific and necessary interactions. These interactions are necessary for the continued persistence of that organic structure in its ambient. External interaction is more flexible and exploratory. The sense halo, brain and epithelial specializations are preserved because they sample and maintain synchronous relationships with a chaotic soup of environmental saliencies. Internal interactions preserve the past and external interactions maintain an adaptive future.

Life is ever complex, but some generalities can be proposed. From an internal point of view the goal of life is to maintain or conserve organization in the form of individual metabolisms and reproduction through descent lines.[132] From an external point of view the goal of life is to find and synchronize new environmental saliencies. This is how systems persist into the future. The process of life is evolution. It is universal. It is the distribution of order over space feeding back into the distribution of order over time (or vice versa). This process can be measured as symmetry, transformation with preservation of structure.

Life is based on positive symmetry. In the biochemical context of early Earth interaction some miniscule percent of those interactions between the outer three layers of primordial Earth was regular (i.e. persistent). Environments that preserve new regularities tend to organize more and more matter-energy within these regular cycles (i.e. inclusion) because emergent regularities broaden the spectrum from which positive interactions can be selected. If maintained in stable environments, regularities propagate. This kind of sustained positive symmetry is variously known as propagating symmetry, positive feedback, inclusion, maintenance or, in the biochemical case of life-on-Earth, reproduction.

Life-on-Earth is a magnified version of the universal pattern of structure construction (i.e. symmetry). The same universal pattern of interaction,

---

[132] This maintenance includes the production of variety (be it mutational or meiotic or otherwise) from which, at populational levels, altered phenotypes can be selected for persistence.

synchrony and propagating symmetry is found in the biological patterns of communication, cooperation and reproduction. Just as all universal structures are the result of persistent interaction selecting for further persistent interactions, all life structures are the result of cooperative biochemical systems being selected for further persistent cooperation (i.e. reproduction). Just as constant, ubiquitous interaction is the conditions from which all universal structures are selected for persistence, so too is constant, ubiquitous communication the condition from which all biological structures are selected for cooperative and reproductive persistence.

Organisms themselves are evidence of the symmetry process in that each level of structural organization within the biological domain was preserved from interaction at lower levels. Each level is an emergent and additional level of constrained release of energy. Prokaryotic cells are cooperative groups of autocatalytic molecules that maintain and reproduce themselves. Their further interaction created the substrate conditions, the interactive background, for the selection of eukaryotes. Eukaryotic interaction was the substrate for the selection of metazoa and so on down the line. Each of the major transformations of life-on-Earth was a result of the cooperative union of previous organisms. The emergence of these cooperative unions as whole organisms required (i.e. embodied or enruled) novel forms of metabolic and reproductive communication whose added levels of controlled release of energy could preserve the union (i.e. maintain and reproduce it). These new biochemical languages emerged from canalizing how the components communicated to select for ever more cooperative unions until the unions were able to cooperate over generations, to reproduce.

Emergent unions are defined not only by the component systems that cooperated to create them, but also by their expanded adjacent possibles. Internally, they are able to accelerate differentiation and specialization due to the, 'gifts of competence designed elsewhere' that individual components brought to the union. Externally, these emergent organisms present a new reactive face to the environment. As they maintain and reproduce themselves, their reactive faces become statistically significant percentages of the total interactive background conditions. Their novel reactive faces created novel interactions in the environment that in turn could be sampled to be selected for persistence. In the end, those interactions that persist longest become the languages (i.e. the means of reproducing the unified system) of the next emerging organisms.

Internally what emerges from this symmetry of life process is whole new rule-sets and product structures for each level of biological cooperation. Prokaryotes, eukaryotes and metazoa all have distinct metabolic rule-sets by which they maintain and reproduce their level of organization. Social metazoa have an additional internal level that is called social behavior. That is, within the group, an additional level of stereotyped interaction exists. This level is constrained by behavioral interaction[133] as well as biochemical interaction. It is still stereotypically reproduced, but this time by the group as well as its individuals. This is to be expected because each level of interaction (i.e. communication) is co-constructed by previous individuals at previous communicative levels.

Externally, what emerges from this symmetry of life process is novel forms of cooperation. In organisms like us this is called novel behavior. Novel behaviors sample the environment for synchronous relationships. Some of these relationships actually capture saliencies and others just couple to them. Most of them die off. Bodies, brains, and by extension, senses, evolved to canalize the successful behaviors organisms needed to maintain and reproduce themselves. These brain-sense systems are selected for what they can recognize and respond adaptively to. Over the history of life 'behavior generation' (i.e. body-brain-sensory environmental coupling) has been in the business of extending this adaptivity through time and space. It is in the business of persisting into the future. Externally, what emerges from the evolution of our descent line is novel ways of generating behavior that can synchronize more and more distant (i.e. spatial and temporal) regularities.

The emergence of these different levels of biological cooperation has maintained organisms organized into nested hierarchies. Organisms are composed of systems within systems within systems. All of these systems together are ultimately nested within universal systems. The human descent line is amongst the most highly intricate of these nested hierarchies, which

---

133 Behavioral interaction is characterized by sensory regularities in the same sense that chemical interaction is characterized by catalytic and bonding regularities,.

means that feedback between the multiple internal and external levels of cooperation is highly complex. This is largely due to the fact that on top of the molecular, cellular, organal, endocrinal, ontogenetic, nervous and instinctual/emotional systems of persistent cooperation, we have an additional behavioral system - symbolic cooperation. Symbolic cooperation is built upon the biochemical, biomechanical and bioelectric systems that all animals have, but like all emergent systems it also produces completely novel behavior. The two sections that follow will build upon the emergence of this new language by showing how it relates to previous 'languages' at previous levels. These sections will also show that the context in which we should imagine the origin of language, and the attendant origin of humanity, is a biological context with biological rules and constraints that are discernible from the history of biological systems.

Many of the terms and specifics introduced in this section may seem rather extraneous or unconnected. The reason they were introduced in this "Life" section is that they vivify the story that is emerging - the human story. Details are important to help us imagine the big picture. The biological domain is a vast and diverse rule-set with a vast and diverse range of structures. Fundamental to this investigation is less the story of each of these diverse organisms and more the story of the major transformations the human descent line has undergone. We must never forget, though, that these major transformations are composed of organisms like you and I, doing what we do: drinking coffee, making love, reading books and talking about them. That is, ancient organisms also ate, sensed each other and reproduced. They also cooperated to live.

# Section 4: Humanity

This section describes first how humanity emerged by coordinating their behavior socially and second how this cooperation resulted in an emergent reproductive unity very similar to life's other major transitions. Like previous emergent cooperative unities in the history of life, this group unity is based on a novel process of communication, cooperation and reproduction. This process is symbolic language and it is just another emergent symmetry.

## Human Emergence

*It is clear that our understanding of human social evolution must be brought within the paradigm of major transitions. We must go beyond kin selection and reciprocity to model a complex regulatory system that binds members of a group into a functional unit.*

*David Sloan Wilson (2002, p. 33)*

**FIGURE 20: HUMAN COOPERATION**

Source: Ritmos Sud Americanos (2013)

The controversy surrounding the details of how ancient primates became modern man is heated, current and changing. While this section is not an in-depth analysis of anthropology, it does seek to put the history of humanity into perspective. This perspective is that modern humans emerged under the same constraints and by the same pathways that all species, organisms and universal structures evolved. Humanity emerges as a typical but infinitesimal branch of universal pattern.

Since Darwin's time, theories of human evolution have tended to focus on the special circumstances that must have existed to create such special organisms as us. This emphasis is due to at least three reasons. First, many scientists presume we are unique and that humanity needs a unique explanation because it is something the natural world has never seen before. This is true but uninteresting. All species alive today could make the same argument. Second, science is in the business of looking for a cause. Because language and thought are arranged into lines of cause and effect, there is the presumption that if we find the right historical moment, and frame the problem carefully enough, we will find the cause of human uniqueness. So far the cause has ranged from opposable thumbs to cooperative hunting, squat foraging, various brain modules, tool making, food sharing, social intelligence and gossip - to name just a few. Third, when theorists look for causes they find the most proximate and unique one, claim a name for it and then quit looking. As soon as they have boxed the problem conveniently enough to justify what they have "discovered" as the cause[134], they look no further back into history. Typically, this means that linguists look no further back than fifty thousand years, anthropologists no further than seven million years and primatologists no further than twenty five million years, and so on.

This investigation attempts to place human evolution in perspective with the history of life-on-Earth, and within the history of the universe. The goal here is to show that universal structure is fractal. The patterns that ordered all matter-energy since the Big Bang are the same patterns that ordered life-on-Earth. At a higher resolution, human emergence is part of that pattern. What we have thought was so special is just another level of the same magnificent pattern.

---

[134] This box is often shaped by the path-dependent history, terminology, and ontological presumptions of their chosen academic domain.

To show this is not to show how human emergence was unique, but how it was typical. Humans emerged just as so many other species emerged, through transformation with preservation of structure. Physicists call this process symmetry. Mathematicians use the equals (=) sign to represent it. Biologists call this process evolution, life, autocatalysis, speciation, metabolism, or reproduction.

## Humans are Animals

Humans are animals. As animals we evolved and like animals we live. We develop, we discern, we move, we eat, we breed, etc. In short, there is nothing unique about us except the totally mundane fact that like all other species on this planet, there is nothing out there quite like us.

To begin understanding how animals change is to first understand that animals in total represent a tiny fraction of the types of life that have evolved on Earth. Animals are one successful branch of the domain Eukaryota. There are at least nine other branches (i.e. kingdoms) each with their own phyla, classes, orders, families, genera and species. This is not to mention the two other entire domains (i.e. Prokaryota and Archea) that have their own entire family trees. Just as the biosphere is an infinitesimally small part of galactic interaction so is humanity an infinitesimally small part of life. Both also seem to represent the highest tip of the range of ordered structure within each context. That is, although both are representative of ordered structure, both are extreme examples from the highly ordered end of the spectrum.

Since we emerged as one tiny branch of the Kingdom Metazoa (also known as 'Animalia'), it seems reasonable to look first at how our brethren emerged, and second, compare this to how we emerged amongst them.

## How Animals Emerge

There is no canonical formula for how animal species evolve, but we could expect they transform like other universal systems. They start from initial conditions that are sorted and ordered by interactive rules (i.e. constraints). Then they interact to be selected for persistence.

The broadest external conditions for the maintenance of the domain of biology are, of course, the history of regular interaction in the universe and conditions here-on-Earth. Internally, the biological domain itself is differentiated into different functional levels by different descent lines with different reproductive mechanisms and cycle times. Bacteria cycle quickly through mechanisms of binary fission (e.g. minutes) and species cycle slowly through mechanisms of speciation (e.g. tens of thousands of years).

From cells up to species, the selection process operates on how well units of selection from that level can sync to their environments. Selective units from each level reproduce differentially largely according to their internal interactions and survive differentially largely according to their external interactions. Bacteria reproduce according to the autocatalytic rules of binary fission and they die according to how successfully they interact with their environments. Herds of buffalo reproduce according to the genetic and epigenetic rules of meiosis, instinct expression and social learning, and they die according to how their group interacts with its environment. Lone buffalo don´t survive. Frame shift up, species reproduce according to the demographic rules of their species and they go extinct according to how they interact with their environments. This point of view corresponds with what is currently often called multi-level selection theory (Wilson D. S., 2002; Wilson E. O., 2012).

Like the evolution of other universal systems, biological systems have many different levels of rule-based interaction that maintain and reproduce many different levels of structure. These structures are sampled by initial conditions sorted by emergent rules of cooperation and they are differentially preserved by how they are reproduced and how they survive interaction with their environments.

## Allopatric Speciation

As mentioned above, species can be seen as units of selection. A unit of selection needs only to maintain or reproduce a characteristic but variable structure, and for this structure to survive differentially in the process of interacting with its environment. Both uranium atoms and species do this. As a result, species maintain, change or go extinct according to how their phenotypes interact with their environments.

In this investigation we are more interested in transformation than stasis or extinction and in this context transformation is called 'speciation.' The favoured theory of speciation today is the theory of allopatric speciation. In as few words as possible, allopatric speciation is understood in terms of populations. Populations inhabit certain ranges. These ranges correspond to the phase space for the phenotype of that population. That is, populations reproduce to inhabit all areas where they can get to and life for them is possible. The edges of these ranges are called margins and often survival at the margins is exceedingly difficult. For example northern margins often represent the coldest climates that a species is able to survive in. As a result, the phenotypes that live at these margins are often small groups of somewhat specialized organisms. This means that for every population in a species there can be as many specialized subspecies as there are distinct marginal environments. It should also be noted that "marginal" subspecies can also exist as physiological not geographical extremes. Any small group that can maintain distinct features is marginal.

Normally (i.e. statistically) environmental changes occur slowly and species are able to adjust their positions within these environments to remain viable. When an ice age comes they move south. However, when conditions change rapidly or adjustment is constrained (say by oceans, desert, disease or competing species) the species can lose its range and go extinct. This happens all the time. Often when this happens new species from different descent lines move in to occupy the range of the extinct species. Another option is that even though the main population goes extinct, there is always a chance that one of the specialized marginal populations was preadapted to whichever environmental change overtook them. In this case, an ice age may make the main population go extinct, but the northern specialists are just able to survive in the warmest regions of the original range. After a period of further phenotypic adaptation to the new conditions this small group of specialists then repopulates the range with its novel, and now reproductively distinct, phenotype. Ian Tattersall gives a fantastically concise description of this process in his book, "Becoming Human" (Tattersall, Becoming Human: evolution and human uniqueness, 1998).

Allopatric speciation is immanently plausible as a source of speciation. As mentioned previously, it is probably not the only way new species are created (e.g. Orgel's second rule), but it does provide a foundation from which to work.

Allopatric speciation depends on a few conditions. Most importantly, environmental conditions must change radically. If they don´t, nothing changes. Some species of bacteria and deep sea thermal vent organisms are believed to have changed very little since the beginning of life-on-Earth, but they are the exceptions, not the rules. Secondly, marginal populations have to exist and usually are somehow preadapted to new conditions. This preadaptation can come as any phenotypic (or social) feature or combination of features that allows them to survive new conditions. Thirdly, for the species to differentiate enough to become a new species (e.g. not be reproductively re-absorbable into its parent species) it needs to pass through a time of high death rates and low populations. High death rates ensure that selection for difference is strong and low populations ensure that gene pools are small enough for distinct advantageous phenotypes to spread.[135] This final point underscores what was previously mentioned, the need for reproductive isolation so the emerging phenotype doesn´t become reabsorbed into the ancestral reproductive pool.

In sum, selection for novel maintainable order happens at many levels of the biological domain. Individuals, groups and species are all units of selection that exist within different frameworks of maintenance, reproduction and selection. At the species level, change is often enabled through allopatric speciation where populations exposed to environmental flux experience death rates so high that only small marginal groups survive by adapting themselves rapidly to the novel conditions. Once the species phenotype finds a new interactive stability within its environment it once again expands to inhabit all areas where life for them is possible.

---

135 Novel, different and advantageous phenotypes are necessary to drive change in the population. However, if the population is too large these novelties are lost in extensive interbreeding. Statistically they can only persist and spread in small, isolated, gene pools.

## Punctuated Equilibrium

Recently we have accepted the fact that the emergence of species most often happens suddenly. For well over a hundred years evolution was thought to change animals little by little and it was through the gradual accumulation of these changes that animals differentiated into species. Like Newtonian physics, this view is less wrong than it is incomplete. Genetic drift and gradual changes in populations do happen. However, the suddenness of the disappearances and appearances of species has been underappreciated. Eldredge and Gould (1972) called attention to this discrepancy by introducing a new phrase for an old process, "punctuated equilibrium."

Punctuated equilibrium is the idea that speciation usually happens quickly not slowly. It used to be thought that the poverty of transitional fossils in the fossil record was due to the poverty of the fossil record in general. Now, with the fossil record more filled out and still so few transitional fossils, punctuated equilibrium is becoming the rule. It says that species appear suddenly in the fossil record because they appear suddenly in nature, not because we don't have enough fossils yet. By this logic, changes within species can still proceed, incrementally extending the degree to which phenotypic features sync interaction, but rarely do these extensions constitute the qualitative change in phenotypic functionality we recognize in different species.[136] Punctuated equilibrium implies that different species are caused by rapid periods of dramatic phenotypic transformation instead of long periods of incremental transformation.

This new viewpoint corresponds nicely with the idea of allopatric speciation. Populations of species go through extinction bottlenecks where they are forced to change or die.[137] Then, once successfully adjusted to the new environment, they statistically "reappear" in the fossil record when they radiate to fill their new phase space (i.e. range). According to punctuated equilibrium, these new species can then go on to enjoy long "life spans" where their phenotypes remain generally the same until again disturbed

[136] Of course, there are exceptions. One exception is termites and ants that are often so similar phenotypically that their species are commonly differentiated behaviorally (Wilson E. O., 1975).

[137] Reference to "bottleneck" speciation is not original. See Stanley (1996).

by environmental conditions. Functionally, this bottleneck transition should have some interesting features.

## Bottleneck Speciation

Speciation is a slippery event to prove. Recent attempts like Jonathan Weiner´s account of Peter and Rosemary Grant´s twenty year study of Galapagos finches have gone a long way to substantiating theories like punctuated equilibrium and allopatric speciation, but to this day, definitive studies of wild speciation remain hard to carry out and even harder to prove (Weiner, 1994). This is only to emphasize that the bottleneck theory we are about to take a close look at is a scientific best guess. Although it almost certainly does take place in nature, the degree, frequency and significance of this kind of speciation has yet to be fleshed out.

If species are regularly created in some kind of population bottleneck, what could be expected of such an event? First of all the environment would have to change rather drastically. This does not necessarily mean large-scale environmental shift. Since different features of environments are weighted differently as to how they contribute to a species' fitness, a relatively tiny change in the overall environment can have catastrophic results for a species that depends on that feature. Catastrophic results can vary from K-T meteorites to the introduction of a tiny species-specific virus or the local loss of one plant species. What is important in generating the bottleneck is that selective pressure, in the form of high death rates, is put on the species.

Selective pressure is a decoupling of the species phenotype from the features of its environment that it is adapted to sync or sync to. This could be a loss of a food source (that it syncs) or a change in water chemistry (that it syncs to) or a disease (that precludes it from syncing to anything). This decoupling is what causes the death rates to rise, the populations to sink and marginal populations, if they are isolated enough to miss the plague or preadapted enough to eat another type of plant, to get lucky.

If the selection pressure is too high, the species goes extinct. If it is just high, one would expect some kind of fragmentation. Fragmentation is the shrinking of the population to leave pocket groups. Using the image of an inverted fitness landscape where low points, not peaks, represent fitness optimalities, one can imagine high death rates affecting the population much like evaporation affects puddles. What was one large puddle (i.e. species) breaks up into a series of smaller puddles as the water (i.e. population) is reduced by evaporation (i.e. death). These smaller puddles are centered on the deep points (i.e. fitness optimalities) that are now reproductively isolated from each other. To continue the metaphor, if the environment rapidly restabilizes and the puddle refills, then the population reintegrates and the species remains the same. Likewise, if the puddle dries up then the species disappears. The case where the species remains fragmented for a substantial period of time is the interesting one. It is here, presumably that many new species are born.

If speciation is to occur then the environment or the species phenotype (e.g. its mode of syncing to its environment) must change. The least interesting case is when a marginal population(s) is lucky enough to be preadapted to survive the new conditions where their mainstream cousins could not. In this case, it goes through a brief bottleneck where the unique aspects of its phenotype that allowed it to survive become fixed in a new, "slightly altered," phenotype that, once it expands to refill its new range, becomes the new species phenotype. The most interesting case is when the species goes through a long and difficult bottleneck from which can appear, as we shall see, some very improbable adaptations.

Long bottlenecks are characterized by alternating periods of lower and higher selective pressure. It may be that the duration of the bottleneck is long because the environment change was extreme or it may be that the species phenotype itself is constrained by its physiology from making the necessary adjustments, forcing it to experiment with numerous improbable adaptations before they find the one or combination of ones that once again function to recouple them. For whatever reason, what seems to be characteristic of this process is an extended period where the phenotype is forced to change rapidly. In the end, this period of searching is only as long as it takes the species to either go extinct or find new ways to survive its new conditions.

'Search,' perhaps, is not the right word. What happens is one of two cases. In the first case the species phenotype (which will for the rest of this chapter be referred to as just the "phenotype") already has a feature that enables it to adjust to the new conditions. This kind of preadaptation is like a 'sleep under the snow' behavioral trait of a group on the northern margin of the species range that suddenly becomes selected for when temperatures drop. In this case, there may be a brief period of strong, 'sleep under the snow' selection pressure where that behavioral trait becomes sufficiently fixed in the phenotype. Then the newly adapted species radiates to fill its new phase space. This kind of phenotypic change is highly predictable because it is the result of highly constrained selection pressures (e.g. only those phenotypes that sleep under the snow and somehow pass this on successfully to their offspring survive). In this case the adaptation really did exist previously and the phenotype just has to fix it behaviorally, epigenetically then genetically (usually in that order) into its population. This case depends on a lucky preadaptation.

The second case is when the phenotype doesn´t have one preexisting feature that allows it to survive. In this case, the phenotype is in a sense cast adrift. Perhaps initially it is just lucky enough to be located in the deepest part of the puddle, or the most outlying. But now, the phenotype finds itself free to vary widely because the conditions that enforced the maintenance of specific environmental couplings have disappeared. Also, the high death rates and small group sizes allow any better variations to spread rapidly and make them especially sensitive to adaptation and change. So now the phenotype is adrift, uncoupled from its ancestral points of sync and very changeable.

Phenotypes in this situation must first of all be durable if they are to survive. One feature that appears to help them is size. Part of Cope´s Rule, which claims that species generally get bigger, is that statistically, it is the smaller species that survive extinction events better (Gould, 1989). In these cases, smaller means fitter and this presumably has something to do with lower energy requirements (e.g. they need less food) and higher representation and reproduction rates (e.g. statistically more of them exist and they reproduce faster and in higher numbers). This kind of size selection is another example of survival by luck.

A different feature would be evolvability, the ease with which the phenotype can adjust itself. One of the prime functions of mechanisms like Gerhart and Kirschner´s (1997) "exploratory processes" as well as Deacon´s (1997) "Darwinian processes" is to enable phenotypic plasticity. Obviously, phenotypes that are able to vary themselves easier are also able to adjust to novel local optimalities quicker. One example Deacon gives is the Hox genes that enable huge structural changes to develop from minimal changes in gene expression - like whole new appendages. Obviously organisms that can make substantial structural changes, like growing new appendages with simple genetic changes and plastic metabolisms (e.g. vascular plasticity), have an advantage when conditions change radically. This is just one of legions of examples, but it serves to illustrate that durability, whether it comes from luck or evolvability, is crucial for speciation.

Small groups under heavy selection pressure can survive by luck or by varying their phenotypes to couple to novel environmental features. In the first case selection pressure is constrained by the preadaptation itself and speciation is just an extension of that preadaptation's preexisting functionality. In the second case selection pressure is unconstrained and the phenotype is left to wander, or explore, its new environment. It alternately fixes on coupling small local optimalities until it discovers one big enough to allow it to fix (i.e. behaviorally, epigenetically or genetically canalize) that coupling and reproductively expand into its new phase space. The second case is interesting because it is a prime candidate for producing radical exaptations in the form of rather large, unexpected phenotypic adaptations.

An "exaptation" is a new word introduced by S. J. Gould and E.S. Vrba that essentially means a preadaptation that changes function. They cite Black Herons that use their wings to cast a shadow in front of themselves in order to better see the fish they are hunting. Black Heron wings were preadapted for flight, but this original function has been exapted to support a completely different function (e.g. shadow hunting). Gould and Vrba (1982) are making the distinction between those adaptations that function to couple the organism the way they were originally designed to and those adaptations that change the functionality of the original design.

In this investigation the functionality of the adaptation will also be important to distinguish. Henceforth, the term 'exaptation' will refer to adaptations resulting from novel couplings while 'preadaptation' will continue to refer to the extension of preexisting couplings.[138] As such, exaptations are the result of the phenotype searching the fitness landscape (i.e. sampling its interactive background) for novel optimalities, novel couplings.

One of the characteristics of this process is an increase in noise. As previously explained, noise refers to "unstructured or irregular interaction." The increase in phenotype-environment noise is characteristic of the loss of coupling and the experimentation necessary to find a new coupling. The interesting thing about this noise is that it can prevent the phenotype from permanently settling on small locally optimal couplings. This property of adaptive systems is noted by Beeby & Kaeser (1990, p. 73), who explain that noise in adaptive systems seems to be a key factor in preventing systems from fixing on small local couplings. By continually displacing these systems off smaller and closer optimalities this noise allows these systems to continue searching for larger, more advantageous couplings even though those couplings are statistically very improbable. Essentially, noisy interaction can push the phenotype to find improbable solutions. The phenotype-environment coupling cannot be too noisy or the phenotype dies. Conversely, it cannot be too synchronous or the phenotype never changes. Yet again, the range of the spectrum of interaction that produces emergent couplings is a tiny, not too hot, not too cold range, suspended somewhere between order and chaos - the goldilocks conditions. However, having found this sweet spot, those solutions that emerged could be highly improbable adaptations to complex phenotype-environment coupling problems. In short, the second case where the bottleneck forces the phenotype to decouple and search for new couplings should be expected to provoke functionally novel relationships (i.e. novel exaptations).

---

138 For a concise and interesting critique of Gould and Vrba´s concept to exaptation see Buss, Haselton, Shackelford, Bleske, & Wakefield (1998).

What this means is that when we are looking at particularly novel and major transitions in biological history, we should expect the contexts that gave birth to them to be to a certain extent noisy. This dovetails nicely with the social solutions that actually characterize the major transitions in life history and with the conditions that gave birth to the three major universal symmetry domains. It is from a background of constant noisy interaction that novel cooperative systems are selectively preserved. Remember, both the ubiquity and the *variety* of interaction are important to establish a spectrum of interaction full enough that statistically improbable interactions can be sampled from it. Noisy environments select for emergent social novelties.

As should be expected, noisy environments do not produce exact fits. As Gould (1993) notes, novel exaptations are often results of "sloppiness," "redundancy," "poor fit" and "quirky design" (p. 97). He cites the case of the evolution of the mammalian ear from the jaw and skull bones of ancestral premammals. His point is that true evolutionary creativity is usually the result of a hap-hazard process of accidental co-option. In this case, the premammalian brain co-opted the jawbone into functioning as a sense organ by the pure accident that the jawbone vibrates and is close to the brain. An important point to understand about the conditions for evolutionary creativity is that the noise produced by the uncoupling of the phenotype in the bottleneck is an ideal condition for finding and co-opting these kinds of 'quirky designs.' The protomammal's jaw would never have become that delicate arrangement of inner ear bones if it had not gone through times of duress. Without these bottlenecks, those marginal populations whose jaw bones conducted sound vibration better would never have differentially survived and the mammalian ear would never have evolved. These extremely improbable solutions to complex problems are discovered under bottleneck conditions. In such events, a noisy environment would be an expected condition for exaptation.

Bottleneck speciation often depends on environmental decoupling. Sometimes that decoupling is brief and results in the functional extension of the ancestral phenotype as represented by extending preadaptations. Under conditions where the phenotype is more seriously decoupled, a highly altered phenotype with novel traits and coupling functions can evolve. History demonstrates that the truly novel transformations have been the result of social unifications. History also indicates that conditions of noisy interaction can result in communication selecting for novel cooperative relationships selecting for novel reproductive relationships. Social speciation (i.e. the emergence of novel cooperative unities) is a rare but powerful version of regular speciation. However, before moving on to summarizing some of the characteristics that can be expected from this kind of speciation it is important to take a quick look at the final phase of the bottleneck process, adaptive radiation.

## Adaptive Radiation

Canonically, adaptive radiation is referred to as the reproductive expansion of new species. This radiation typically follows extinction events, however, the term can also refer to any expansion of species that follows the discovery of new phase space. This includes the possibility that a species that discovers a new trick, like echolocation in bats or water mammals, can also experience periods of relatively uncontrolled expansion and speciation.[139]

What is especially interesting is that where the bottleneck produces constrained or unconstrained development of new species traits so does the process that follows it. In the bottleneck paradigm, adaptive radiation occurs when the new species begins to successfully couple (e.g. survive and reproduce) after a period of duress. This expansion period is characterized by extremely low selection pressure. This period of low selection pressure can be relatively short and constrained resulting in a single new species with a functionally extended ancestral trait or it can be long and unconstrained, with unpredictable results.

Long periods of successful expansion can be due to conditions in which the new species expands into a preexisting but vacant phase space, like the discovery of the Galapagos islands by Darwin´s Finches, or the post Permian extinction expansion of Therapsids (i.e. mammalian precursors) to fill the

139 The reverse is also possible. The discovery of a new physical phase space like air for bats or water for marine mammals could also lead to sampling of new exaptations like wings, fins or echolocation. Once again, structure and process merge.

under populated ranges of extinct clades. It can also be due to the discovery of a new trick through exaptation as evident in the adaptive radiation of echolocating bats or multicellular animals in the Cambrian Explosion.

What is interesting is that long and/or extreme periods of any kind of adaptive radiation can create conditions for phenotypic novelty and speciation. Basically, expansion into an open phase space can relax selection pressure in much the same way that severe decoupling in the bottleneck does. Terrence Deacon (2010a) calls this "relaxed selection" and interestingly enough also indicates that this relaxed selection often results in traits that are, "more variable, more complicated and more influenced by social experience" (para. 18). Essentially, reproductive success for these animals no longer depends on very specific phenotype -environment couplings. His example is domestication, but it could equally be any situation where old couplings are relaxed. Wolves become dogs by the de-emphasis of feral traits (i.e. couplings) in reproductive success and the new emphasis on the varied domestic traits given to them by human breeders. Interestingly enough, he also notes that the relaxation of selection introduces noise, presumably in the form of increased noncoding introns, into domestic genotypes. Once again, noisy interaction seems to be the background from which novel structures are selected for persistence.[140]

In any case, he claims that the result of relaxed selection is increased genotypic noise and phenotypic variation. Obviously, this too can result in those kinds of radically novel phenotype-environment couplings that are characteristic of exaptation. Interestingly, these exaptations are often highly complex and as Deacon notes, "more influenced by social [read: 'cooperative'] experience."

Ultimately, relaxed selection as evidenced in post extinction radiations, post bottleneck exaptations and domestication contexts[141] can all lead to speciation. In extinction events, surviving phenotypes expand, relaxing their previous coupling constraints and presenting more phenotypic

[140] Interestingly, this kind of genotypic noise can be favourably compared to the view that genetic mutations drive organismic change. It is a good example of the traditional view that change comes from inside the organism. This investigation challenges this view and proposes that change comes from outside the organism, from how it survives interaction with its environment.

[141] These are not mutually exclusive options.

variation to the new environment. Speciation in this scenario is constrained by the amount of phenotypic variation and the amount of couplable environmental features. Likewise, exaptations that emerge from bottleneck speciation present new traits to new environments and their power to create new species is limited by the variability of the potential couplings inherent in both. Domestication happens on a smaller scale, but still represents a good example of how uncoupling can facilitate speciation and other jumps in phenotypic functionality.

As a final note, it would also seem reasonable that the length and complexity of the sorting process itself be proportional to the degree of novelty of the phenotypic solutions. For example, the amount of time spent in the bottleneck is an indication of the amount of time the phenotype remained uncoupled. The amount of time spent uncoupled could be statistically connected to the amount of novelty the new species exhibits. That is, the more time spent sampling the adjacent possible, the larger the spectrum of interaction is and the statistically more likely that the solution actually discovered will be unlikely and thus highly novel. Additionally, the more novel the phenotypic solutions are, the more unspecifiable is the phase space they eventually expand into. This would imply that species resulting from long complicated bottlenecks would have a greater chance to develop highly unique and complicated exaptations that are more likely to open large new phase spaces.[142]

## Characteristics of Emergent Species

What emerges from the bottleneck is a whole world of specially constrained interactions and expectations. As per Orgel's Second Rule, "Evolution is cleverer than you are," it should be imagined that in real speciation events, periods of stress and relaxation are to be mixed and matched over inconsistent time periods. The number of species spun off and the features of their individual traits would most likely depend on the frequency, duration and extremity of environmental change as well as the adaptability of the

[142] The opening of new phase spaces is analogous to the concept of "niche construction" by Deacon (2010a).

phenotype and, of course, the history of broken symmetries. As such, the products of the intermittent drying and then partial refilling of the puddle (so that novelties in one puddle become mixed with novelties in another only to separate under different conditions later) should be expected to be formidably variable.

However, this process would produce some reasonable expectations. First, any processes that enabled rapid phenotypic change in the sense of coupling potentials would be preferentially preserved. The aforementioned Darwinian and exploratory processes presented by Deacon, Gerhart and Kirschner would be increasingly represented in species emerging from these bottlenecks. These internal mechanisms for adaptability make any phenotype that has them more labile and thus more likely to adapt and survive.

Processes that help species survive significant duress in the form of bottlenecks should be preferentially represented in new species. This means that along with lucky features like small size or antifreeze sap, features that allow for behavioral flexibility should be differentially preserved. Like the evolution of the senses, the evolution of genetic, epigenetic and embryological flexibility helps the phenotype to rapidly search for, find and then sync external saliencies in difficult times. The only real difference is that the "senses" search is conducted in the immediate timescale of phenotypic interaction and the "genetic-epigenetic" search is conducted over population and generation timescales. Both can be expected to use different methods of preservation, but both are essentially examples of selection for the extension of flexible interaction with noisy external features.

This last point is especially interesting because it is generally behavior that leads the landscape search and coupling process for animals. As Stanley (1996) notes, it was the eminent Ernst Mayer who,

> suggested that a change in basic behavior within a group of animals has often preceded a major evolutionary shift of bodily proportions. Indeed, this generalization probably holds, because behavioral changes, even genetically

> programmed ones, tend to be easier for evolution to achieve than wholesale anatomical restructuring (p. 167).

It is behavior that leads the change. What this means is that, in animals like us, features like more complex nervous systems (i.e. brains) should be continually selected for, at least in periods between massive extinctions where smaller, less complex phenotypes are chosen for. Animal behavior is the most flexible tool available to the phenotype for finding and coupling to novel environmental features. Behavior maintains the largest spectrum of interaction. It is the most likely place for the positive feedback, in the form of differential selection, that results in novel functional relationships becoming canalized in the phenotypic internal.[143] If this feedback can be stably maintained behaviorally, it can begin to be differentially preserved at epigenetic, ontogenetic, and finally genetic levels of interaction. To the degree that animals are capable of it, behavior comes first in this cascade of novelty preservation. Behavior finds the novelty, other processes help to preserve it. Communication, cooperation and reproduction: the process starts with communication and in animals this is behavioral.

Curiously, this external to internal process of novelty canalization continues on down the line. Various parts of the body itself are the ontogenetic and epigenetic 'external' for Hox genes and other exploratory processes. Hox genes must be regularly activated under novel cellular (i.e. external) conditions for new appendages to appear. Neural nets that over-connect and then die back are culled by externally induced signal constraints - or lack thereof (Deacon, 1997, 1998; Gerhart & Kirschner, 1997). At each level of novelty production the internal is syncing external interaction. The internal

[143] To a large extent behavior is the boundary interaction that differentiates the internal from the external. It is the reactive face that negotiates the historic internal's future external relationships. As such, one would have to presume that most external interaction in the form of behavior results in either neutral or negative symmetry. However, this reactive face has been historically selected to find and maintain enough positive symmetry relationships to survive long enough to reproduce itself. As mentioned above, this suggests that even though a majority of interactions at the reactive face boundary may incur negative feedback, the phenotype itself is an instance of the temporary displacement of that negative feedback in favor of positive feedback. In short, behavior samples lots of interactions, but selects, and is selected for, the positive interactions.

is taking external order (i.e. saliencies) and converting it to higher degrees of internal order - canalizing it.

In just this sense, one could expect that any boundary dampening processes that reduced noise in the form of converting unstructured external interaction to structured internal interaction to survive speciation differentially. The classic example here is the big brain. Bigger brains are behaviorally flexible enough to act as an inclusion machine, allowing the animal to couple to a wider variety of more complex environmental features. Less classic examples would be the genetic process of splicing where genetic noise in the form of introns is systematically cut out leaving only the protein coding exons, or the example of autoimmune system evolution where incoming noise in the form of harmful bacteria or viruses are identified and synchronized (in this case sync means neutralized - i.e. recognized and destroyed). These dampening processes survive bottlenecks because they help their organisms remain alive long enough to capture and couple a wider variety of external features. In fact, they enable their phenotype to convert more unstructured external interaction into structured internal interaction than others can. As a result, we should expect to see an increase in the representation of these noise dampening processes. Once again, the selection for noise dampening mechanisms like brains and autoimmune systems would be tempered by the number, frequency and severity of extinctions that, as Gould (1989) notes, have a tendency to make luck the determining factor in survival.

A final point on what we should expect to emerge from speciation is an extension of the argument for long bottlenecks to produce more unique novelties. What long bottlenecks create is an extended fitness landscape search facilitated by noise, uncoupling and survival mechanism adaptations. More searching means the phenotype is more likely to find highly improbable, sloppy exaptations that extend the animal´s ability to exploit 'competence designed elsewhere.' This can happen internally or externally. In the evolution of the ear this was jawbone-brain cooperation. In the evolution of ants this was ant to ant pheromonal cooperation. Cooperation is key in the process of exploiting competence designed elsewhere, whether that 'elsewhere' is in the neighboring bones and cells or in the neighboring organisms. It is these cooperative alliances, wherever, whenever and however they appear that, once they find a stable reproduction platform, go on to produce some of the most novel and successful features of nature. Witness eyes. Witness eukaryotes. Witness social insects. Witness brains. Witness humans.

In the end, this searching process is ideal for generating large exaptations. When we compare Gould´s and Vrba´s exaptive process to Dennett´s concept of the biological co-opting of "competence designed elsewhere" and Margulis´ concept of symbiogenesis we find they all look suspiciously similar. They all look like versions of the process of inclusion found in all universal systems. Also, each represents the rather extreme end of the inclusive processes in that they involve the co-opting not of simple environmental features (e.g. food capture or seasonal coupling) but of features of complex, interactive functionality. In short, they represent the co-option of the functionality of other living systems.

Historically, the two characteristics that define the most radical emergent species are: 1) social cooperation and; 2) the canalization of a reproductive method that preserves that cooperation as a unified whole over time (e.g. autocatalysis, binary fission, mitosis, meiosis, stereotypy, learning, etc.). Together, the picture of bottleneck speciation as characterized by allopatry, punctuated equilibrium and adaptive radiation presents a series of characteristic expectations for those species that are spun off from it. The first would be a fossil record that is significantly muddied by closely related subspecies and rapid, indeterminable regional fluctuations caused by the intermittent mixing of surviving 'puddles.' The human fossil record seems to fit this pattern very well. The second would be an increase in evolvability, noise dampening mechanisms and exploratory processes. As we shall see in upcoming chapters, both increased brain size and cooperative capacity in early Hominins meet this expectation. The last expectation would be that depending on the length and severity of the bottleneck, the probability of extreme and unlikely solutions increases. The most extreme solutions in the history of life are Smith and Szathmáry's major transitions, all of which are social unifications characterized by 1) increased cooperation resulting in 2) canalized cooperation in the form of reproduction. The importance of this kind of emergent cooperative unification in Hominins will be demonstrated presently.

## Presymbolic Human Evolution

> *"Ice age climate is not uniformly frigid but extremely variable. Evidence from the last glaciated era suggests the dramatic swings occurred within the course of a human lifetime."*
>
> *National Geographic (2010)*

Richard Leakey′s diagram (Figure 21) shows a typical if somewhat controversial outline of the timeline of human evolution. The major problem with finding a better one is that there are almost as many timelines as there are paleoanthropologists. This makes each timeline controversial. That said, we are dealing with generalities here, not specifics. The generalities of human evolution are dauntingly general. So general in fact, and so contentious, that we will skip most of the well-known details and summarize them. It is also useful to divide human evolution into two epochs, pre- and post-symbolic language. This division is justified by the sudden and massive appearance of symbolic culture starting about fifty thousand years ago in East Africa, the Middle East and Europe - the so called 'great leap forward.' It is also justified by convenience. Since "Mind" is the title of the next section, we need to find a point where "mind" became part of the human adjacent possible and this will be facilitated by the emergence of language.

**FIGURE 21: TIME LINE OF HUMAN EVOLUTION**

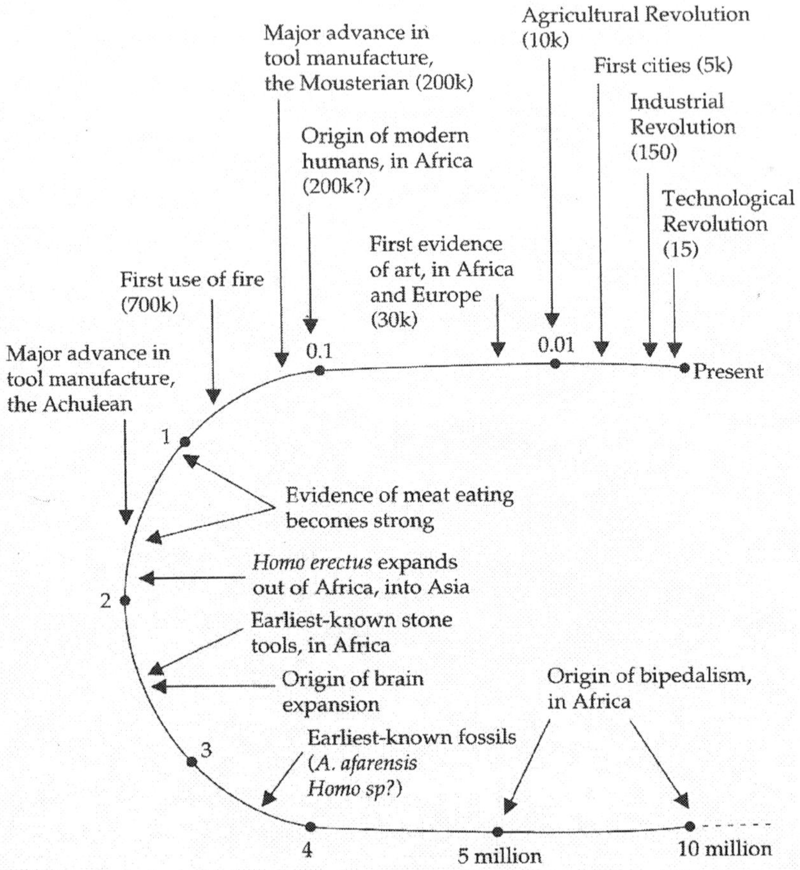


Source: Leakey (1994)

Consensus is that the human decent line split from our closest living relatives, the chimpanzee line, about seven million years ago. For the next five million years there is a bewildering array of potential homo ancestors and parallel hominid species who seem to wink in and then out of existence. Just over two million years ago one of these species, *Homo erectus*, appears in the fossil record as one of the only relatively uncontroversial Homo ancestors and spreads out from eastern Africa to populate much of Asia. This kind of rapid and extensive spread is indicative of adaptive radiations. Since that diaspora, the Erectus descent lines have split into an even more bewildering and controversial array of fossil hominins[144] throughout Asia, Europe and Africa. Notably, however, the majority of the more recent potential proto-human fossil species have also been found in eastern Africa.

Eastern Africa, specifically the Rift Valley, is generally thought of as the home and hotbed of human evolution. Wills (1994) describes it as follows:

> The valley has through all its existence, provided a rich and complex environment for our ancestors. This environment, instead of being some unchanging Garden of Eden, was marked by fluctuations over periods of thousands or tens of thousands of years from wet to dry and back again. … Although the planetary pulse to which the Rift Valley's climate responds remains unclear, the result has been large changes in both weather and vegetation. During the cooler periods vast forests stretched the length of the valley, while grasslands and deserts marked the drier and hotter periods. … It is far more likely that our species was formed through the ceaseless migrations, separations, and mingling of hominids as they moved north and south along the perpetually well-watered valley itself (p. 100).

Other factors also characterize this area of Africa. Some of the most influential were the ice ages that had complex and often dramatic effects on this valley. One example was the repeated aridification of the northern part of the valley to create intermittent periods of connection to the Asian continent through fertile passageways. Complicating this were the preexisting geophysical changes in this area. During the period leading up to the last 2 million years there was the general drying of the African continent that led to the loss of woodlands and the invasion of grasses. With the grasses

144 The term 'Hominid' is used in the modern sense of the larger 'Hominidae' group of great apes. The term 'Hominin' is used for the sub-group of humans and their ancestors.

came invading grassland species and the infamous savannah environment. But these changes were further complicated by the overall raising of the African continent that created a rain shadow and even more pronounced differences between the eastern and western sides of the valley (Diamond, 1992; Leakey, 1994; Leakey & Lewin, 1992; Stanley, 1996; Tattersall, 1998, 2002). According to Spier (2010, p. 119) Dutch ethologist Adriaan Kortlandt even proposed that the dramatic differences between the dry and wet sides of this massive cleft in the continent divided the other primate lineages from the homo lineages. This is not even to mention the extreme mosaic of interlaced environments that must have been represented within the valley itself as arid/grassy heights plunged to tropical depths while straddling the equator and connecting the volcanoes and deserts of the north with fertile temperate zones to the south. Put all of this in the context of radical and rapid ice age climatic fluctuations and the chaotic environment from which our species was born begins to come to life in the imagination.

The background of hominid evolution in eastern Africa is one of massive and frequent climatic and ecological variation. Droughts, floods, volcanoes, sporadic Asian connectivity, lakes disappearing and then filling the valley, invading grasses, invading grass-eating species, geological and climatic upheaval. It all points to a cauldron, a hot pot of evolutionary flux. It is not hard at all to imagine our ancestors winking in and then out of existence, radiating out sporadically to other African and Asian environments and then disappearing almost completely, leaving pockets of uniquely adapted hominids over great distances and then overrunning and interbreeding with them in the next diaspora. This mosaic environment was the background interaction from which we emerged. It was the long and complex bottleneck that alternately forged, rended and then re-forged our ancestors to fit conditions that refused to stabilize.[145]

The two great success stories of the Rift Valley species were *Homo erectus* and *Homo sapiens*. Both expanded their ancestral ranges by constructing and populating niches very different from their home niches. We know this

[145] The conclusion that climatic instability contributed significantly to human evolution and to brain evolution has long been championed by both Potts (1998a, 1998b) and Stanley (1996). Recent surveys of physical evidence have added significantly to these points of view (Ash & Gallup, 2007; Bobe & Behrensmeyer, 2004).

because both species also radiated extensively from East Africa to populate the rest of Africa and Eurasia. Interestingly, the pictures that emerge of these species have striking similarities. First of all, as species phenotypes, they are both large for their environments, large brained and large bodied. They are also aggressive, both as hunters/foragers and as niche expanders. They are both increasingly carnivorous. They are both highly behaviorally labile (i.e. versatile), or intelligent. And finally, they are both highly cooperative, or social. Also notable is the environmental context from which both of these species emerged. They both appear amidst a background of dramatic ecological and climatic flux and they both appear suddenly and then radiate geographically very rapidly.

This picture of Erectus and Sapiens is largely accepted amongst anthropologists. Even if the different qualities are usually weighted differently, it is important to note that in the last two million years, through innumerable changes, it is basically the same package of features that both lineages display, to their obvious advantage. As such, any description of the evolution of 'humanness' in the sense of tracing our immediate ancestors should take into account that it is describing increasingly large, aggressive, carnivorous, intelligent, versatile and social animals.

As an aside, this rather cynical estimation of the fundamental human character was driven home to me when I attended the "Safari zum Urmenschen" installation at the Senckenberg Museum in Frankfurt (9/10/2009 - 18/4/2010). This installation re-created the heads and faces of twenty to thirty of our hominid cousins ("Urmenschen") from seven million years of fossil Hominid skulls. Even though the anthropologist-physiologist artists went to obvious pains to keep emotion from the faces of these, our brothers, I was shocked at how unsettled they made me feel. I couldn´t help the overwhelming gut feeling that this was a group I wouldn´t want to meet on a lonely road anywhere. I have no problem imagining that they felt as I did about each other in their own times.

The second point is about the conditions under which these fossil species appear. Hopefully it is enough to say that the lack of a clear human phylogeny after almost two centuries of some of the most intense and encompassing fieldwork in scientific history definitely favours the hypothesis that human evolution is a complex story. It is no facile search for a missing link either in the form of a single causative environmental feature or a single ape man phenotype. So many potential missing links have been found in the form of distinct phenotypes (i.e. presumed 'species') that the real missing links might never be definitively sorted out. This, combined with the ecological and climatic evidence seems to confirm recent human evolution as a case of an extended and complex speciation event.

Essentially, the last two million years have been one long bottleneck for the emergence of humanity. Unraveling the complicated series of convergences and divergences of factors extending from the genetic to the geological changes that distinguish the human descent line from other failed descent lines is the description of this bottleneck event. It is a daunting job and this investigation will skip it. Not because it is unimportant, but because so much headway has already been made that proto-humans and proto-human environments can be nicely summarized by a fantastic Walt Whitman word, 'multitudinous.'[146]

What is essential to understand in the evolution of modern humans has already been said. We know how systems change and we know how species change. We also know the broad strokes of what our ancestor's main features were and the environment that conditioned their interactions. To understand how language evolved as a result of all this, we should start considering this information with a special eye for not overstepping our bounds.

[146] For surveys of human evolution see Diamond (1992), Leakey (1994), Leakey & Lewin (1992), Stanley (1996), Tattersall (1998, 2002) and Wills (1994).

# Language Emergence

## What a Good Theory of Human Language Would Look Like

First of all, this investigation proposes to explain human emergence as synonymous with the emergence of symbolic language. This means that anatomically modern humans dating back to between one hundred thousand and two hundred thousand B.P. are considered just anatomically modern. Truly modern humans, 'like us,' don't appear until evidence for symbolic thought appears on the walls of their caves or in the designs for their artifacts about fifty thousand years ago. This division, like most historical divisions, is both useful and somewhat arbitrary. Take it for what you will, it will not be further dickered about here. What this does is identify symbolic language as the feature we are trying to explain as our defining human trait. We are looking for Deacon´s (1998), "Homo Symbolicus." But before we start in earnest, perhaps we should take a moment to think about what a good theory of language would look like?

What level of specificity about the emergence of language can we safely assume considering the evidence at hand? Not much. Speciation is very hard to see, precisely because it happens to very small groups under extreme duress over long periods of time. We should expect bottlenecks to be fragmented and messy affairs. For exactly this reason the hard anthropological and archeological evidence we have to go on is scanty and largely in need of human interpretation. The pitfalls of this kind of broad speculation from limited evidence are manifest in the history of these disciplines. To name just two well-known instances one only has to look as far as the historical debates on the meaning of European cave art[147] or

[147] Gould (1998) beautifully illustrates the effect of sociopolitical mindsets on scientific theory as per the changing historical interpretations of prehistoric cave art.

the cultural significance of the fertility goddess figurines.[148] In both cases, massive unverifiable assumptions (like hunting magic or matriarchal societies in these cases) have led to wildly differing viewpoints over time. All of these viewpoints seem defensible and all of them are equally susceptible to the sociopolitical mindsets of their defenders.

At first glance, the evolutionary origin of language seems to be embroiled in very similar circumstances. As previously mentioned, the history of proposed 'causes' of the evolution of symbolic humans has a long and coloured past. Was it upright posture, opposable thumbs, food sharing or mirror neurons that cemented our evolution of enlanguaged brains? Was it all of them together? ...To what degree? ... When? Surely they all had *something* to do with it? So is the list even complete yet? How will we know when it is complete? What can we really say about the evolution of language in humans?

Here are a few suggestions on what we shouldn´t say. Any theory that depends heavily on one specific cause is suspect. Language appears to be a unique exaptation that comes from a long and complicated bottleneck process. It does not appear to be the result of a simple extension of one ancestral function, what was previously called a 'preadaptation.' There is no question that all new features come from ancestral features. Ear bones come from jawbones. The question is how different is the function of the ancestral feature from the evolved feature? Ear bones represent a massive change in functionality from jaw bones and human language functions very differently from vervet monkey warning calls. This change of functionality is the result of a protracted coupling and uncoupling of different features over varying landscapes and time periods. This will be discussed in more detail presently, but for now the more specific the cause, the more suspect the theory.

In this case, causal specificity isn´t necessarily wrong, just improbable. At best, language theories based on very specific causes disappear as their specific conditions locate them in smaller and smaller portions of this long,

[148] Eisler (1987) proposed that matriarchal cultures dominated Europe in prehistoric times. She largely based this presumption on fertility goddess artifacts found throughout Europe. Since then, this presumption has been widely questioned. The assumption I am making here is not that she was wrong, just that her interpretation of the evidence was overzealous largely due to the growing cultural importance of academic feminism.

complex and multilevel puzzle. At best they become one of the perhaps thousands of necessary, but not sufficient pieces of the language puzzle.

Another suggestion is that any serious language theory should make sense in the context of the history of life. This means that we cannot afford to propose theories that cannot be identified as 'the kind of thing organisms have always been doing.' 'Novelty' in this sense, is novel in specific structure only. The process by which that novel structure is created needs to be an identifiably organic process. This means it should have a history of comparable instances in the history of life. Proposing miraculous 'one offs' is, to paraphrase Terrence Deacon, 'not the kind of answers science should be looking for´ (Deacon, 2010b). Once again, this doesn´t mean that most theories are wrong, only that they are not inclusive enough. Most theories only choose one or two levels of interaction from which they contextualize their causes of language. Psychologists choose behavioral contexts, anthropologists choose cultural contexts and neurologists choose brain contexts. If life is a history of interaction at many levels, then any one change should be evident within many different interactive levels. A language theory that makes sense neurochemically and behaviorally, but doesn´t make sense ontogenetically is worse than one that includes all three. Specifically, this means that better language theories should be supported by as much evidence from the physical, chemical, and behavioral studies of the history of life as possible.

A final suggestion is kind of an extension of the previous one. Language is not mysteriously unique. It is no more mysterious or unique than snake venom proteins. The difference in complexity between snake venom proteins and other organic proteins is so vast as to be comparable to the difference between human language and animal communication. It would seem it is only our hubris that elevates our own unique features above those unique features that other animals have evolved. Like snake venom proteins, language must have come about like all the other major exaptations in the history of life. As such, too much attention has been given to explaining language as a result of its unique features. The majority of theories still focus on explaining the unique, never before seen, aspects of language, like syntax, recursion, discrete infinity or special language modules of the brain, instead of focusing on how language is part of what life has always been doing.[149] A good theory of language would include evidence for the similarities that make language just another interaction, just another extension of behavior, along with evidence for the differences that make it unique.

That said, the following chapter introduces a few of the main contenders for the theory of language evolution.

## Problems with Language Origin Theories

The problem with language origin theories is that they all look reasonable. Different theories identify different aspects as being operative to different degrees and in different sequences and combinations under different conditions at different times. Frederick Neumeyer claims that linguists have traditionally neglected the study of language origins for a similar reason. Language specialists cannot even agree on what the important features of language are (Christiansen & Kirby, 2003, p. 4).

To give an idea of the fantastic depth and intricacies of this debate, this investigation presents a series of quotes from Morten Christiansen's and Simon Kirby's superb summarization of the different states of research into language origins. After each quote will follow some comments on how that quote relates to the approach that will be presented here. Please keep in mind that the aim here is to bridge these theories at a metalevel not divide them in details. Very little presented within the context of these theories is wrong as much as it is insufficient. The overall failure of these theories to come up with one compelling account of language origins is largely due to four shortcomings. The first is the failure to contextualize language broadly enough. As a result of this, the second shortcoming is that group selection is not seriously considered. Thirdly, no historical comparative analysis is built that properly contextualizes language within biological history. The fourth and final shortcoming is the persistent scientific illusion that effects have unitary causes, that there is a key or a condition that caused language. The first three shortcomings are all related to the inability or unwillingness to

[149] An interesting exception to this generalization can be found in Studdert-Kennedy and Goldstein (2003) where they deny Chomsky´s claim that discrete infinity is unique to human language. For more see also Christiansen & Kirby (2003, pp. 235-255).

contextualize language as part of the natural history of the universe. The last shortcoming is a result of our language scissors artificially constraining reality to serial and digital effects with serial and digital causes (see chapter "Thought").

To be fair, many of these theories are not 'origin theories'; they only profess to make arguments for parts of an overarching theory. This makes them ideal contributors to a higher level, more general explanation for language like the one offered here. The theory of language origins to be offered here is that language is just another level of constraint in the universal process of selection for persistence of cooperative interaction. In many ways language is just like some of the major transformations that preceded and enabled it. As such, most of these theories are developed to a level of specificity not present in the theory introduced here. What these researchers have found is incredibly important within the limits of cause and effect they have chosen to frame their questions. For this reason, comparison with this investigation's theory is restricted to the broad outlines of their theories. Only the theory summaries are presented here.

Readers who have little interest in current state of the art linguistic and origin of language theories are encouraged to skip to the conclusion of this rather long chapter. For those who want even more detail, feel free to refer to the individual works of each of the researchers that are featured in Christiansen and Kirby (2003). The following excerpts are from that book:

> The landmark paper 'Natural Language and Natural Selection,' published in 1990 by Steven Pinker and Paul Bloom … proposed the theory that the human ability for language is a complex biological adaptation evolved by way of natural selection. On this account, language has evolved as an innate specialization to code propositional information (such as who did what to whom, when, where and why) for the purpose of social information-gathering (p. 3).

The argument for language as a complex biological adaptation is now a given, in part thanks to Pinker and Bloom. In fact, it is the tracing of complex systemic transformations from the Big Bang forward that gives this investigation its unique foundation. Additionally, Pinker and Bloom identify the idea of an, "innate specialization to code propositional information," as fundamental to what language is. The phrase "innate specialization" is a reference to the fact that although no one knows how or why language is acquired, it seems to be automatic and it seems to come from the inside (i.e. genetic). Presently this assumption, either in its Pinkerian genocentric or Chomskian hand waving form, will be challenged.[150] Notice also that "propositional information" is exemplified by, "who did what to whom, when, where and why," questions. All of these concepts refer directly to the uniquely human capacity to identify, and react to, not-here and not-now information (i.e. information not present in the immediate spatial and temporal environment).

> James Hurford ... argues that language evolution needs to be understood as a combination of both biological pre-adaptations ... and learning based linguistic adaptations over generations. He points to several possible ... preadaptations for the production of speech sounds (phonetics), for organizing the sounds into complex sequences (syntax), for forming basic and complex concepts with them (semantics), for complex social interaction (pragmatics), and for an elementary ability to link sounds to concepts (symbolic capacity). Once humans were language-ready with these pre-adaptations in place, language systems would have grown increasingly complex due to the process of transmitting language across generations through the narrow filter of children's learning mechanisms (Christiansen and Kirby, 2003, p. 4).

The issue of pre-adaptations is important. Primate studies imply that our ancestors likely had a range of features that pre-adapted them for language

[150] Chomsky's Universal Grammar theory originally focused on a unique brain module that coded language's deep structure. More recently Chomsky has changed his focus to genes as the deep structure that codes for language. The 'hand waving' referred to here is in reference to Chomsky's original LAD brain module. See the chapter "The Poverty of Stimulus Fallacy" in the "Mind" section for more.

(Bell, et al., 1977; Savage-Rumbaugh, 1997; Seyfarth & Cheney, 1997). However, it is important to remember that pre-adaptations are most likely only part of the story. Many theorists rightly stress that most of the changes that brought about language are best imagined as co-evolution (Deacon, 1998) or multilevel evolution (Christiansen & Kirby, 2003). That is, environment, communication structure, brain structure and body structure are all feeding back into each other all the time. The most underappreciated facet of this complex issue is that novel adaptive structures are usually created by novel environmental interaction. It is communication that is selected for cooperation that is then selected for reproduction. Canonically, new adaptations are thought to appear in reverse; reproductive mutation is traditionally thought to select for cooperation, which is further selected for emergent communication.

As we have seen, the study of the major transitions in evolution encourages a different view. Truly novel adaptations almost always develop from the outside in. We should look first to how our ancestors interacted to see how they cooperated. We should expect to find evidence of that cooperation starting to move into the internal physiological systems as cooperation was routinely selected for over long periods of transformation (as it was differentially reproduced). In this context, pre-adaptations can be seen as internal physiological constraints that are eventually co-opted to preserve successful novel external interactive processes. The jawbone structure limited the inner ear to bone components, but only through continual selection for external sound-brain interaction.

It is also interesting to note the separate categories that Hurford identifies as parts of the puzzle; phonetics, syntax, semantics, pragmatics and symbolic capacity. This investigation proposes that the sequence of development that most resembles the natural process of novel system emergence is something like pragmatics itself maintaining a kind of social semantics and, eventually, a syntax that was slowly canalized by individual human bodies into physiological processes like phonetics and neural architecture.

Note as well Hurford's mention of the importance of how language would be constrained in form by how and what children could learn. It must be remembered that the "narrow filter of children's learning mechanisms" is not so narrow as to be exclusive. Even now, a lot of important information is not deemed learnable by children until they are quite old. Symbolic information is passed horizontally as well as vertically and thus it need not be unduly constrained by what children can or cannot learn. Another way to imagine this is to realize that language, like DNA, is an information system that has been maintained, unbroken, since its inception. Children may re-embody language, but they do not reinvent language. One can easily imagine complex religious concepts being passed down from adult to adult through rituals that are restricted to the adult members of a group. Children have evolved complex relationships with language. Like all highly adaptive behaviors, language has undoubtedly shaped and been shaped by human ontogeny, but children still inherit a continuously unified and pre-formed system that is at least fifty thousand years old.[151]

> Frederick ... Neumeyer suggests that a more measured approach to uniformitarianism is needed because there may have been differences in the use of language across language evolution. For example, language may originally have been used as a tool for conceptualization rather than communication (Christiansen and Kirby, 2003, p. 4).

It is crucial to realize that, "there may have been differences in the use of language across language evolution." This would be especially likely if the selection process was long and complex as characterized by the ice age bottleneck. In the case of the theory to be proposed here, what functions originally to convey just locative information (e.g. "I'm over here") can begin to function as a global response mechanism when that locative information is potentiated by social networks surviving in complex environments. This will be elaborated shortly.

[151] See Deacon (1998) for an in depth treatment of the co-adaption of language and humans. Pigeon languages may be cited here as an exception except that children growing up in mixed language environments already know one language and pigeons are just a way of mixing down and then mixing up a fusion, not a true invention. There are no examples of groups of children in languageless environments spontaneously evolving a truly new language.

The gestural theories especially depend on the functional transformation of language. On some level they have to explain how gestures became subordinated to vocalizations. As for the implication that "conceptualization" preceded "communication," there is little evidence for natural symbolic concepts in primate studies and plenty of evidence for communication. Although concepts could have spontaneously appeared in a mutated individual's brain it seems unlikely. Language is social, not individual behavior. Concepts in symbolic form are maintained and reproduced only because they bridge meaning between at least two individuals. Individuals die, concepts are passed from person to person, from generation to generation. Concepts are a feature of social networks.

> Derek Bickerton ... suggests that when approaching language from an evolutionary perspective it is important to look at language ... as the coming together of three things: modality, symbols and structure. He argues that a largely cultural emergence of symbolic representation combined with a biological adaptation of brain circuitry capable of encoding syntactic structure were the two distinct evolutionary sources that gave rise to human language. Only later would a preference for the spoken modality have evolved, and then entirely contingent on the prior existence of the symbolic and structural components of language. From this perspective, the evolutionary dissociation of symbols and structure are reflected in ape language studies, where learning of symbolic relations approaches a near-human level of performance but where only a limited grasp of syntax has been demonstrated. Bickerton concludes that a capacity for structural manipulations of symbols may be the key adaptation that gives us, but no other species, language in all its intricate complexity (Christiansen and Kirby, 2003, p. 5).

"Modality, symbols and structure," is another useful way to simplify the problem. However, Bickerton is presuming that brain structure capable of encoding syntactic structure is at all remarkable. Neural nets evolve to be trained by external patterns (e.g. either environmental patterns, body patterns and/or other neuronal patterns). Brains are exceptionally plastic and that makes them one of the first places where external patterns leave an impression. If we consider Philip Lieberman's hypothesis that syntax co-opted the motor sequencing patterns of the basal ganglia we can see that complex sequencing skills for important adaptive behaviors (e.g. running, throwing, etc.) not only pre-existed in the brain, but also have a long history of canalizing newly adaptive behaviors (e.g. upright running, throwing spears, etc.). Syntax is just another sequencing pattern (albeit highly complex) for a new behavior - talking.

Certainly a brain is necessary, but had we not learned to sync to language patterns the way we did (and vice versa- e.g. language synced to us), we would have learned another way. Brain structures are more often the first evidence of selected patterns of interaction not reasons for the origin of that interaction.[152] Our basal ganglia learned to order our feet into running not because it wanted to, or was built to, but because we kept co-opting it to help us run and those ganglia that helped organize running better, survived.

The vertebrate brain is actually highly conserved. The differences in vertebrate brains are more the result of specializing an old brain bauplan than of adding new parts. This means that important new brain functions (like new behaviors or senses) are usually dependent on hypertrophied, or otherwise specialized, areas of an ancient brain-plan (Campbell, 2008). Ontogenetically and genetically, it is not hard to 'add more volume' to a certain part of the brain and then that new brain matter (read, 'neocortex' for humans) is available for novel potentiation by its environment. Syntax is the rules of organizing symbolic representations within complex interactive routines. At some point it was heavily selected for. When it became heavily selected for it co-opted extant brain structures and thereby differentially preserved those

[152] This argument is a simplification. It gives the impression that the brain is inert without external stimuli, that the brain is like play dough just waiting for the imprint of external experience to give it a functional form. Closer to the truth is that the brain maintains certain historic structural and rhythmic patterns that are disturbed in various ways by external stimuli to produce responses that can be selected for adaptability. Philip Lieberman describes such an intercourse when he theorizes that syntax emerged as a co-option of the basal ganglia's preexistent function of coordinating complex motor skill.

brain areas that maintained syntactic patterns, not the other way around. The neocortex didn't 'cause' syntax, the selective pressure for syntax helped 'differentially preserve' the neocortex.

Another way to see this argument is that if the brain ran the show, it would run the show like an efficient engineer. It would wait for mutations that encouraged new functions and then develop these mutations into new modules with new functions. This would be a neat, static, modular brain. Keep what works and add what you need. If the environment ran the show, the brain would be sloppy. The environment would present a fluctuating suite of constraints and let the brain sort itself out. Instead of waiting for mutations to create special new parts, this kind of brain would preserve whatever worked, whenever it occurred. It would be characterized by brain areas that are often co-opted by new functions and create shifty alliances of 'kludged' brain parts that are full of the ghosts of ancient functions waiting to be recaptured by unpredictably odd futures. This brain would be a sloppy, dynamic and contingent brain.

Bickerton presents an excellent argument for the centrality of syntax to language evolution. He also mentions the necessity of understanding language as a cultural emergence. Unfortunately, how language culturally emerged is not explored.

> Michael Tomasello … shares the view that it was the separate evolution of capacities for using symbols and grammar (syntax) that distinguishes human communication from the communication of other primates. In contrast to Pinker and Bickerton, he suggests that there was no specific biological adaptation for linguistic communication. Rather, Tomasello argues that there was an adaptation for a broader kind of complex social cognition that enabled human culture and, as a special case of that, human symbolic communication. A crucial part of this adaptation was an evolved ability to recognize other individuals as intentional agents whose attention and behavior could be shared and manipulated (Christiansen and Kirby, 2003, p. 5).

Tomasello's theory is very similar to the one presented here, and vice versa. Language is the result of an, "adaptation for a broader kind of complex social cognition that enabled human culture and, as a special case of that, human symbolic communication." In general, Tomasello is very interested in the use of gesture to create shared contexts of intentionality, both in pre-linguistic children and in apes. Both of these interests focus on cooperative activities being selected from communicative backgrounds (Tomasello, 2008). In the upcoming chapters we will revisit the spirit of Tomasello's argument for the centrality of shared contexts of intentionality in a different framework. We will look at the unique way that humans use each other to 'sense' each other's environments and expand a communal sense halo within which cooperative arrangements can be sampled and potentiated to maintain an extended range of group behaviors. Even though Tomasello principally restricts his interest to individuals and goes no further back than primates and children to make his point, his kind of theory is best supported by the history of major emergences in the universe and of life here on Earth.

> Arbib suggests that the mirror system forms the evolutionary basis for a link between a sender of a message and the perceiver of that message. The same subset of neurons appears to be active in the mirror system both when generating a particular motor act and when observing others producing the very same motor act. Following the evolution of a unique human ability for complex imitation, Arbib proposes that language originated in a system of manual gestures, and only later evolved into a primarily spoken form. Finally, Arbib joins Hurford, Tomasello and Davidson in arguing that syntax emerged as the result of subsequent cultural evolution (Christiansen and Kirby, 2003, p. 8).

Arbib is interested in mirror neurons as a means for motivating symbolic language evolution. The problem with mirror neurons is not that they undoubtedly played a part in training our brains for language, but that they do not explain the behavior in the first place. They function to imitate, they are a classic "coupling" mechanism like pair bonding or heightened aggression in mothers. Like the basal ganglia, mirror neurons were maintained by our ancestors because they were extremely adaptive for a number of behaviors long before symbolic language co-opted them. Obviously, a feature such as mirror neurons would be very helpful in preserving novel adaptive behaviors (like call and response) once they were initiated. Unfortunately, they do not explain the novel functionality of symbolic language in the sense of what made it selective, only how it was maintained. Like the basal ganglia, mirror neurons were co-opted by language behaviors. They didn't cause language behaviors.

Both Tomasello and Arbib praise mirror neurons, but to imply that, "the evolution of a unique human ability for complex imitation," is a necessary precursor to language evolution is overstating its importance. We didn't need specially augmented mirror neurons to call and respond and to begin to develop symbolic group reactivity routines. Ants and bees have highly coordinated, differentiated and specialized adaptive routines and yet, presumably, no mirror neurons. Of course, once call and response became the selective background for an especially powerful form of behavior, selection would begin to transform apelike mirror neurons into humanlike versions supporting increasingly specialized symbolic behaviors.

> "Michael Corballis also sees language as originating with a system of manual gestures, ... he argues that whereas non-human primates tend to gesture only when others are looking, their vocalizations are not necessarily directed at others" (Christiansen and Kirby, 2003, p. 8).

Once again, there is little doubt that gestures were a large part of many protolanguages. Anybody who has been to Mexico knows that gesture is still a large part of modern languages. However, despite hundreds of millions of years of gestural communication, it is hard to imagine any aspect of the symbolic thought process for which physical gesture is a sufficient explanation. Gesture was probably necessary for language origins, but not sufficient. Gestures may have bridged some important developmental gaps during tough times in the bottleneck. Like mirror neurons, they are so clearly part of our communicative history that we shouldn't rule them out.

The thirteen billion year history of emergent universal systems suggests that language is the result of a preservation process, specifically selection for persistence eventually resulting in inclusion, differentiation and specialization of what begins as a group activity. If language marks the emergence of a truly novel phenomenon then it should be considered in light of the other major emergences. This means that it should be considered a social phenomenon produced by some kind of group selection logic like the others. Language and thought are the preservation of a shared symbolic map of not-necessarily-here and not-necessarily-now saliencies selected for group coordination.

Where gesture seems highly effective in one-to-one communication when plenty of time is available, it rapidly becomes inefficient for one-to-many communication. Both religion and politics define humanity and it is hard to see how they could evolve without a one-to-many communication system. Such one-to-many communication events are crucial to team building and often include different instructions for specialized members within time constraints. Even with the help of a gestural language derived from modern verbal symbolism (e.g. ASL), deaf basketball and football teams surely take more time, training and/or have less dynamic playbooks than verbal teams.

> Robin Dunbar argues … that the arguments in favour of a gestural origin of language are largely circumstantial. He moreover contends that gestural language suffers from two major disadvantages in comparison to spoken language: it requires direct line of sight, and it cannot be used at night (Christiansen and Kirby, 2003, p. 9).

Robin Dunbar is a great theorist. He is absolutely bang on with identifying the glaring faults of gestural theories of language origins. He also is one of the few theorists that recognize the environment as being the only context within which novelty, in this case symbolic system origins, can be explained. He too is looking for the kind of interactions that could initially maintain a social protolanguage and then later, with selective feedback, fix it into the individual brain. In fact, the idea of one-to-many communication is the backbone of his 'gossip' theory of language origins. However, for reasons that we will explore later, gossip (or social grooming) is an insufficient cause of language.

> Michael Studdert-Kennedy and Louis Goldstein ... focus on the mechanics involved in producing the sounds of human languages. They propose that a key pre-adaptation for language was the evolution of a system in which a limited set of discrete elements could be combined into an unlimited number of different larger units ... They suggest that the ability for vocal language draws on ancient mammalian oral capacities for sucking, licking, chewing, and swallowing. Subsequent evolutionary pressures for more intelligible information exchanges through vocalizations would then have led to a further differentiation of the vocal tract. ... [The] combination of phonetic gestures into larger complex structures would have occurred through processes of cultural evolution involving attunement among speakers through vocal mimicry (Christiansen and Kirby, 2003, pp. 9-10).

These "Ancient mammalian oral capacities," will figure in comparisons of the mammalian predilection for the radical aural/vocal sensory adaptations we find in bats, whales and humans. Also, the evolution of 'combinable discrete elements' is part of what will be referred to as the 'digitalization' of language, and it has an important impact on human language beyond recursion and discrete infinity. Studdert-Kennedy and Goldstein (2003) also mention the long history of 'combinable discrete elements.' They argue that combinable discrete elements characterize all physical, chemical and genetic systems and that this contradicts Chomsky's (2000) claim that "discrete infinity" is a "biologically isolated" property (p. 3).[153] They see linguistic discrete

[153] Chomsky's idea of discrete infinity is a set of finite variables (like words) that can be linked (syntactically) in an infinite number of ways. 'Discrete' refers to the finite set of variables and 'infinity' refers to the endless number of word combinations this set can produce. A result

infinity as a generic natural process. Where their theory fails is in focusing on recursive discrete infinity, with a heavy dependence on gestures, to explain language development. It could easily be argued that all reproductive languages (e.g. autocatalysis, binary fission, mitosis and meiosis) expand their precursor's spectrum of recursivity.

Expanding the recursivity of substrate interaction (e.g. through discrete infinity) is not unique to human language. In fact, discrete infinity and recursivity appear to be predictable features of systems emerging from selection for symmetric persistence from an expanding spectrum of interaction. The addition of new units of selection with new reactive faces to an environment that continuously feeds these discrete units back into the environment in cycles of maintenance and reproduction is itself a form of recursive discrete infinity. All emergent reproductive rule-sets should include these features. In a mundane way, recursivity is just reproduction, and vice versa.

This implies that recursivity and discrete infinity are not the missing link between ape communication and human language. They are simply a convergent feature of any novel rule-set that maintains emergent structures. In fact, the transformation to symbolic language was so complex it should not be thought of in terms of having one missing link. Studdert-Kennedy and Goldstein do an admirable job of explaining how language may have evolved. They describe one possible route after something like linguistic discrete infinity mysteriously emerged. In this sense it is once again a theory of what may have been necessary for symbolic language to emerge, but it is not sufficient to explain language, much less its emergence.

> Philip Lieberman ... argues that upright walking would have resulted in biological adaptations of basal ganglia - a collection of subcortical brain structures - for the learning and sequencing of more complex movements. These changes to basal ganglia formed the key adaptation en route to language... A consequence of this view is that language has a rather long evolutionary history, with simple symbol use (in

---

of a system of discrete variables endlessly interacting according to a rule-set (i.e. universal grammar) is 'recursion,' a repetitive pattern.

> the form of naming) and rudimentary syntax dating back to some of the earliest hominids. Lieberman notes, however, that modern speech would have emerged considerably later in human evolution, given his interpretation of the fossil record and comparisons with the vocalization abilities of extant apes (Christiansen and Kirby, 2003, p. 10).

Once again, everyone wants to propose "key adaptations." Undoubtedly, there were many essential physiological preadaptations that put language into the human adjacent possible; however, they speak little to the actual process that selected for and maintained language. In a trivial sense, every feature of proto-humans was a potential preadaptation for language because only that suite of features was available for co-option in the series of adaptations that did preserve language.

Lieberman is recognized for courageously opposing the simplistic approach of Noam Chomsky by asserting that language wasn't a magic "universal grammar" module added on to the human brain. He was one of the first to demand a specific evolutionary explanation for language. His work depended largely on establishing that the basal ganglia, an ancient part of the vertebrate brain that helps coordinate motor skill sequences, was essential for the human ability to build syntax (i.e. word sequencing rules). Lieberman argues that syntax is based on ancient motor skill adaptations like learning to walk. In a sense his syntax argument parallels George Lakoff's (1987) upcoming "cognitive semantics" argument. They both provide good evidence to believe that language co-opted ancient systems of adaptive behavior. Syntax co-opted an already functional motor sequencing skill. Semantics co-opted an already functional behavioral system. Like mammalian hearing systems (ears) co-opted existing jawbones and brains to maintain an exaptive coupling to novel external saliencies, so did syntax and semantics (human language systems) co-opt existing motor sequencing and behavioral systems to maintain an exaptive coupling with novel external saliencies. Both Lieberman's and Lakoff's theories also evidence the internalization of what begins as adaptive *external* interactions. We will revisit George Lakoff's cognitive semantics in the next section's "Women, Fire and Dangerous Things" chapter.

To be clear, given the history of emergent systems, language was most likely not a result of novel internal interactions. Internally, systems are conservative; they preserve successful external interactions. Only in a very limited sense do they invent them. It is external interaction that finds maintainable novel interactions. This means that it is unlikely that changes to basal ganglia or mirror neurons caused language to evolve. More likely is that the basal ganglia changed to maintain syntax because complex symbolic sequencing was preferentially selected for within the proto-human environment. Both basal ganglia and mirror neurons certainly played important roles in preserving and internalizing symbolic behavior, but they did not incite that behavior themselves.

Lieberman also hypothesizes that symbolic behavior has been around for much longer than syntax. He argues that fully functional syntax is probably fifty thousand years old while 'sloppy' symbolic behavior is much older. This argument is based on solid physiological and anthropological evidence and will be revisited in the epilogue. As for Lieberman's historical analysis, there is no doubt as to the, "rather long history" of language. As another novel system of maintainable interaction, language is as old as life, older. Transformation with preservation of *symbolic* structure is merely the most recent emergent symmetry in the thirteen billion year old story being told here.

> Lieberman and Corballis both point to the evolution of more complex sequential learning and processing abilities as forming part of the foundation for the origin of language... Simon Kirby and Morten Christiansen similarly relate general properties of sequential learning to the structure of language. Specifically, they propose that many language universals - that is, invariant sub-patterns of language - may derive from underlying constraints on the way we learn and process sequential structure, rather than from an innate biological adaptation for grammar. ...specific language universals can be explained by sequential learning constraints. This perspective

> further suggests that languages themselves can be viewed as evolving systems, adapting to the innate constraints of the human learning and processing mechanisms. … They argue that the broader consequences of this work are that language evolution must be understood through processes that work on three different, but partially overlapping timescales: the individual timescale (through learning and development), the cultural timescale (through iterated learning across generations), and the biological timescale (through natural selection of the species) (Christiansen and Kirby, 2003, pp. 10-11).

One of the most rewarding and encouraging events in researching emergent processes through different classes of natural systems is to see professionals on one type of system echo professionals on another. Nothing sounds sweeter. In this case Lieberman, Corballis, Christiansen and Kirby are describing language processes much like Goodwin (1994) describes ontogenetic processes. Goodwin goes to great lengths to show how the physics of chemical and mechanical relationships between molecules, cells and other regularities below or above the genome level maintain generic patterns that are responsible for a great deal of the organization in animals. Traditionally, genes have been considered the only source of order in organisms. He points out that a vast amount of the order in organisms comes from non-genetic regularities (e.g. quantum patterns, chemical patterns, astronomic patterns, geometric patterns, cellular patterns, maternal patterns, etc.).

In most cases, genes take advantage of this 'generic' order to initiate and sequence how their information is activated and structured. You could say that this generic order is stored in the gene's environment. Genes are one pattern buried within many other patterns, one sequence within many other sequences. These other patterns are absolutely necessary for the successful reproduction and persistence of those genes. Genes are only one level of order upon which many higher and lower level processes of constrained release of energy operate to maintain pattern. The above theorists are noticing much the same thing. They all want to explain the, "structure of language," by referring to the, "underlying constraints," in the system generation process.

Goodwin's developmental processes are constrained by physical and mechanical rules, what he calls "morphogenetic fields." Lieberman's syntax processes are constrained by the motor sequencing rules of the basal ganglia. Christiansen's and Kirby's information processing rules are constrained by network, logic and informational constraints. In all cases, the resulting patterns are maintained by differential preservation of free and natural, lower level (internal) and higher level (external) symmetric transformations.

Another fascinating element of Christiansen and Kirby's work is the identification of the interaction of three distinct levels of maintained systems. They call them "timescales" but one could equally think of them as 'change scales' or 'symmetry scales.' At each level the product structures are maintained by the interactions of their component structures as well as the conditions of their external interactions. They identify the maintenance process of organismic systems (e.g. individuals) as "learning and development," of social systems (e.g. groups) as, "iterated learning across generations," and of species (e.g. species) as, "natural selection of the species." Upon examination you will find that order at each scale level depends on the reproduction of informational relationships (i.e. synchronic maintenance) be they mechanical, chemical or behavioral. More to the point, emergent novelties like language start with interaction on the individual timescale (e.g. communicative level) and need to be selected for persistence on the cultural (e.g. cooperative/social level) and biological (e.g. reproductive/species level) timescales. However, regardless of where one identifies the starting point, the language pattern is *maintained* by differential preservation of free and natural, lower level (personal) and higher level (social, species) symmetric transformations. To complement this idea try replacing the word "timescale" with "transformation-scale" or "change-scale."

Most importantly, they are entirely correct in claiming that language must be explicable on all levels. Single level explanations are not only incomplete, but the more levels the emergence of language patterns are explicable on, the better the explanation.

> Ted Briscoe focuses on the possible emergence of biological adaptations for grammar. … he suggests that we may be able to understand how language-specific learning biases could have arisen in our evolutionary history by exploring how learning itself may impact our ability to procreate. The assumption is that aspects of language, which were previously learned, would gradually become genetically encoded through 'genetic assimilation' - that is through genetic adaptations for language selected to increase reproductive fitness … On this account, language started out relying on general-purpose learning mechanisms, but through biological adaptations learning gradually became language-specific (Christiansen and Kirby, 2003, p. 11).

Ted Briscoe's process of internalizing language is right. Adaptive behaviors are preserved by moving them into the internal milieu of conserved relationships. This process can be called the Baldwin Effect, genetic assimilation, inclusion or internalization. The terms 'inclusion' or 'internalization' seem better because both the Baldwin effect and genetic assimilation are restricted to genes and to living systems. Not only can behaviors be preserved through external interaction (e.g. non-genetic interactions), but they never necessarily need to move into the genome at all. In fact, non-living systems also internalize each other in cycles of maintenance and reproduction: witness autocatalytic cycles, binary stars and stellar nucleosynthesis.

One unsatisfactory aspect of Briscoe's "Grammatical Assimilation" argument is that it seems to treat the evolution of language as an incremental extension of functionality, like a slightly longer neck on giraffes or slightly thicker fur on northern wolves. This investigation argues that these kinds of extensions of pre-adaptions are by far the most common evolutionary emergences, but they are not characteristic of the major transitions of evolution. Language is a whole new way to relate to the environment. It opens up an immeasurable phase space for human communication, cooperation, reproduction and, ultimately, radiation. From a historical point of view our symbolic transition doesn´t look like an incremental extension of extant capabilities as Briscoe´s theory implies.[154] It looks more like a major exaptive transition.

The principle problems with language origin theories are related to not considering language in a broad enough context. To start with, these theories generally do not recognize the directionality of biological novelty. As previously mentioned, emergent systems start with external interactions not internal interactions. Theories that imply preadaptations, mutations or other internal systems as language 'causes' are missing the point that language most likely began outside the proto-human body. All of these internal systems probably played an important role in capturing, coupling and eventually canalizing protolanguage behavior, but none should be considered as causal at the beginning except as they represented adaptations selected in preexisting contexts for preexisting functions.

Secondly, with a few notable exceptions, group selection is largely overlooked (or vaguely assumed) in these theories. Many of them give lip service to the importance of social or cultural contexts, but none of them take the opportunity to envision language pragmatically (in terms of social relationships). Those that do, still use the context of how individuals benefit from language proficiencies within the group to motivate their origin story. This stands in stark contrast to how human groups as units of selection benefited from language-like proficiencies within complex environments.

Thirdly, there is a dearth of historical comparative analysis in these theories. To interpret brain evolution, scientists take endocranial casts and then apply a process of comparative anatomy. Endocranial casts allow them to deduce how a specific brain was organized by how its surface shaped the inside of the skull. Comparative anatomy allows them to deduce how brains changed in relation to those brains that came before and after them. The first method establishes how brains are structured and the second establishes how brains structures transformed over history. Together they create a complementary, dynamic, multilevel understanding of the trajectory of brain evolution.

[154] Even major transitions are composed of incremental changes. The difference is in the functional results of these incremental changes. The incremental extension of giraffe necks seems functionally different than the incremental extensions of proto-mammalian jaws to become ears.

The overwhelming majority of these language origin theories use a carefully contextualized theoretical framework that deduces language origins from either the structure of language itself (linguistics) or from the structure of the brain (neuroscience). The conclusions drawn in these theories are void of any serious consideration of language within a comparative history of systems. Language seems to be considered totally and wholly unrelated to the historical development of the communicative-cooperative-reproductive systems that form its evolutionary substrate. It's like language has no historical context. Maybe Eve made language from God's rib?

This last fault is hard to explain other than in the context of an age-old human misconception, that we are unique. The only reason not to look for related transformations in the history of universal transformations is the presumption that nothing with any language-like features ever happened before. This is rather like only considering one endocranial cast to determine how brains are organized. No attempt is made to develop a context within which the trajectory of transformation makes sense. Language origin theories suffer from shortsightedness. A lot of evidence from the proximal sources of psychology, neurology, anthropology and linguistics has been gathered. The results of these investigations are crucial to the solution. However, even more crucial is the overall context of language emergence. This context is the history of emergent systems, particularly the history of how life here on Earth has selected for communicative mediums (e.g. 'languages') that enable emergent cooperative unities that reproductively persist, thereby growing new spectrums of interaction from which new cooperative unities can be sampled and maintained. Over the last four billion years many cooperative reproductive codes, or 'languages,' have evolved to reproduce new cooperative unities: autocatalysis, binary fission, mitosis, and meiosis. Each of these aforementioned languages radically altered the face of the Earth in its time. None are considered here.

Finally, there is the consistent and pervasive failure to recognize that language is not an effect in need of a cause. Language is not a thing, produced by other things. It is a pattern of transformation selected from the interaction of other patterns of transformations. It is an added way to constrain interaction. In this way it is much more like other reproductive processes than like a one-off Big Bang. It is a pattern of persistent interaction maintained by how its components, in this case humans, interact.

## General Differences in Language Theories

> *However, the issue essentially boils down to a choice between two possible pathways to language as we have it. Both assume that, in the here and now, language can be used for both social and technological information exchange: the real question is which came first. The traditional view is that language evolved to subserve a technological function and, once we had it in place, its intrinsic capacities allowed it to be used for other more trivial purposes (such as social information exchange); the alternative view is that language evolved to subserve the social function, but its intrinsic properties allowed it eventually to be used for other purposes (e.g. technology) that were themselves valuable. The issue then, is which was cause and which consequence?*
>
> *Robin Dunbar (2003, p. 220)*

When it comes to theories of language origins, it is a jungle out there. Chomskian nativist theories claim, one way or another, that language is innate. That is, a hardwired ability for symbolic language somehow recently coalesced out of a complex brain doing a complex job (Chomsky, 2006, p. 184). Although, this is not impossible, it seems lazy, unspecific and miraculous. Genes are the current favorite for what holds this hardwiring. Unfortunately, the closer we look at genes the more we realize they need cellular environments around them and molecular and quantum substrates within them. There is no lone 'language gene' and moving the innate magic of language from LAD brain modules to genes solves nothing. Genes are just one level of pattern whose survival and reproduction depends on many internal and external levels of pattern. In general, nativist theories are altogether too vague. Often they seem contrived to conveniently avoid the issue so their authors can get on with more important issues of language.

In the end, there are a lot of different theories with different kinds and levels of evidence. This has been just a small sample of some theories and which ones seem better. However, before we move on to develop a new theory it would be useful to take a look at reducing the complexity of the issue. In a way, the diversity of these theories can be seen as a reflection of the diversity of the language adaptation itself. The possibilities for adaptation amongst such richly interactive hominids over such a long period of speciation and extinction in such an unstable environment should be expected to be numerous and complex. Perhaps it will be helpful to identify how this puzzle can be most profitably approached given the current state of evidence and theory.

## A Deductive or Inductive Search for Language

There are two ways to approach the origin of language problem. The first is inductive. This approach is based on accumulating evidence from sources like fossils, comparative anatomy or neuroscience. Induction combines the evidence to produce a solution. It moves from specifics to the general. However, in this case the glut of evidence from all sides is muddying the water more than clarifying it. It is precisely the labyrinthine permutations of how the evidence can be strung together that keeps us from finding the answer. In the end, this is the complaint about theories that are too specific. There is so little conclusive evidence and so much circumstantial evidence that, like the Bible, any statement can be supported. Eventually, if we keep collecting evidence and comparing it in this manner we will find the solution, but after how long?

The other approach to the problem is deductive. Deduction moves from the general to the specifics. A deductive approach would be the following. If language looks like this, what evidence can we find that supports such a view? This approach is theoretical and speculative. It takes the result and then looks for the evidence that supports it. It calls for generating a "Just so story" and then comparing it to the evidence. The danger of the deductive

---

*So Stories* which is a collection of fanciful stories about how animals got the features they have (e.g. how the elephant got his trunk and how the leopard got his spots).

approach is that it can invent or co-opt the evidence it needs to make itself work. To protect against this kind of fabulation, deductive reasoning needs to be tested and supported by as many diverse and credible sources of evidence as possible.

To use inductive reasoning for the origin of language is hopelessly complex. At least with deductive reasoning we have a sure place to start from, a theory of language.

## A Group Selection Theory of Language Origins

At this point we have an overview of how systems evolved, how life evolved and how presymbolic *Homo sapiens* evolved. The net cannot be cast any wider in the search for human origins. However, the catch has yet to be hauled in. What this investigation proposes is that language, thought and humanity can be shown as part of the universe doing what it has always done. We are no more special, and perhaps considerably less so, than the emergence of molecules, solar systems or brains. To find the true outlines of our origins we cannot fail to consider previous origins. Emergent systems have a long history of emergence in this universe. For thirteen billion years systems have been maintaining certain communicative, cooperative and reproductive mediums to preserve and re-create themselves. To consider the human mind outside this history is hubris, plain and simple.

In the interests of progress this investigation will not dally with further fascinating and important linguistic, psychological, neuroscientific, anthropological, paleontological, etc. evidence for language origins except when they pertain to the theory that will be promoted. This is truly unfortunate. Perhaps the future will afford the luxury of giving other theories and their authors the attention they deserve. They are the shoulders we all stand on.

True to the proposed deductive line of reasoning, the "Group Selection Theory of Language" will first be proposed, and then supported. Like each of the other major transformations in the history of life, *language evolved to cue complex coordinated social behavior.* It "cued" because protosymbols, like vervet monkey warning calls, evolved to initiate adaptive reactions in others by broadcasting stereotyped behavior. It was complex because the adaptive reactions necessary for survival in such varied social and environmental conditions rewarded plasticity. It was coordinated because it maximized behavioral flexibility (e.g. groups that differentiated and specialized their members survived preferentially). It was social because language is primarily a social not an individual phenomenon. And finally, it was behavior because that's what big brained proto-humans specialized in, flexible behavior.

Many theorists will see the ghost of their own syntheses in this synthesis. This is true. However, it is not a ghost as much as a reconstructed model, at the best resolution possible, of the spotty and disparate body of evidence we have. Unfortunately, many theorists have gotten so specific in their theories of language origins that they have forgotten the utility of a functional one. This theory does no more than state that groups of proto-humans that could maintain a greater variety of vocally (or gesturally) coordinated adaptive reactions survived better. The question is, how did they do this?

## Flexing your Behavior: how to beat the bottleneck

> *Social behavior, like all other forms of biological response, is a set of devices for tracking changes in the environment. No organism is ever perfectly adapted. Nearly all the relevant parameters of its environment shift constantly … the parameters change at different rates and often according to independent patterns.*
>
> *E. O. Wilson (1975, p. 144)*

Historically, the best way to beat the bottleneck of extinction in times of upheaval is luck. As Gould (1989) so eloquently argued, luck is critical to survival. Often the luckiest ones are the small ones, the ones whose modest metabolic needs can be easily met in times of scarcity. However, although there is evidence for significant geological and ecological upheaval around the time of the emergence of both *Homo erectus* and modern *Homo sapiens* (Dennett, 1991; Leakey, 1994; Leakey & Lewin, 1992), these times seem to have been less times of overall scarcity, like those imagined after the K-T meteorite that killed the dinosaurs, and more times of rapid and complicated change. In these times, resources didn´t disappear as much as they were constantly in climatic and ecological flux. They were the kind of times that reward flexibility as well as luck.

Flexibility in this instance refers to three levels of interaction. These three levels are comparable to the levels previously referred to by Christiansen and Kirby (i.e. individual, group and species level interactions) and they also reflect the levels of behavioral response E. O. Wilson outlines in "Sociobiology" (see Figure 16). First we will take a look at the species and individual levels and then we will move on to the crucial group selection level.

The individual level of interaction describes each animal´s ability to recognize and capture, or couple, variable environmental saliencies over its lifespan. Flexibility starts at the shortest timescale with individual behavior and then gets preserved over generations in descent lines or social arrangements. Ultimately, it becomes stored in the various internal systems of the individuals that constitute the species phenotype. As Gerhart and Kirschner (1997) note, "behavior, as a form of phenotypic plasticity, can generate the most immediate adaptations but it also creates new selective opportunities for evolutionary change as well" (p. 608).

On the individual level, much has been made of our lineage´s steady increase in brain size. Mammals are brainier than reptiles, primates are brainier than mammals, apes are brainier than primates and *Homo sapiens* are brainier than apes. As researchers have noted, a bigger brain is good for generating larger arsenals of adaptive behaviors (Deacon, 1997; Gerhart & Kirschner, 1997). Essentially individual organisms have two main paths to success. The first is specialization. Both clams and trees concentrate on building impregnable defenses (e.g. massive fibrous trunks or hard shells) to the exclusion of much else. Nature is replete with these niche specialists. The second path is flexibility. Build big brains that can sense, navigate and adaptively respond to increasingly varied and extended environments. This is one of the key sources of mammalian success. It is also a hallmark of our lineage. Bigger brains also seem to persist in times of flux. Ultimately, bigger brains can be seen as evidence of selection for individual behavioral flexibility.

On the species level, flexibility is built into our species phenotypes on many levels other than the brain level. We have immune systems that adapt to whatever microbial dangers our immediate environments present. We have gene pools that present a safe variety of expressive features to be selected by the conditions its individuals find themselves in. Epigenetically and 'epi-neurally' our species has a fantastic range of semi-independent systems at various levels that all work to present variation for selection of those combinations of flexible phenotypic features that have historically survived under ancestral conditions. The species itself represents these different genotypes and phenotypes spread out over populations to maintain a variable reactive face to its macro environment. Thus species can maximize their potential for surviving the inevitable environmental fluctuations.

Species level flexibility can be compared to Gerhart's and Kirschner's idea of evolvability. In short, certain adaptations like big brains, exploratory processes and Hox genes function to enable the species phenotype to track environmental changes by enabling phenotypic plasticity over generations. This plasticity allows the marginal populations to make the relatively quick adjustments necessary to survive extinction events. Evolvability enables speciation. Another way to say this is that species flexibility is often embodied in demographically distributed individual evolvability.

Flexibility on the group level has further implications. On the group level, members can adjust their behavioral connectivity through social learning to present novel group behaviors to their changing landscapes. Behaviors that maintain new group level social patterns can also be selected for survival. Lion prides can learn swamp hunting techniques. Dolphin pods can learn to swarm up fish in mud shallows. Social solutions have a long history in biology and are based on the concept of cooperation, of canalizing beneficial communication. Essentially, where one fails, many can succeed. One buffalo can´t protect her young from wolf packs as well as many can. One ant can´t forage and reproduce as well as many can.

Social flexibility is another specialty of mammals and particularly of apes. For behaviorally complex animals, sociality presents another level of opportunity for differentiation and specialization. Buffalo young survive because they get behind their mothers and stay there when danger is present. In this case the differentiation and specialization is within the life span of each buffalo. When young they get behind the adults and when old they get in front, shoulder to shoulder. At a more canalized level ants begin differentiating for their "lifelong" special jobs when they are just larvae. In this case the differentiation and specialization is within the life span of the colony itself. An important part of colony survival is adjusting the numbers of different specialists it breeds as conditions change. In both cases, at different levels and time scales, social differentiation and specialization helped these species coordinate individual behavior on a global level. For ants, buffalos, lions and dolphins, it is coordinated behavior at a social level that enables them to persist.

By all accounts proto-humans were social. They just weren´t built for individual survival. In fact, the examination of fossil hominids seems to show members of our lineage, like Erectus and Sapiens themselves, as being decidedly less physically robust than some of their cousins.[157] Even so, it was they who survived. It seems the meek did inherit the Earth. It also seems they did it in cooperative groups.

Our apelike ancestors weren´t small and they didn´t have simple needs. Growing a hard shell or fibrous body was not in their physiological adjacent possibles. Most likely, they used the tools they had at hand to survive the last seven million years. This meant preserving any mechanisms that improved their abilities to capture and couple to a greater and greater variety of shifting environmental saliencies. In our case (i.e. most Hominin species) this meant the differential preservation of large brained individuals. Those individuals that recognized, captured and coupled new opportunities represented the successful descent lines. Generation of flexible behavior, in the form of differentiated and specialized adaptive reactions, was how we beat the bottleneck. Frame shift up and flexible behavior became so important that *groups* that socially differentiated and specialized their *individuals* began surviving preferentially.

## Human Group Selection

> *"Thinking of social groups as like organisms has been out of fashion for thirty-five years, and now it turns out that organisms are themselves social groups! Moreover, each transition requires group-level selection, exactly as Darwin proposed."*
>
> *David Sloan Wilson (2002, p. 18)*

[157] For more on robust hominids see Leakey (1994), Tattersall (1998) and Wills (1994).

As David Sloan Wilson says above, group selection has only recently, and reluctantly, become accepted (in this case reaccepted) as an authentic feature of evolution. Smith and Szathmáry themselves did not want to accept group selection, but their major transitions in the history of life clearly implies that all truly novel classes of life are the result of group selection events (Smith & Szathmáry, 1995). Obviously, group selection should be considered a rare, but momentous force for natural change. Group selection may be unusual, but when it happens, it can have extremely surprising results.

Bacteria are still a major part of the biosphere, but when a few got together symbiotically to form the original eukaryotic organism they revolutionized the biosphere. Likewise, cooperative eukaryotes, in the form of multicellular organisms, revolutionized the biosphere. What happens when a group of multicellular organisms gets together? This is the question. And if you want to know how our group unified you cannot overlook how these other groups unified. Many would argue that humans could not be group selected, but then they would be forced to consider humans as the exception in life's major transitions. *All* of the previous major transitions are transitions from reproductive individuals to reproductively unified groups.

Historically, groups unify by canalizing communication to maintain cooperative interaction and then internalize those cooperative functions to a degree of cooperative specificity we call reproduction. All group level selection events that ended up in the emergence of novel unified organisms started out with the canalization of cooperative interaction. The coordinated interaction these organisms engaged in is called behavior. Animal behavior became interesting when groups of specialized cooperative cells called neurons began to differentiate and specialize into brains and nervous systems that could further differentiate environments and specialize responses. Animal behavior became especially interesting when social cooperation became the catalyst for reproductive canalization.

Symbolic language is essentially a new form of group cooperation. Autocatalytic interaction between molecules drove molecular cooperation (i.e. autocatalysis) in protocells and ultimately, it canalized into the form of reproduction we call genetic. Cellular interaction drove prokaryotic cooperation and ultimately, it canalized into the form of reproduction we call mitosis.

Eukaryotic cell interaction drove protist cooperation and ultimately, it canalized into meiotic sex. In each case, the group selection event incrementally preserved processes by which the group organization could be reproduced faithfully enough for continued existence, for continued representation in the biosphere. In each case, a new form of group cooperation had to be canalized by a new reproductive system to preserve the coordinated, cooperative and unified features of the group organization. For a social group to become an autonomous unit of selection, it must become autopoietic - it must reproduce itself.

If we do indeed succeed in becoming the next in this rare but ongoing process of group selection, symbolic language will be the communication medium that will have been canalized to produce the cultural reproduction our groups exhibit. Symbolic language will be the way we interact to reproduce our cultures.[158]

Proto-humans were selected for how they organized themselves cooperatively. How they organized themselves cooperatively depended on how they communicated. Ultimately, language is an extremely unlikely form of reproduction, but is it more unlikely than binary fission? Or mitosis? Or even meiosis?

## Why did Protolanguage Evolve?

> *"Human cooperative communication emerged phylogenetically as part of a broader adaptation for collaborative activity and cultural life in general."*
>
> *Michael Tomasello (2008, p. 324)*

This chapter's question is one hot mess. Unfortunately, this approach will answer fewer questions than it provokes, but it is the highest level of specificity we can afford at this moment. Protolanguage evolved as the form of

[158] Bacteria, protists and metazoa all successfully radiated and have maintained themselves in the order of billions of years. I speak of language enabled cultural reproduction as yet to join this group simply because although we have radiated to populate most of the world, we have yet to differentiate or survive much more than a hundred thousand years. Although I remain positive, I reserve congratulations until we have made it a few more hundreds of thousands of years.

communication that was 1) available for selection in the human adjacent possible to 2) best serve to differentiate and specialize how proto-human groups could react adaptively to their environments.

Just as each individual animal has a sense halo, each social group of animals has a sense halo. Part of this group sense halo is the combined set of all distinctions its members can communicate to each other under adaptive reaction circumstances. These distinctions are every change each communicating member can sense (e.g. internally and externally) and how changes in these states inform each other's behavior. Individual behavioral reactions to these changes are what constitute the components of the social system. How this system reacts to its environment is the group reactive face.

Obviously proto-human (pre-language) individuals communicated. It is a ubiquitous feature of the apes. It seems equally apparent that these proto-humans lived and died in social groups. On some level, group selection was part of the considerable selection pressures they were under. Even more importantly, we shouldn't underestimate the complexity of the conditions these groups had to adapt to over the last two million years (vis-a-vis Orgel's second rule).

The picture they form is groups of large, aggressive, carnivorous, intelligent, versatile and social cooperators adapting to rapidly changing environments. How did they do this? By extending their distinctive trait, flexible behavior. And how do you maximize the flexibility of your behavior? By growing the biggest brain possible *and* by extending your repertoire of behavioral response routines to the level of coordinated group responses.

Each proto-human group lived under unique conditions that offered unique opportunities for group response. Like individual behavior, group behavior generally starts with recognition of external saliencies (e.g. opportunities or dangers). Unfortunately, a 'group brain' does not exist that can automatically sort the information that is coming in from each channel (e.g. each individual in this case) and reduce it to coordinated behavior. What does happen is that each individual senses his/her own environment and produces his/her own behavior. That individual behavior then becomes part of the environment for the rest of the group. Individuals reacting to each other's behavior is a large part of how that group behaves. Individual behavior is as much part of the group phenotype (e.g. how it is connectively potentiated and how its individuals are differentiated and specialized within the group) as cellular behavior is part of the individual phenotype (e.g. how its cells are connectively potentiated and differentiated and specialized within the individual). Both of these social entities present a unified reactive face to their external environments. This supposes that if you presented a similar challenge, again and again, to a group of *Homo erectus* on the savannah, they would likely do the same thing, or set of things, every time.[159] From a different point of view, small monkeys, pigs or antelopes probably see a foraging group of male chimps as a *unified* predator-entity.

One big difference between successful and unsuccessful groups is the degree to which they are synchronized. Individual organisms are composed of highly synchronized parts (e.g. cells, regulatory systems, etc.), groups are less so. Within the group, each component's interaction is loosely constrained by the actions of the others. Within the individual, each component's interaction is highly constrained by the others. Within groups of social proto-humans, coordinated behavior was a primitive communication network. To some degree, sensitivity to conspecific environments exists in all social animals. In a way, being a social animal is defined by being sensitive to, and constrained by, social partners. Wolves are social because their behavior is constantly influenced by pack-mates, tigers are not social because their behavior is rarely influenced by other tigers. In fact, higher dimensional degrees of synchronic interaction, as evinced by increasingly mutually constraining behavior, could be seen as a definition of sociality. Increasingly synchronized behaviors, like those proposed by social intelligence theories of language, are just one instance of a characteristic of a far more extensive class of universal phenomena.

Socially, language provides adaptive benefits for its members. One of these benefits is the extension of the individual sense halo by maintaining a group sense halo. As E. O. Wilson (1975) states, "Defensive superiority is the adaptive advantage of cooperative behavior reported most frequently

159 At least they would do one of a small set of stereotypical responses. There have also been reports of experiments where a group of chimps was exposed to the same simulated attack in forest and in open plain conditions and they spontaneously chose defense strategies consistent to the terrain.

in field studies, and it is the one that occurs in the greatest diversity of organisms" (p. 37). One of the basic functions of groups is improved protection. In groups that are foraging and omnivorous, like ourselves, a group sense halo can also improve the ability to recognize food or other 'offensive' capture and coupling opportunities as well. The group sense halo augments individual opportunities by having members signal to others. Individual reactions to others' signals can then create positive feedback situations, where adaptive reactions to conspecific behavior are selectively preserved.

Carnivorous pack animals like lions and hyenas express this kind of instinctual and/or learned coordinated behavior when hunting or protecting each other. Individuals react adaptively to the behavior of others. For example the sound of an attack on one pack member often incites an aggressive reaction in others. Evidence for the internal canalization of these responses can be found in the evolution of mirror neurons, mother aggression and other hardwired empathetic instincts. As this feedback progresses group members can become increasingly conditioned to respond in certain ways. Over time, social adaptive reactions can be preferentially preserved. Functionally, a social communication system connects the members of the group and extends the individual sense halo by proxy.[160]

Another word for flexible behavior is opportunism and that is basically what proto-humans were, opportunists.[161] Opportunism works best by maximizing both the number of opportunities recognizable and the number of ways to take advantage of those opportunities. One fundamental way to recognize more opportunities is to extend your sense halo. Where this is costly and time consuming for individuals (e.g. better senses and bigger brains take time to evolve genetically) it is easier to do by co-opting others,

[160] Traditionally group selection has been discounted because it was thought that noise at the lower, individual level would disrupt the formation of group coordination. In this context the individual reactive instincts of each member constantly disrupt this communal coordination causing 'noise' in the system. However, as we have seen noise can also have a beneficial result in keeping the phenotype searching for more beneficial couplings. In this case the phenotype that is searching is a 'group phenotype' not an 'individual phenotype.' This noise can also be superseded by high selection pressure; essentially the members must cooperate or go extinct. Both of these conditions seem applicable to human evolution.

[161] Three sources of evidence support the claim that we were opportunists. We are currently opportunists. Most of our closest ape relatives are opportunists and our bodies are adapted to eat a wide variety of foods, like most other opportunistic species (e.g. rats, dolphins, cockroaches, etc.).

especially if your group comprises extremely behaviorally flexible individuals capable of imitation and learning.

What happened in our case was that under heavy selection pressure, groups were selected for how well they communicated cooperative opportunities and coordinated the capture/coupling of those opportunities. Groups whose members could recognize and communicate more types of danger or food opportunities over greater distances to produce more adaptive reactions, survived differentially. It was communication to cue complex coordinated social behavior.

As previously mentioned, this type of opportunism depends not only on extending systems of recognition, but also on taking advantage of what is found. Individuals have brains and pre-evolved reaction routines (i.e. instincts) that initiate behavioral responses, but groups must evolve them. Groups that are connected differently, react differently to environmental saliencies. Those groups whose connectivity results in better adaptive reactions survive differentially. The feedback loop in this case is on how individuals cooperate to produce adaptive social behavior. If the selective pressure is in the form of predation then perhaps the group learns to stand shoulder to shoulder in front of their children. If the pressure remains stable, the mechanisms that preserve the behavior transform from learned behavior within a lifetime, to iterative learning over generations to internal epigenetic (i.e. stereotypy) or genetic mechanisms (i.e. genetic assimilation). If the selective pressure is varied and shifting then the capacity for multiple variable response routines is slowly internalized by expanding flexibility mechanisms like brains, immune systems, exploratory processes or Hox genes. Fluctuating environmental conditions especially encourage the age old process of differentiation and specialization. Increased capacity for different responses can search the fitness landscape faster and convert (i.e. canalize) found opportunities into different specialties faster.

What proto-humans were doing to survive was extending their sense halos by learning to respond adaptively to conspecific behavior. As coordinated groups they could do what uncoordinated groups and individuals couldn't: hunt-and-forage simultaneously, defend kills, coordinate babysitters, food share, etc. By differentiating and specializing communicative connectivity they could specialize in organizing the variety of adaptive reactions necessary for the variety of saliencies they found. There is no reason why all of the proposed social and technological sources for language should not be considered to various degrees, and at various times, valid. Our specialty was behavioral flexibility and our language comes from the pressures and constraints of preserving systems of communication that can support complex, coordinated group behavior.

Ultimately, language is a feature of the evolution of human cooperative behavior. As with the emergence of other cooperative organisms throughout the history of life; communication, cooperation, differentiation/specialization and selection play pivotal roles.

So far this has been a description of the kind of relationships that would have encouraged the evolution of a new differentiated and specialized mode of communication to support the differentiation and specialization of group adaptive reactions. At this point this investigation will switch over to describe what the steps to this process could have actually looked like.

## How did Protolanguage Evolve? steps to symbolism

> *"How can organisms move to a truly novel environment, with needs imposed for functions simply absent before?"*
>
> *Stephen Jay Gould (1993, p. 101)*

Presumably all animals use representational associations to behave. Terrence Deacon interprets Charles Sanders Peirce's different levels of representational associations as a kind of ascending hierarchy of complex behavior (Deacon, 1998). The first to evolve were iconic associations. Iconic associations directly link saliencies to behaviors. Paramecia feel sugar gradients and move towards the higher concentrate. For paramecium, sugar molecules directly incite sugar behavior. With iconic associations organisms are constrained to react to saliencies immediately present in both time and space. They can react, but they cannot predict.

More complex animals can maintain indexical associations. Indexical associations are the ability to learn to associate extended physically dissimilar causal relationships. Pavlov's dogs learned to associate food with a bell sound even though the bell sound did not look, smell, sound or feel, in any physical sense, like food. In this case, the food need not be present for the behavior to be incited (e.g. the dog salivates when the bell is rung whether food comes or not). Indexical associations depend on iconic capabilities (the dog needs to be able to directly recognize both food taste and bell sounds consistently) but they allow for a kind of second-degree causal relationship to be learned. Animals capable of indexical associations are able to extend the power of iconic behavior and anticipate immediate future events (e.g. 'the bell *means* food is coming'[162]). With indexical association capacity, organisms can react to saliencies extended in both time and space. They can predict limitedly.

Symbolic associations extend adaptivity even further, but they too depend on a maintained substrate of iconic and indexical associations. Where iconic associations require the direct presence of the saliency, indexical associations can learn to expect the presence of the saliency in the near future. Symbolic associations can socially maintain these adaptive links without the saliency ever being present. Symbolic associations extend future adaptivity to a degree never before seen. With symbolic associations we can react to saliencies vastly distant from us in both time and space. We can predict prodigiously.

Each emergent level of this hierarchy of representational associations is a super-ordinate rule-set for the complexification of adaptive reactions, a new rule-set for behavior, an extension of behavior into more distant and cryptic adaptive futures. Only humans are able to behave symbolically, but this capacity was selected from an organismic history of communication substrates (i.e. iconic and indexical representational associations). Like each of the major transitions in evolution, the transition to human symbolic behavior added a

[162] In the context of 'meaning' as 'information that maintains adaptive futures' this implies that sugar molecules *mean* 'go that way' to paramecium, bells *mean* 'expect food' to Pavlov's dogs and 'the great wall of china' *means* a set of adaptive futures to humans. These adaptive futures can range from simple individual advantages; a trip to that famed and far off landmark will up my social status, to complex social advantages; massive walls are effective short term but ineffective long term solutions to unwanted invasions (see chapter "Meaning as Adaptive Futures").

new level of symmetric interaction that could include more extensive relationships within its cycle of maintenance and reproduction.

Proto-humans survived by extending their power of behavioral flexibility to the group level. They did this by communicating. The greatest difference between ape communication and human language is symbolic representation. Where ape communication is limited to responding to changes in their immediate emotional and environmental states (e.g. iconic and indexical associations), human language is somehow not so limited. Human language can somehow dissociate its users from reacting to their immediate physical environment in a way apes cannot. Presumably this dissociation is grounded in the way language represents the human environment with symbols. As such, humans seem to be able to respond to their own, or another's, symbolic representational environment as well as those iconic and indexical representational associations in their immediate physical environment. A human can choose to constrain their immediate desire to eat a chocolate bar because they have an extended desire for social utility (appearing desirable at future social environments). Other apes do not 'see' that far into the future. They eat the chocolate bar. This topic will be addressed in detail at the beginning of the "Mind" section, but for now let us assume that we will need a reason for the evolution of the dissociation of symbolic interaction from iconic and indexical interaction.

First of all there is plenty of evidence for the primate capacity for symbolization. The canonical vervet monkeys, as well as language studies on other captive primates, have demonstrated that under certain conditions, primates can symbolize (Cheney & Seyfarth, 1990; Savage-Rumbaugh, 1997; Seyfarth & Cheney, 1997). That is to say, the capacity for symbolic behavior was presumably in our ancestor's adjacent possible.

Another fact about our proto-human ancestors is that they had significantly larger brains than vervets or chimps with the presumed attendant increased capacity for flexible behavior. Our ancestors had an unprecedented ability to learn behaviors that helped find and capture/couple novel saliencies. One of these tricks was using each other to form cooperative networks to perform complex or specialized jobs. They learned to manipulate conspecifics like tools to extend their adaptive efficacy in time and space. Thus, as our ancestors began to differentiate and specialize the ways they organized themselves they would have needed a way to cue each other's behavior. Currently, the two significant contenders for the medium that conducted this cueing are gestures and vocalizations. It may have been either, but seeing as we communicate using both now it can be assumed that we did so then too. However, for convenience both will be called 'vocal' simply because in the end it was the vocal medium that carried the symbolic load.[163]

So, as groups encountered the plethora of environmental problems that surrounded them, the groups that could learn specialized responses survived more often. They did this by individuals learning to respond stereotypically to certain gestures or vocalizations. Notice that we no longer need to posit reasons like gossip or mirror neurons or tool making. Each of these reasons was likely to have encouraged language evolution to different degrees under different conditions.

Our presymbolic ancestors were both vocal and social. They survived in groups whose organizational and behavioral flexibility was important to group survival. This made other group members consistently key elements in each member's environment. Essentially, other members were exceptionally rich indices of extended behavioral associations. At the same time, each member was still locked into reacting to his or her own immediate environment. How did they extend their own environments to include other's environments? This is Gould's question at the beginning of this chapter, "How can organisms move to a truly novel environment, with needs imposed for functions simply absent before?" He goes on to talk about how the jawbone evolved into the bone parts of the inner ear, but the question is equally valid here. How did individual "grunts, coos and trills" (Seyfarth & Cheney, 1997) evolve into distinct vocal cues for coordinating group reactivity routines?

---

163 The degrees to which each is used now are probably different than the degrees to which they were used then. It also seems that over time vocal cueing would be more likely to increase due to its lower energy requirements, greater range of differentiation and its superior functionality in representing past or future emotional and symbolic features where gestures are largely restricted to communicating immediate emotions. I choose not to go into depth on these reasons mostly because they do not affect my argument. Even if language was purely gestural at the beginning, it ended up largely vocal, most likely because vocalizations coordinated group behavior better.

Before proposing an answer, we will take a quick detour into the vocal/aural histories of some of our cousins.

## Vocal/Aural Adaptations: bats and whales

> *"But bat sonar, though clearly a form of perception, is not similar in its operation to any sense that we possess."*
>
> *Thomas Nagel (1974)*

Whales (i.e. *Cetaceans* - including dolphins and porpoises) and bats (i.e. *Chiroptera*) do something very unique in the animal world, they echo locate.[164] Underwater, this sensory mechanism is often called sonar and more generally it is referred to as echolocation. Echolocation is the use of sound to see. Essentially, the animal emits a sound that bounces off of dense objects in its ambient and then the animal interprets the echoes to determine the characteristics of the object, which in turn determine how it should react.

What makes echolocation interesting is first, that whales and bats are mammals like us. As Michael Studdert-Kennedy and Louis Goldstein propose, perhaps there is something about being a mammal that encourages radical vocal/aural adaptations (Christiansen & Kirby, 2003). Despite birds flying longer than mammals, the Asian cave swiftlet and the South American oilbird are rather poor echolocators compared to bats. Likewise, teleost fish have populated the oceans much longer than dolphins and whales, yet they have failed to take sonar to the functional level of success that these recently arrived mammals have. In both cases an entirely new class of organism (mammal) moved into highly competitive and established niches previously dominated by different classes: *Actinopterygii* (teleost fish) and *Aves* (birds). This is unusual in the sense that it is generally very difficult for new organisms to move into established, highly competitive niches.[165] Something gave them an edge. Interestingly, both cases also evince that there is something about

[164] In fact, only the *Microchiropteran* suborder of bats are typical echolocators. The *Megachiropteran* suborder doesn't typically echolocate with the exception of the Egyptian fruit bat.

[165] For more information see Smith and Szathmáry (1995).

mammals that makes them very adaptable, especially when the adaptation involves the extension of the vocal and aural senses.[166]

Secondly, all of these mammals have relatively recently moved into functionally unique evolutionary niches. This second point suggests that radical adaptations often accompany radical changes in evolutionary niches. Echolocation was evolved by land mammals learning to function in water or in air. These are extreme changes of environment, probably even more radical than our ancestors moving out of tropical forests and into woodlands, savannahs, freezing climates, islands, deserts, and, in fact, into pretty much every land based eco-niche on this planet. The functional challenges of these kinds of phase space expansions are extreme and, like in bats and marine mammals, require rather extreme remodeling of interactive capabilities.

Thirdly, they have each modified vocal and aural sense capacities in order to take advantage of the niches they founded. Phylogenetically, bats, humans and whales are not particularly close. Functionally however, it seems that each has taken advantage of innate mammalian propensities for vocal/aural sensory differentiation and specialization to solve different problems in unique ways by modifying similar sensory systems to navigate radical environments - classic convergence.

Of course, the problems each faced were very limitedly similar. Bats had to learn how to differentiate aural saliencies at night so that they could use them to their advantage. Whales had to learn how to differentiate aural saliencies in opaque water so that they could use them to their advantage. Humans had to learn how to differentiate aural saliencies of social partners so that they too could use them to their advantage. Imperceptible water, imperceptible air and imperceptible partner space are not the same environments. However, they all share the definitional trait of lacking sensory feedback by which an organism can constrain its behavior. Radiation into 'imperceptible' environments requires the evolution of novel modes of perception. In this case we are examining echolocation, sonar and proto-language as novel modes of perception. Like echolocation, language is a vocal/aural sense extension. Symbolic language enabled humans to co-opt

166 Interesting genetic work has been done in the study of the human FOX P2 gene which has been associated with human vocal capabilities as well as bat vocal capabilities (Li, et al., 2007; Yong, 2009).

each other to vocally extend their sense of foresight into an 'unforeseeable' ice age environment.

Bats evolved sixty-five to fifty million years ago. The change from a visually oriented tree forager to a highly accomplished flier and echo-locator must have been quite a radical change, but not without its benefits. Modern nocturnal insectivores like bush babies and lorises hunt from the trees they live in.[167] Many insects are capable of flight, so the air in which these insects live adjacent to the trees represents a very real adjacent possible for these insectivores. The problem for bat ancestors was how to find and catch insects in the air with little to no usable light.[168]

The solution was almost unbelievable. Somehow, bat ancestors not only learned to fly, a significant adaptation in and of itself, but they learned to navigate and make their livings at night, where they faced little to no competition from birds (Speakman, 2001, pp. 123-124). How was this unprecedented and complex move into the adjacent possible accomplished? It was accomplished by modifying vocal and aural capabilities. First of all they had to discover the locative potential of echoing and then, generation by generation, extend it. Bat hearing became bat seeing much like the chewing apparatus of early mammals became the hearing apparatuses of their descendants (i.e. their jawbones became their ear bones). Where jawbones began to be selected for how they conducted hearing-like information to the proto-mammalian brain, vocal echoes began to conduct seeing-like information to bat brains.[169]

Daniel Dennett talks eloquently of a transition like this when speaking of Odmar Neumann's "all hands on deck" theory of the evolution of nervous systems. Basically the theory holds that nervous systems evolved as emergency orienting responses. Once the capacity to orient responses to saliencies

---

167 Neither bush babies nor lorises are exclusive insectivores, but insects do compose a large part of their diets.

168 Obviously this presumes that the first bats were *Microchiropteran* echolocators, not like the fruit or vampire *Megachiropteran* bats.

169 On a similar note, there has been success in training blind people to see (navigate) by translating visual information from cameras to tactile information on their skin. Apparently, certain technologies can enable a blind person to learn to see by feel (See the work of Paul Bach-y-Rita in sensory substitution). For an even more interesting case of a blind boy who can echolocate look up information on Ben Underwood (Ben Underwood Foundation, 2009).

became possible, then there are advantages to triggering this orienting response more and more often because being oriented is adaptive (Dennett, 1991, p. 180). The same logic can be applied to echolocation. Once echoes were recognized as carrying information that helped bats recognize, capture and/or couple important features of their environments, what began as something akin to passive listening changed into active vocal cueing combined with listening.

Bats learned to use their aural sense by activating it with vocal cueing. This new sense was just a functional combination of older capabilities to realize new types of synchronic interactions with their environments. Ultimately, they were able to move into a whole new adjacent possible and open a whole new phase space by functionally extending their sense halo into a previously 'unforeseeable' environment.

Whales also extended their sense halos by changing the adaptive behavior they could vocally and aurally cue.[170] Theirs is a unique story, but like the bats, it goes to show that vocal/aural cueing has a history of producing highly improbable and highly powerful behavioral adaptations for mammals in extreme environments. These adaptations are essentially ways of extending their ability to sense distant and cryptic saliencies. Saliencies that were previously unforeseeable. These adaptations helped them open niches (i.e. phase spaces) in which they functioned extremely differently than their ancestors did. Even though the problems bats, whales and humans all faced were very different. Each was solved by modifying the same sensory package so as to enable adaptive reactions previously unavailable to those species. In each case it was the extension of the sense halo into what was foreign environments that brought about shockingly powerful adaptations that changed forever how these species interact with their environments. Bats, Whales and Humans are some of the most successful and highly radiated mammals in the world.

There may even be neuroscientific evidence that links the convergent evolution of supra-sensory vocal/aural exaptations in bats, whales and humans. A study by Samuel Wang, Damon Clark and Partha Mitra noted that

170 It seems that the development of echo locative perception in *Cetaceans* was linked to their increasingly deep-water search for their squid food source (UC Berkeley, 2007).

both bats and whales exhibit abnormally large cerebellums and that this result could be attributed to their adaptation for echolocation (Lundmark, 2001). In unrelated studies paleoanthropologists are finding abnormal increases in human cerebellums correlated to increasing "cultural pressures" (Weaver, 2004) and "higher cognitive functions" (Balsters, et al., 2009). As a purely speculative suggestion, it should be possible to compare the neuroanatomical history of bats, whales and humans to see if the brain adaptations for echolocation compare interestingly with the brain adaptations for symbolic language. Cerebellums are traditionally associated with coordination of motor control. Getting good at a new sensory modality like echolocation would require the coordination of vocal/aural motor control. Presumably the same would be true of symbolic language. The question posed would be how much does language look like a sensory adaptation? Do the changes in the human brain as a result of symbolic language resemble the changes that bat and whale brains made in response to echolocative sensing?

The comparison of three such obviously different types of mammals in this chapter is intentional and made specifically to present another loose kind of prediction about human origins. In the case of both *Chiropteran* and *Cetacean* origins the prototype ancestors first encountered a new niche and then developed senses that extended their sensory capabilities into that niche. Bats learned to fly first and then developed echolocation (Simmons, et al., 2011). *Cetaceans* learned to swim first and then developed sonar (UC Berkeley, 2007). In both cases an environment in which different kinds of information were positively reinforced by selection eventually drove those phenotypes to develop and preserve new sensory modalities inside themselves. In both cases selection for external behavioral couplings later became physiologically internalized (e.g. perhaps in similar cerebellar adaptations). If humans were to follow the bat and whale model they would first have to find themselves in a significantly novel environment and then selection would power a new vocal/aural adaptation that 'read' that environment well enough to consistently produce adaptive futures. The proposal made here is that the ice ages environment that drove the symbolic language adaptation was characterized by extreme fluctuations on many levels of selection. Our ancestors found themselves in different environments regularly. To survive they evolved a new sensory modality, symbolic language.

In the 90's philosopher Thomas Nagel wrote an influential article called, "What is it like to be a bat?" in which he argued that despite both humans and bat's being mammals, we cannot experience the world like a bat does. The quote at the beginning of the chapter comes from this article. In one sense Nagel is entirely correct. You and I cannot sense unseeable physical objects with our voices and ears the way bats do.[171] What we can do is sense unseeable physical objects with our voices and ears the way humans do. I can tell you there is a thing called the Great Wall of China and 'sensing' this, you may get curious and react by going to visit it.

When I was a child we used to go to the edge of the ravine at dusk and throw pebbles into the air. The bats would often swerve and follow the rocks down. Apparently, bats are curious about what they hear is in their environments too?

## Vocal/Aural Adaptations: humans

Humans are as different from bats as bats are from whales. Despite some convergence in their adaptive histories we should expect significant differences that correspond to the specific problems and environments each came from and survives in.

The biggest difference is in the environments each species adapted to. For bats and whales their new environments were the well-known physical conditions of nighttime and dark water where vision was limited. To extend their sense halos and capacities for adaptive reaction into these environments they had to develop nonvisual senses for an otherwise stable environment. That is, the problem they had to solve was how to sense an environment that remained similar except for its lack of light. So what was the foreign environment that humans adapted to? What was the problem that vocal/aural cueing solved that conventional ape senses couldn't?

The answer can be found in the mercurial and kaleidoscopic ice age environments. The problem proto-humans faced was how to track an

171 This is not strictly true. See the case of Ben Underwood, a boy who can echolocate (Ben Underwood Foundation, 2009).

environment that didn't stabilize. Unlike, bats or whales, humans were moving into an environment characterized by instability and flux on many levels. The problem was not, 'How do I sense what is in front of me without light?' It was rather, 'How do I sense what is in front of me tomorrow?' - 'How do I sense the future?' The human predicament was based on the problem of sensing changes through time as well as through space. [172]

This brings us to the major difference between the human solution and the bat solution. The human solution was social. What bats needed to do was to develop a sense that helped them order a new environment, to synchronize themselves to it. Echolocation gave them a way for individuals to make sense in the dark. Humans also needed to sense their environments, but how do you sync to something that keeps changing its interactive face to you? How do you make sense of something that is altogether beyond your individual capacity to predict? You learn to co-opt your family and friends.

Notice that the moves to echolocation for bats and whales are typical examples of individual level selection. They are very successful examples of phenotypes coupling to non-conspecific aspects of their environments. The human move to symbolism is an example of group level selection. It is a very successful example of individuals coupling to conspecific aspects of their environments. The human move can be best compared to Smith's and Szathmáry's other major transitions in evolution for two reasons. The first reason is because the human exaptation was social, like all the others. Second, the human exaptation is radically altering the face of the Earth, like all the others. The first prokaryotic cells, the first eukaryotic cells and the first metazoa all radically revolutionized the biosphere in their own times.

---

[172] Some may argue that the time-scale (i.e. change-scale) difference between the human life-cycle and ice age environments is too great for the former to be shaped by the latter. There are two faults with this thought. First, ice age climatic flux was far more severe than most of us can imagine. National Geographic notes that the, "Ice age climate is not uniformly frigid but extremely variable. Evidence from the last glaciated era suggests the dramatic swings occurred within the course of a human lifetime" (National Geographic, 2011). The second fault is to compare individual life cycles to the climate not group life cycles. Groups evolve characteristic relationships with their environments. Inuit groups maintain and reproduce their differentiated and specialized cultural structures differently from South-Sea Islander groups. Group cultures actually change on much longer time-scales than individuals and thus can arguably become susceptible to differential preservation on climatic time-scales.

Social solutions to complex environmental problems have a long history that has already been noted and will be revisited in the upcoming chapters. For now, suffice it to say that behavioral feedback between conspecifics can produce cooperative adaptive reactions that are truly amazing. Where bats and whales learned to take advantage of the audible qualities of solid objects to extend their capacities for adaptive reaction, our ancestors took advantage of the audible qualities of their neighbors to extend the group capacity for adaptive reaction. Pheromonally, ants did a very similar thing hundreds of millions of years earlier.

To use a sonar term, the advantage of pinging a night sky is knowing the locations of, "things to avoid" and, "things to seek out." The advantage of pinging neighbors is knowing who is where and how they might be useful. The environment that humans moved into was social. Where the vocal and aural characteristics of dark water and dark sky interaction constrained the new sensory modalities that bats and cetaceans evolved, the vocal and aural characteristics of conspecific interaction constrained the new sensory modalities that humans evolved.

This brings up the problem that bats, marine mammals and humans had at the beginning, how to tell important information from noise. Presumably, the first bats had very poor echolocation abilities. The problem was separating the vibrations that were adaptively important to their survival from all the other vibrations in the air. More than likely, proto-humans had a similar problem.

What ape vocalizations do well is communicate the emotional states and the locations of members (Seyfarth & Cheney, 1997). They convey the signals, "do not come near me now, I'm very pissed off" or, "over here" very well. This is the starting point for human language. For apes, and presumably for our ancestors, communicable information was about emotional states and physical locations. This information was signal, and everything else was noise.

A simple improvement of the signal to noise ratio is not what happened because if this were so we would still be limited to communicating emotively. This is the greatest weakness of all of the social theories of language evolution, they fail to describe how primarily emotive information transformed into symbolic information.

A good example is Robin Dunbar's Gossip Theory which asserts that language evolved to take the place of grooming, a kind of grooming at a distance. He argues that grooming is a type of bonding that is necessary for large groups to stay together. According to Dunbar (1996) the problem language solves, is that groups couldn't get bigger unless individuals could bond at distances. Ignoring the improbability of consistent selection for larger groups throughout the bottleneck, grooming seems to bond by reinforcing emotional, not symbolic relationships.[173] Presumably, it works something like the somatic markers described by Damasio (1999) which use emotional feedback to train adaptive behavior. The problem with this kind of theory is that there is no way, and no reason, to invoke abstract symbols into this kind of mix without relying on questionably bounded confabulations of probabilities and statistics (and, if you remember, these are exactly the kind of magical explanations we should avoid). In fact, as the art of poetry attests, emotional bonds are some of the most difficult relationships to put into words. Symbols just weren't invented to represent emotional states, a non-symbolic sigh does the job much better.

That leaves us with locative information. Baboons use vocalizations to stay together when moving through brush.[174] In a sense those vocalizations represent, "I am here." Much like the bat pings the dark, apes ping each other. For Baboons this function is limited to maintaining group cohesion under specific conditions, the rest of the time they communicate like other primates, emotionally. However, there is no reason that highly intelligent, behaviorally flexible and exceptionally social proto-humans under extreme environmental duress needed to stop there.

---

[173] Remember that larger groups are also resistant to phenotypic adaptations while it is the smaller groups that are especially likely to adapt. Dunbar is proposing that language allowed proto-humans to expand the upper limit for primate group size. Apparently primate groups become dysfunctional when they number over about 50 individuals so, presumably, Dunbar is proposing that human groups were regularly selected to be over 50 individuals. This seems to be rather large for proto-human groups, but small for population genetics. It remains to be seen how many is too many for speciation to occur in this case.

[174] Although baboons are not our closest cousins they do make their livings in the same kind of open savannah that proto-humans did. Baboon environmental pressures, and thus baboon behaviors, could be closer to proto-human behavior than that of our closer cousins, the forest living chimpanzees.

## Locative Sensing: imagination

> *"When calling to one another, monkeys seem to lack one of the essential requirements of language: the ability to take into account another's mental state."*
>
> *Robert Seyfarth & Dorothy Cheney (1997, p. 38)*

What this investigation suggests is that the best evidence we have for apes symbolizing in the wild is, well, apes actually making use of vocal representations to produce adaptive behavior in others in the wild. In addition to apes using vocalizations to localize troop members, vervets have been seen vocalizing the incorrect warning call to provoke the correct mode of escape in troop mates (Cheney & Seyfarth, 1990). Likewise baby chimps have been documented deceiving their mothers into chasing rival chimps away from a prized tuber by vocally representing a false attack and having their mother chase the rival off.[175]

It is no secret that apes use vocalizations to communicate with each other. As mentioned, however, these vocalizations overwhelmingly represent emotional here-and-now states that incite emotional here-and-now responses. Instances of symbolic vocalizations where the signaler has to 'imagine' the response of others, like those mentioned above, are rare. 'Imagination' in this case is referring to what many philosophers call "a theory of mind" (Seyfarth & Cheney, 1997). A theory of mind is the ability to attribute mental states to others and adjust one's behavior according to the listener's presumed (imagined) mental state. This investigation calls this process imagination because proto-imagination can be seen as a colloquial term for 'attributing mental states to others.' There is no evidence that animals imagine anything other than what they themselves experience. The proposal here is that the first

175 See Leakey & Lewin (1992). In fact, ethologists have for a long time been reporting on primate behavior which resembles the kind of behavior humans, as opposed to animals, engage in. My favorite is a very believable account of a 19th century Dutch ethnologist studying chimps in South Africa. He watched while three young male chimps ambushed a panther which was stalking their troop. Two of the chimps died from the attack, but so did the panther.

and most fundamental kind of imagination that humans experienced was the 'attribution of mental states to others.'

The problem with imagining does not appear to be in generating such instances. As previously mentioned, both vervets and chimps have been witnessed imagining others' mental states. The problem seems to be in functionally maintaining imagination once generated. How do you teach your friends and family to use the eagle warning call except when it will cause them to pause just as the eagle is about to strike, in that case use the jaguar call which incites immediate flight? How do you reproduce that kind of empathetic cooperation in others? You do not. You can learn to do it yourself, but you cannot teach it. Even if it is occasionally sampled from the spectrum of environmental interaction, how do you canalize that cooperative instant to make it reproductively persistent? The problem is how to teach imagination?

What we need is a way to introduce maintainable (i.e. re-creatable, recursive or reproductive) imagination to ape communication. The way to do this is through call and response. Bats send out sounds and interpret the response. How their sound changes is how the world changes for them. The great thing about sound is that it is absolutely consistent. Bounce the same sound off the same thing at the same angle and it produces the same echo. Bats and whales use the predictable properties of sound to teach themselves how to react to their environments. The physics of sound is the same 'language' everywhere (i.e. sound transforms by the same rule-set everywhere). If a chimp baby gets the idea to run a rival off a tuber by crying for help from its mother; it calls, but the response varies. Maybe its mother is out of range now? Maybe the rival has become dominant to the mother in the meantime? In the end each chimp has to learn its own tricks for getting others to cooperate with it because behavioral responses vary. In the end, each chimp's bag of tricks dies with that chimp. The cooperation has no platform for reproduction.

Locative communication is different. When I cannot see you and want to know where you are I can call to you and you respond. Like an echo, the response carries certain infallible information like how far away and which direction you are facing. Under all conditions this information stays the same. Now we are speaking the same language. When I vocalize you know something about how I am disposed and when you respond I know something about how you are disposed. I ping you, and you ping back. The difference between this and what the baby chimp or vervets do is the response.

Responses carry important information about how the responder is oriented. A simple grunt can tell the sender the direction the responder is moving, the direction she is facing, if she is in the open or in dense bush, her level of arousal and/or attentiveness, her mood and a host of other more or less cryptic information about the responder's behavior (i.e. reactive face). All of these saliencies are information in that they reduce the sender's uncertainty of the responder's behavioral adjacent possible. The sender now has knowledge about what the responder will do. If the sender has a large plastic brain able to build vast libraries of indexical behavioral routines then the signal from the responder becomes an especially rich index of potential adaptive reaction. Responders are essentially broadcasting a suite of potential cooperative utilities to those with the brainpower to differentiate and specialize them, to those who can read their partner's reactive face. Cooperation is evolution's most powerful tool. It is selected for persistence from a rich spectrum of interaction. Large powerful brains trained to differentiate external cues in order to identify beneficial couplings would find a rich spectrum of potential coupling in their partner's reactive face the moment this reactive face was broadcasted. The more often it was broadcasted, the more often the represented saliencies could be selected for persistent novel synchronies.

Bats and whales did not wait for their new sound environment to inform them. They learned to incite the sound information themselves. They called and their environment responded. Humans too had to learn to incite emergent co-operative couplings from their environment. The difference was that bats and whales interacted with anything that reflected sound. Humans interacted with each other - with anything that reflected human behavior sound. In a sense the limits of the broadcasted information became the boundary limit of a new cooperative unit of selection - of the emergent symbolic group. Within that boundary limit, information in the form of behavioral broadcasting informed our ancestors about each other's reactive faces and thus allowed their brains to sample and maintain synchronous behaviors by doing what brains do best, extending prediction to include more within their cycles of maintenance and reproduction. Child care, foraging, food sharing, reproduction and defending kills are just a few of the activities in which knowing the physical and emotive disposition of those near you is adaptive. Once the call and response becomes behaviorally canalized, the group as a whole has a communicative medium upon which it can quickly, and consistently, rearrange itself.

Call and response is already a feature of primate behavior, which locates it right in our ancestor's adjacent possible needing only variation and selective feedback to drive differentiation and specialization. There are few doubts that many hominins experimented with the capacity for vocally cued coordinated social behavior over the last two million years. The benefit of this kind of experimentation is in the capacity for vocal information to be modified. Unlike gesture, vocalizing is especially functional in signaling changes in emotional states. Literally, the internal state of the sender is transformed into airwaves inside them. Those airwaves are then 'felt' by the internal state of the receiver's inner ear and brain. As such, vocal locative information would be especially rich in emotional modalities. Immediately, a wealth of emotional information about the internal state of the vocalizer is available once his vocalization is received. As we now know, those internal states are stereotypical reactions to his or her own ambient. In a sense, broadcasting your own internal state contains cryptic information about your immediate environment.

Undoubtedly, nonhuman apes react to the emotional vocalizations of other apes. What nonhuman apes lack is a regular medium to feed other's reactions back into their own reactive routines. They lack a consistent 'pingback.' In a sense, locative vocalizations (i.e. call and response vocalizations) provided a medium that consistently transmitted valuable environmental information that was emotionally modifiable. Our ancestors learned to use regular updating of each other's locations to begin reacting to each other's local ambients. They learned to read not only locations in vocalizations, but the environmental conditions at those locations. Metaphorically, you could say that primitive vocalizations broadcasted localized emotional weather reports. Each of the previous major transitions in evolution marked an extension of behavioral capacity. Our ancestors extended their behavioral capacity by learning to react adaptively to not-here ambients, not-here conditions.

We are just starting to experimentally recognize that our own bodies do this all the time. The study of body maps is realizing how the brain represents different parts of the body. Surprisingly, scientists have found that the brain inherently maps the space around our bodies as well as the space within it. University of Arizona neuroscientist Michael Miller (personal communication, December 22, 2012) stated that rats with electrode array crowns surgically implanted onto their heads have to relearn how to navigate their enlarged heads through their mazes and cages. People with hats learn to duck when they go through doors. People in cars learn to incorporate its dimensions into their driving movements (Blakeslee, 2007). The proposal here is that we learned to incorporate other's behaviors through their broadcasts. Remember, our bodies are actually very coordinated *societies* of cells that have learned to incorporate (i.e. cooperatively predict) each other's behaviors.

In just such a way proto-humans learned to incorporate each other locatively. The group itself can become a functional unit in this way. It can coordinate complex defensive, offensive, foraging or caregiving processes all by modifying individual responses to information about each other's ambients. The power of emotive responses to differentiate and specialize cooperative behavior is functionally accessed by locative call and response updating.

The first symbols represented remote behavior. One could also say they represent displaced or not-here behavior. This investigation uses the term 'locative' because besides indicating that the behavior represented is remote, or not-here, locative also implies a connection with echolocation and the baboon-like tendency to maintain an environment of aural information that is used to extend its participant's ability to sense and include previously cryptic environmental information (bats in the dark and troop mates in the bush, respectively).

The term 'locative' means representing not-here opportunities for adaptive reaction. Specifically, it refers to representing the behavioral state of conspecifics in the form of a map of nearby loci of potentially adaptive cooperative responses. This "representing" is in the form of individual reactions to saliencies in other's ambients. In a sense it is a new form of extended indexical association where the stimuli associated is your neighbor and the feedback created between you two is cooperative (i.e. synchronous). The reality of other people's saliencies must be maintained virtually, through group vocal updating. Call and response vocal updating is exactly the kind of behavioral trait that other mammals like bats and marine mammals have canalized to extend their own sense halos, to extend their own adaptability in extreme environments. The process of locative sensing is a proto-system for attributing mental states to others. It is a crude sort of imagination. What else would you call reacting through another's process of reacting?

## Locative Symbols

> *"For an observer an entity is an entity when he can describe it. To describe it is to enumerate the actual or potential interaction and relations of the described entity."*
>
> *Humberto Maturana & Francisco Varela (1980, p. 8)*

At this point our ancestors were experimenting with speeding up adaptation to variable environments by learning to react to how other group members broadcasted their own ambients. In a way they developed a sense for imagining themselves as others, a sense of not-here saliencies. The question now is how these "grunts, coos and trills" become words, symbols?

To start with, this situation is once again very similar to Dennett's portrayal of Odmar Neumann's "All Hands on Deck" theory of consciousness. Once updating is possible it becomes adaptive to trigger it more and more often. Instead of wanting to know what is around you so you can react, our ancestors wanted to know what was around their friends and family so that they could react. They did this by 'calling' them to get a sense of what they would do from their 'callback.' What begins as the development of a largely passive sense for reading others becomes an active tool for inciting them. When this feedback loop is established a new medium for the coordination of group activities becomes possible. Vocal interaction now presents an environment of ubiquitous and variable vocalizations for selection. From this interactive background successful coordinated group responses can be selected for persistence. Vocalizations are now differentiable. What needs to be differentiated? Which grunts and coos incite which responses?

The groups that survived the bottleneck were those groups with the most behavioral flexibility to find and take advantage of the changing saliencies around them. This behavioral flexibility included individual intelligence and behavioral flexibility. But where individuals within a species do not vary much in these qualities, they can vary significantly in how they approach these problems in groups. Different cooperative solutions to complex environmental problems are what differentiated our ancestors. The sense of how to react to other's senses is only as discerning as the number of distinctions it can make. Equally, the number of different adaptive reactions you can incite and partake of is constrained by the number of different signals you can produce and respond to. If coordinated behavior is selected for, selection will preserve those groups whose individuals can differentiate conditions, and their adaptive reactions to those conditions, better.

Evolution starts with ancestral structures. The first symbols were ordinary animal-like indexical signs that referenced behavior. They were vocalizations that cued adaptive behavior. In a sense, the first humans developed a range of stereotyped vocalizations to incite a range of stereotyped reactions in their neighbors. Grunts, coos and trills automatically carry emotive information, but that information cannot be easily differentiated. Emotions take a certain amount of time to incite and a further amount of time to clear before new emotions can be incited. Also, emotions have no natural divisions that can be exploited to differentiate responses. Anger, for example, often fades into melancholy with no consistent and decisive behavioral division between the two. To differentiate this communicative system and make it fast enough to react adaptively, our ancestors had to modify something other than their emotions. They had to vary a different aspect of their reactive face. They had to modify the communicative media itself - their vocalizations.[176] Over time it was the vocal modifications that were differentially

[176] In fact, their vocalizations, as a communication media, may have provided enough variation just within the sender-media-receiver system for differentiation to occur on its own. That is, there may be so much variation (noise) in just sending and receiving a message that 'modifying' words was more like preserving novel words when they occurred than inventing them.

preserved to support more and more of the distinctions that initiated specialized cooperative actions, not the emotional modifications.

The first symbols represented remote behavior. They were differentially preserved vocal representations of the sender's reactivity to her environment. They were maintained by feeding back into the group's ability to consistently coordinate multi-individual responses to complex environmental problems. Seeing as these first humans probably had no pre-evolved random access memory systems to internally maintain large numbers of distinct vocal representations it is also likely that these vocally cued complex coordinated social behavioral routines were to some extent externally maintained, by the group itself. Of course, individuals needed to remember which cues represented which behavioral saliencies, but each individual could only store a limited number of symbols. However, a group that maintained a constant symbolic information environment could distribute the memory of different symbols in different individuals. The group itself could maintain a greater 'vocabulary' than any one individual could. This regular cultural practice of constant use, imitation and teaching were necessary to keep emergent symbols in existence long enough for basic learning circuits to fix them into individual brains. It's a good thing those brains were preadapted to be plastic.

The anthropological evidence for the evolution of our vocal tract, in the form of lowering the larynx and shaping of the skull base, goes back over two million years ago (Leakey & Lewin, 1992; Lieberman, 2001). From that point on there appears to be some kind of selective pressure to differentiate the sounds we could make. As a back of the envelope kind of calculation, this could be roughly how long the locative sense has been evolving.

With the creation of a social system that maintains symbols, the first humans had a nascent capacity to codify environmental saliencies. This codification is the appearance of a novel sense for the capture/coupling of extended saliencies. Proto-humans began to be selected within a social environment of symbols that triggered and constrained each other's behavior. They could sense how others would react by what sounds they made and they imitated those same sounds when they sensed their own reactions to the same triggers. Saliencies in their environments got vocal names so they could all react in a coordinated fashion.

By symbolically codifying their environments humans created another buffer zone within which they were safe to experiment, feedback and extend connectivity. This process is strikingly similar to the first multicellular organisms that engulfed extracellular space to enable a level of cooperative differentiation and specialization between cells that the world had never seen before. Essentially humans extended their group sense halo by symbolically stereotyping everything that was important to group survival. The salient features of the environment that each of them was sensing had a name that identified how the group should capture/couple it. If the feature was named 'baby' they coordinated their behavior to ensure it was cared for and taught. If the feature was 'enemy' they either defended their community or escaped.

By extending and canalizing their vocal connectivity proto-humans were able to coordinate a group reactive face that could accomplish much more than any one of them could. In the very same way, cooperative cells extended and defended their own group borders as the first multicellular animals. The communicative network multicellular cells used, was a novel kind of language. This language wasn't the language of ionic, covalent or gravitational synchrony; although it included and depended on the maintenance of all of these. It was the systems of biochemical cooperation called signal transduction. As connectivity grew, cells began to differentiate and specialize for the jobs that were important to group functionality instead of individual functionality. It was group functionality that now insured individual survival. Epithelial specializations like eyes, skin and brains began to establish the extended sense and behavioral halos that took over the individual cell's external negotiations and consolidated them within an emergent cooperative and reproductive boundary.

The same thing happened for proto-human groups. Like all emergent systems, the language of cooperation that was selected to maintain them was unique. In this case it was behavioral connectivity based on a vocal medium that preserved the group. By maintaining cooperative group behavior they extended individual internal buffer zones to the limits of their communicative community, to the limits of their communal symbolic senses and behavioral halos. Like other major transitions in life, they expanded what was conserved internally by extending their sense and behavioral halos. Within that new symbolic boundary all important features were communally 'known.' They could be cooperatively 'metabolized.' Outside this boundary was the unknown. Outside what the group had names and behavioral routines for, were those features that they needed to learn to sense if they were to radiate. Like all emergent organisms, the proto-human group created its adaptive future by extending what it could sense and react to. For our ancestors, this meant giving everything names, locating them symbolically within the social context. Symbolically it was the *social* context that created meaning by differentiating and specializing what could be done by who to promote communal survival.

Language is a social phenomenon. It preserved groups of our ancestors by coordinating group reactivity. It did this by stereotyping (i.e. canalizing) vocal cues that coordinate cooperative behavior. The connective media proto-humans used to maintain this cooperative unity was vocal and symbolic. Obviously, this was not the first time cooperative unity through novel forms of interaction (i.e. communication) had happened. Witness protons, witness atoms, witness molecules, witness solar systems, witness DNA.

## Language Origins

We are finally in a position to put this information to work for us. What makes humans special is language. In other respects we are just another species. The origin of language in this sense is the origin of humanity as we know it. When Morowitz (2002) states, "that metabolism recapitulates biogenesis" (p. 78), he refers to how the organizational relationships extant in metabolic processes are physical evidence of the organizational relationships that were captured in the first metabolisms. How metabolisms function now, reflects how the molecular components of the first cells functioned together. In a similar manner, language origins can be deduced from their current functional organization. Deacon (1998) brilliantly describes the systematics of language thus:

Because symbols do not directly refer to things in the world, but indirectly refer to them by virtue of referring to other symbols, they are implicitly combinatorial entities whose referential powers are derived by virtue of occupying determinate positions in an organized system of other symbols. Both their initial acquisition and their later use require a combinatorial analysis. The structure of the whole system has a definite semantic topology that determines the ways symbols modify each other's referential functions in different combinations. Because of this systematic relational basis of symbolic reference, no collection of signs can function symbolically unless the entire collection conforms to certain overall principles of organization. Symbolic reference emerges from a ground of non-symbolic referential processes only because the indexical relationships between symbols are organized so as to form a logically closed group of mappings from symbol to symbol. This determinate character allows the higher-order system of associations to supplant the individual referential support previously invested in each component symbol. This system of relationships between symbols determines a definite and distinctive topology that all operations involving those symbols must respect in order to retain referential power. The structure implicit in the symbol-symbol mapping is not present before symbolic reference, but comes into being and affects symbol combinations from the moment it is first constructed. The rules of combination that are implicit in this structure are discovered as novel combinations are progressively sampled. As a result, new rules may be discovered to be emergent requirements of encountering novel combinatorial problems, in much the same way as new mathematical laws are discovered to be implicit in novel manipulations of known operations (pp. 99-100).

Fascinatingly, Deacon's explanation of how language works mirrors the kind of process that Gerhart and Kirschner describe in the cooperative origin of multicellular animals. Multicellular animals are closed systems of interaction between cells. These cells, like symbols, do not form direct relationships with external saliencies. Principally, they form relationships with each other. They refer to each other and only within that organizational "topology" of internal relationships can they be maintained and reproduced (only so do they 'exist'). Animal cells and symbols are both elements within, "logically closed group[s] of mappings." In the one case the mappings are, "from symbol to symbol," and in the other they are, 'from cell to cell.' Read by a cellular biologist, Deacon's first sentence could easily be: 'Because animal cells do not directly interact to things in the world, but indirectly interact to them by virtue of interacting to other cells, they are implicitly combinatorial entities whose interactive powers are derived by virtue of occupying determinate positions in an organized system of other cells.'

Notice however, that neither the cellular nor the symbolic systems can be adequately described as the simple occupation of, "determinate positions in an organized system of other symbols [i.e. cells]." In both cases what we are describing is more than just relative positions of components, be they cells or symbols. This is a dynamic process not a static snapshot. We are describing how components within the system regularly constrain the transformation of other components. It is a symmetry rule-set. These systems transform, and they transform by preserving structure. They maintain an organizational "topology" of enruled transformations between themselves. In cellular societies (e.g. animals) these symmetry systems are called metabolism and reproduction. In human societies they are called language, thought and culture.

Let us presume that, just as cellular cooperation gave those first multicelled animals a power to sense and react to their environment's more adaptively, symbolic cooperation in the form of some kind of protolanguage gave the first humans the power to sense and react to their environments more adaptively. In this context, we are not looking for the origin of symbols themselves, but for the origin of a system of symbolic interaction. This system needs to not only create, but to maintain and reproduce a, "logically closed group of mappings," that operate as a, "higher order system of relationships." Specifically, those relationships are not indexical in the sense that they are causally related to real external saliencies. The word 'banana' has no inherent causal relationship (i.e. iconic or indexical relationship) to the oblong yellow fruit. Symbolic relationships are a different kind of relationship - a super-ordinate relationship. Their rule-set is not constrained by 'what is,' but by how we react as a group to 'what is.' Symbols refer to each other first and to the thing they represent second.

When Deacon says, "this determinate character allows the higher-order system of associations to supplant the individual [indexical] referential support previously invested in each component symbol," he is talking about language alone. However, this line of reasoning implies that we need an explanation for the subordination of the old indexical relationships that we, like all animals, individually had with salient features of our immediate environments. This kind of immediate, learned and/or instinctual reaction to immediate environmental features creates a meaning system that reinforces meaning mappings only within the here-and-now context. I see a yellow oblong fruit and I move to grab it because it is yummy.

Conversely, symbolic relationships map to each other. They are their own environment. They, "do not directly refer to things in the world, but indirectly refer to them by virtue of referring to other symbols." Symbolic components must primarily react to each other within the rules of the, "distinctive topology" of the symbol system itself. Symbolic language is a metasystem, or added system, of constraint on human interactive behavior. The question is how such a metasystem can be maintained in the face of constant disruptive noise from each member's indexical system of interaction? What kind of process could change an ape's fascination with whatever happens to be around him physically to a human's capacity to be fascinated with whatever ideas happen to be around him conceptually?

What we need in this case is a platform that supports the maintenance of a system of self-referential interactions. Our ancestors did this by vocally stereotyping their own direct experiences. In doing so they created a shared local topography of sound information, an environment of stereotyped vocalizations in which the components are represented experiences, but the global function is a coordinated group reactive face. In this proposal the, "higher order system of relationships," is the vocal stereotyping of each other's ambients. The systematic interactions of their vocalizations themselves were selected to maintain the first order relationships necessary to create a metasystem of group interaction. It was maintained because it was selected for on the group level, coordinated group reaction was selected for. Only, secondarily did those symbols map to specific features of each individual's environment. It didn't matter what word everyone used for banana, as long as everyone knew to initiate coordinated (therefore specialized and differentiated) behavioral routines when it was represented within the group system.

In this way one can imagine the slow and steady subordination of each individual's personal adaptive reactions to a super-ordinate group system of adaptive reactions. Imagine the growth of a self-referential socially maintained system of symbol to symbol relationships coming to exist on top of the ever present individual animal system of iconic and indexical relationships to specific here-and-now features. Our ancestors could iconically see, smell, feel or taste an actual banana. They could also learn to recognize indexical referents that mean that bananas are not far off, like banana trees, banana tree habitats, the sound of someone eating a banana and oncoming banana season. But what became more and more important as time passed, was using the banana symbol socially to constrain cooperative behavior to a degree that allowed access to and control of more and more banana patches. Many can accomplish what one cannot thorough the power of cooperation.

Individuals were now adapting to a second environment, a *socio-symbolic* environment. The mappings that naturally described each individual's relationships to their own immediate sensory world became conditionally subordinated to a super-ordinate rule-set that was the group meaning mappings of language. This proposal walks us through a plausible scenario. The "higher order system of relationships" necessary for language creation and maintenance is represented by the group itself. What potentiates the group reactive face? The environment does, in the form of selection for coordinated group reactivity. Those arrays that produce adaptive group responses are selected for. Thus, the system of relationships our individual ancestors had with their immediate ambients was subordinated to the system of relationships they had to each other - to their socio-symbolic ambients. Obviously, relationships to things in their own immediate ambients remained as triggers for personal adaptive reactions, but now a new type of relationship, relationships to each other, began to trigger specialized cooperation routines in group adaptive situations. A metasystem of meaning was born.

Just as Deacon can say that, "no collection of signs can function symbolically unless the entire collection conforms to certain overall principles of organization," we can say that, 'no collection of humans can function symbolically unless the entire collection conforms to certain overall principles of organization.' Those principles are the principles of language - syntax and semantics. Even more conclusively we can say that, 'no collection of cells can function metabolically unless the entire collection conforms to certain overall principles of organization.' The language that cells use to function metabolically is one of signal transduction where "logically closed group of mappings," represented by chemical, electromagnetic and mechanical transformations create the, "higher order system of relationships," that constitute the, homeostatic, ontogenetic, metabolic and reproductive coordination of cellular components within multicellular animals.

Similarly, the language that atoms use to function atomically is one of quantum interaction where, "logically closed groups of mappings," represented by electromagnetism in coordination with strong and weak nuclear force transformations create the, "higher order system of relationships," that constitute the atomic elements of the periodic table of elements. Language is not special. It is just another system resulting from selection for persistent positive symmetry interaction. In biological parlance, it is just another example of 'selection of communication for cooperation' and 'selection of cooperation for reproduction.' In fact, Deacon (1998) himself alludes to the symmetry selection process:

> The structure implicit in the symbol-symbol mapping is not present before symbolic reference, but comes into being and affects symbol combinations from the moment it is first constructed. The rules of combination that are implicit in this structure are discovered as novel combinations are progressively sampled. As a result, new rules may be discovered to be emergent

> requirements of encountering novel combinatorial problems, in much the same way as new mathematical laws are discovered to be implicit in novel manipulations of known operations (p. 100).

Sound familiar? The novel system emerges through components sampling their interactive background for novel synchronies. In Deacon's example above the interactive background is the phase space of mathematical logic. The structure implicit in the emergent system is not present before interaction, but comes into being and affects how that novel structure further interacts the moment the novel structure is first constructed (and more importantly 'maintained'). Deacon is describing positive symmetry. The rules of combination (i.e. the rule-set) that are implicit in this structure are discovered as interaction continues (i.e. as interaction is "progressively sampled"). As a result, novel structures interacting novelly with their environments discover new rule-sets for persistent interaction that in turn create new product structures. In fact, "new mathematical laws are discovered to be implicit in novel manipulations of known operations," in exactly the same way that new laws for persistent interaction (e.g. emergent symmetry rule-sets) are discovered by the constant sampling of environments of the same old elementary particles. Interestingly, Deacon's reference to math here is prophetic seeing as, if you'll remember, symmetry is defined as transformation with preservation of structure. This is a mathematical definition. Symmetry has been the guiding principle of mathematics for millennia.

Language can be seen as just another emergent system, just another persistent level of constrained release of energy. The system logic of language is unique, just as all emergent system logics are, but the process that gave birth to it is the same. If, "metabolism recapitulates biogenesis," then language recapitulates language origins. The system that gave birth to language is a system of relationships to relationships, not relationships to things. Such a metasystem needs an explanation outside the conservative processes of internal interaction, outside the brain, outside the body. Group selection provides such an explanation. The group, distributed over time and space, provides a physical platform upon which a metasystem can be potentiated. Thus, symbolic meaning can not only be plausibly and organically discovered, but even more importantly, it can be maintained as a higher order system of self-referential relationships. It is a system founded on the indexical and iconic relationships of animals, but emerging, like each of the other major biological transformations, as a group phenomenon. Furthermore, this group phenomenon is co-constructed on a substrate of communication that both coordinates and reproduces itself. Symbolic language is an emergent symmetry rule-set that reproduces an emergent organic unity - like binary fission, like mitosis and like meiosis.

Essentially what we have here are interactive components that canalize cooperation into an emergent reproductive unity. We have seen this before. Just as Morowitz notes that metabolic systems still evidence their original autocatalytic relationships, symbolic languages should still evidence their original social-behavioral relationships. As we shall see in the next section, one of the world's foremost linguists, George Lakoff, has done just that. According to Lakoff, language is based on our original relationships with those external saliencies that fundamentally organize our existence and behavior. Basically, language is derived from behavior (see chapter "'Women, Fire and Dangerous Things': animating concepts").

## Discourses on Humanity

### Meaning as Adaptive Futures

*"Meaning is adaptive relationship to the environment."*

*Warren Brown (2009, min. 39:15)*

It is worth mentioning that a definition of 'meaning' is emerging from this account of language origins. This definition is very similar to the definition neuropsychologist Warren Brown gives of meaning in dynamical systems studies. As noted in the quote above, he says that in the mathematical study of dynamical systems, "meaning is adaptive relationship to the environment" (Brown, 2009, min. 39:15). What are, 'adaptive relationships to the environment?' They are those synchronous relationships with the environment that allow an organism to survive into the future. They are those relationships that allow it to exist. He further mentions that in interacting with its environment the organism, "reorganizes in such a way to meet that environmental challenge and ... if that [reorganization] can be used in the future, then they [dynamical systems theorists] talk about that as 'meaning'" (Brown, 2009, min. 39:12). By such a definition then, meaning is captured material or coupled information (or vice versa) that allows the organism to persist in the future. Meaning is in the positive symmetries selected to maintain and reproduce the system.

Dynamical systems theory can apply to biological systems only in its nonlinear form. Organisms are not closed ideal transformations in a closed ideal phase space (e.g. linear mathematical phase space). Organisms live in noisy and entropic environments. They can only be modeled as *nonlinear* dynamical systems and even then the comparison is not direct but indirect.

Organisms seem to act similar to nonlinear dynamical systems under many conditions. That said, this definition of human symbolic meaning as it corresponds to meaning in dynamical systems is only suggestive.

However, what is being suggested is that biological, symbolic and mathematical meaning are related. They all represent an extension of adaptivity. They all represent a function of preserved structure over transformation. The environment in which mathematical structure is selected for persistence is different from the environment in which biological structure is selected for persistence. However, both kinds of transformation, both kinds of meaning, are based on symmetry, on the preservation of structure in future transformations.

This is only a seed of an idea at this time, but it does warrant attention. It suggests that symbolic meaning of the type that we humans use to think and talk is based on survival. Our bodies are meaningful because they use their naturally selected organization to help themselves maintain and reproduce their lineages into the future. Language was derived from behavior that worked to help us maintain and reproduce not only ourselves but our social groups into the future. What was meaningful to us was extending ways to capture and couple our environment and we did this by co-opting each other's signals. What is meaningful to us now, perhaps millions of years later, is an exotic derivative of what was originally meaningful, survival. We maintained what allowed us to persist and in our case that happened to be symbolic communication.

Brown said that the dynamical system "reorganizes in such a way to meet that environmental challenge and … if that [reorganization] can be used in the future, then they [dynamical systems theorists] talk about that as 'meaning'" (Brown, 2009, min. 39:12). If this is so, we should ask if language 'reorganized us in such a way to meet environmental challenges and we preserved or canalized this reorganization so that we could use it to survive in the future.' And, if we use that symbolic information to maintain and reproduce ourselves into the future, can we talk about that as 'meaning?'

A curious language fossil in modern German illustrates the connection between words and meaning beautifully. The German verb *heißen* means both 'to name' and 'to mean.' A German brain sees no difference between the sentences, 'X *means* the treasure is buried here' and 'X *is named* the treasure is buried here.' To them, naming and meaning are often the exact same idea.

This connection between dynamical systems meaning, biological meaning and symbolic meaning makes language a socially maintained repository of adaptive futures. Like DNA or memory systems, language is a code, a rule-set that preserves what has worked before in order to maintain what will persist. It equates 'adaptive futures' with 'meaning.' What is meaningful then is whatever the system maintains for its ability to contribute to adaptive futures. This makes human symbolic meaning a fractal subset of the same meaning that maintains all universal systems.

What is meaningful to solar systems are those astronomic, chemical and quantum systems that maintain and reproduce it. Ultimately, though, gravity is the symmetry rule set that is the most meaningful to solar system persistence. What is meaningful to humans are all those symmetries that contribute to the many levels of human persistence. Ultimately, though, much of that meaning is tied up in the symbolic communication system that both maintains and reproduces human groups on this Earth. Human meaning is tied up in the words and concepts we use to build communal adaptive futures. The Germans were right to note that meaning for humans is fundamentally tied to naming.

'Meaning' won't be further discussed in this investigation because it is a divisive word. Too many academic traditions are invested in different versions of it for it to be useful. It distracts from the principle idea here, which is that, like all other rule-sets, language was selected for its capacity to maintain persistent relationships. We know this because the only kind of relationships available for selection are those that maintain adaptive futures.

## Cooperative Unity: multicellularity and humanity

This investigation is essentially about a certain class of scale free natural processes. On examination, they appear to be self-similar enough to give them a single name, symmetry. It must be stressed again that positive symmetry is, by itself, highly improbable. The likelihood that any two or more systems will maintain a synchronous relationship when they interact is most likely low. However, we humans are constituted of the very outlying edge of improbably ordered systems. We are surrounded by nested hierarchies of improbable order, starting with our solar system and moving on to our planet, and the immeasurable nested phylogenies of ordered interaction we call, life-on-Earth. If there is a consistent thread of pattern (e.g. positive symmetry) starting at the Big Bang and leading right up to mind, then we should be able to see the different manifestations of positive symmetry throughout the history of the universe, the history of life and the history of thought. The transformations at each scale level should exhibit recognizable characteristics despite each level being constituted of functionally novel (i.e. emergent) structures. These characteristics are processual. Like Per Bak's process of self-organized criticality, we should see common characteristics of transformation in all manner of interactions, between all manner of things and at all time and size scale levels.

This is an immense goal. Within the scope of this investigation it is sufficient to touch upon a few examples of what appears to be the same kind of processes operating on vastly different and traditionally unconnected scale levels. The diversity of these systems combined with the similarity of their transformations should be enough for the scientific community to consider the proposition seriously. This is all that is being asked.

Unfortunately, research for this kind of investigation has no precedent. Traditional universities do not enable such investigations. It would mean co-ordinated instruction in Astronomy, Physics, Chemistry, Biology, Paleontology, History, Philosophy, Psychology, Anthropology, Sociology and Linguistics to mention just a few principle departments. Up until now, each of these professions has its own language that divides the phenomenological components of its field into a unique learnable system. To see past the names in these time honored phenomenal taxonomies is to see the underlying patterns of relationships that form the kinds of systems that each profession deals with.

Researching this investigation was an amazing experience for the sheer number of 'Eureka!' moments where I recognized that a concept or process in one professional jargon had numerous analogues in other professions. For example, I do not know this, but I would be willing to bet there is a concept in economics that corresponds to the Baldwin Effect in evolutionary biology. It would look something like one company finding an innovation that opens a successful new niche and then its fellow companies being selected over time for their ability to couple to this new market saliency until most companies in that market embody the new 'adaptation.'

These trans-profession patterns seem to happen all the time. They were part of the inspiration for starting this investigation. It is like finding pieces of a pattern that keeps repeating itself. The fractal-like form of this process is the overriding metaphor of this investigation. What looks like is happening in galaxy formation is also happening in organismic formation. What looks like it was happening at the beginning of universal expansion is still happening in everyday interpersonal interaction. And the self-similarity of these patterns seem to exist far beyond static instances of spatial relationships. They are dynamic through both time and space. It is a process that continually bifurcates and complexifies, yet always remains tied-in to universal structures as expressions of the same symmetry process.

Once again, to give an exhaustive report on all the levels at which positive symmetry operations are significant is beyond this investigation. However, a taste will hopefully whet the appetite. What follows is one particularly fruitful set of comparisons between the origins of multicellularity and the origins of language. Most of these examples come from John Gerhart and Marc Kirschner's excellent textbook on multicellular animals, *Cells, Embryos and Evolution*. These examples show the parallels that exist between how the cells of the body came to cooperate and reproduce as whole animals and how we should see ourselves as coming to cooperate and reproduce our thought and languages as whole cultures. This is not an attempt to claim that the transformation from protists to multicellular metazoa was the *same* as the transformation from primates to "Homo Symbolicus" (Deacon, 1998). It is an attempt to show the similarities in development and to thereby strengthen the inference of an underlying logic that constrained both processes.

There is no evidence for these multicellular animals before roughly seven hundred million years ago. There is no evidence for symbolic thought or language before roughly one hundred thousand years ago. However, both emerged and both are very real parts of our lives. For economy several quotes from Gerhart & Kirschner (1997) will be given, each followed by a few comments:

> "The evolvability of metazoan populations depends on the sophistication by which cells can generate, sense, and respond to the extracellular environment" (p. 595).

Metazoan refers to multicellular animals, or 'composed of a society of cells.' Read with a frame shift up into the human evolution context, the same quote could be read, 'The evolvability of human groups depends on the sophistication by which humans can generate, sense, and respond to each other and their environment.'

> "The evolutionary breakthrough metazoa had was the loss of the communicative barrier of the cell wall and the biochemical creation of a system of cell adhesion" (Gerhart & Kirschner, 1997, p. 595).

The loss of a communicative barrier is functionally the same as the discovery of a novel method of communication. They both connote an increased connective functionality. In brain, blood vessel and immune system development this was expressed as the increased connectivity necessary for exploratory process to functionally feedback. In our case, one instance was symbolic communication. The, "biochemical creation of a system of cell adhesion," is analogous to the idea of human group selection with behavioral forces of adhesion replacing biochemical ones. Humans had to survive in groups and groups were defined by their connectivity and their boundaries (i.e. constraints). Both of these concepts are inherent in the idea of "adhesion." With a little imagination you can read the above quote substituting 'human social groups' for "metazoa," 'human' for "cell," and 'symbolic' for "biochemical." Try it.

> The lack of a cell wall ... permitted the ancestors of animal cells to interact directly with each other, to crawl on surfaces, to differentiate into complex shapes, to engulf other cells by

> phagocytosis, and to engage in junctional communication with other cells. Cell adhesion and junctional communication are characteristics of epithelia and the segregation of an internal milieu, which are found in all metazoa (Gerhart & Kirschner, 1997, p. 11).

Once a sufficiently sensitive communication system is established all kinds of emergent cooperative functions can be achieved. Proto-humans could maintain functional structures in travel. They could kill and/or co-opt complex ambient saliencies like large prey, domestic plants and animals or other proto-humans. Symbolic communication is the "junctional communication" that both creates and defines the human group. In a sense it is the systemic interrelationships of language that bind the group and give it a unified reactive face to its environment, its 'world view.'

In Metazoa, the ectoderm both protects the internal metabolic-reproductive cycles and embodies the capture-coupling mechanisms that mediate the animal's relationships with its external ambient. The ectoderm, divided into the brain and epithelia, is the embodiment of the human reactive face. In individuals this reactive face is bounded by our epithelia. In human groups, capture/coupling functions are often achieved through another, extended layer of symbolic cooperative behavior. Like an epithelial boundary, the boundary of symbolically coordinated cooperation is how proto-human groups would have externally defended themselves, hunted and foraged for each other. Internally, that same socio-symbolic epithelia would have protected the young and sick until they were able to do so for themselves, it would have encouraged the differentiation and specialization of labor essential to better tool making and nutrition and it would have encouraged specialized child care and enculturation (e.g. allowing the group the luxury of investing decades of instruction on young members).

Functionally, "the segregation of an internal milieu" is analogous to social differentiation and specialization. Where multicellular organisms differentiate cell groups into organs and regulatory systems, human social groups differentiate individuals into butchers, bakers and candlestick makers.

All cooperative behavior is enabled by the plasticity of, "junctional communication." In our case, this, "junctional communication" is turning out to be symbolic language. In the case of metazoa it turned out to be what Gerhart and Kirschner (1997) refer to as 'signal transduction.' In solar systems it is 'gravity.' In molecules it is 'electromagnetic' bonding ... and so on.

> Multicellular lifestyles [entail] ... morphological variety ... this may in part be due to the development of the capacity for intimate contact between cells and their environment in eukaryotes, or between cells and other cells, leading to the enclosure of extracellular space. Both feeding and sexual conjugation make use of complex discrimination of the outside world, a different response to different stimuli, and some degree of cell differentiation (Gerhart & Kirschner, 1997, pp. 14-15).

First of all, the assumption that Gerhart and Kirschner make here is that "multicellular lifestyles" and "morphological variety" are analogous. Once again structure is folded into process. What an animal is built like *is* how it transforms itself and its environment. Structure and process are two sides of the same coin.

Also, Gerhart and Kirschner (1997) call attention to two features of this process. The first is the, "intimate contact between cells (read: 'proto-humans') and their environment." The second is the cellular co-option of other cells that ultimately leads, "to the enclosure of extracellular space." Proto-humans had a suite of sensory relationships with their environments that included interpersonal communication. When this interpersonal communication evolved sufficient sensitivity (read: "intimate contact") then those proto-humans were able to create a communal domain of organized capture and coupling of otherwise unattainable resources. To maintain this communal domain they also had to begin to reproduce their communal structure through generations. The enclosed "extracellular space" for multicellular organisms was the communal domain enclosed by the epithelia. The enclosed 'extra-personal space' for humans was the communal domain enclosed by the socio-symbolic network. This communal domain is instantiated in communities of differentiated and specialized individuals who begin to maintain and reproduce their own lifecycles so as to maintain and reproduce the group lifecycle through more and more symbolic communication, cooperation and reproduction. Like multicellular cells, humans began to refer primarily to each other symbolically and secondarily to their environment. Like signal transduction, language preserved an emergent level of social self-reference - of autopoiesis.

If one is able to frame shift to a different vocabulary, one can begin to see some very suggestive parallels. According to the evidence, this is our current best guess for how multicellular animals evolved. Given this evidence, we should ask ourselves if human origins are really that unique.

Finally, it is interesting that Gerhart and Kirschner (1997) mention two of the historically prevalent triggers for language evolution, "feeding and sexual conjugation," in correspondence with what can only be the functional analogue for the human senses: the cell's "complex discrimination of the outside world." Only through developing these systems of complex discrimination can cooperation to the degree of reproduction of the social whole be achieved and the individuals within that whole can begin to specialize into, "some degree of cell [read: human] differentiation." These are two completely separate emergence events here, multicellular animals and human society. Both emerged by selection for persistence from an interactive background: communication selected for cooperation selected for the reproduction of an emergent social unity. Like bats and whales, both also required new senses to discern those "complex discrimination[s] of the outside world" that facilitated their emergences.

> "Cell types may be considered to be compartments of conserved reactions with different combinations and amounts of proteins, a compartmentalization of the cell's repertoire of gene expression" (Gerhart & Kirschner, 1997, p. 588).

Different cells in your body express the same genes in different ways to create different protein components that allow that cell to interact in different ways with surrounding cells. This is how the same genetic information coordinates the different cellular functions of your body (e.g. immunological, vascular, endocrinal, neural, etc.). Different humans in symbolically organized groups express the same cultural heritage in different ways to create different skill and personality components that allow those humans to react in different ways to others. As with genes, differential individual expression of the same social information package is what coordinates the different functions of the group components (e.g. care givers, hunters, hockey goalies, etc.).

Coordinated groups of humans are similar to coordinated groups of cells. Symbolic sociality in this case becomes just another level of constrained release of energy in a thirteen billion year old story of additional levels of constrained release of energy leading to social unities. Note please, that each level of social unities (e.g. prokaryote, eukaryote, metazoa) has a distinct medium of communication by which its components cooperate to maintain and reproduce distinct structures.

The following idea is attributed to Carl Woese by Wills & Bada (2000). It is interesting because it lets us see that the same kind of process that formed animals (e.g. groups of cells) and cultures (e.g. groups of humans) is also presumed to have formed the first cells from inanimate chemicals.

> "The earliest cells, he [Carl Woese] suggests, were actually aggregated communities of very simple proto-organisms, each having some of the properties needed to allow the entire community to survive ... giving rise to cooperative groups of proto-organisms that had very different properties" (Gerhart & Kirschner, 1997, p. 209).

Cooperative unity is a fact of existence since the dawn of the universe. From the firmament of post-bang matter-energy interaction, nuclei were selected for persistence. From the firmament of nuclei and elementary particles, atoms were selected for persistence. From the firmament of atoms and molecules, galaxies were selected. From the firmament of our planetary system, bacterial cells were selected. From the firmament of bacterial cells, protists were selected. From the firmament of protists, animals were selected and from the firmament of animals, humanity was selected. Each of these fundamental transformations of the universe was an instance of cooperative unity. Particularly in the last few cases, "organisms that could enrich their extracellular signals could further enrich their evolutionary options" (Gerhart & Kirschner, 1997, p. 595). In our case, 'proto-humans that could enrich their cooperative communication could further enrich their evolutionary options.' In each case, 'systems that could enrich their positive symmetry interactions could further enrich their evolutionary options.' This statement could introduce every chapter in this book.

## Human Speciation

> *[Alan] Wilson supposed that behavioral changes brought about by burgeoning brainpower must have increased the rate of morphological change. Increased brainpower increased the range of behaviors open to animals and birds, which in turn caused some of them to move into new ecological niches. The different selective pressures in these new niches, by bringing about rapid changes in appearance in their descendants, gave rise to new species.*
>
> *Christopher Wills (1994, pp. 245-246)*

*Homo sapiens* is a new species. It is almost certainly the product of something like an extended bottleneck speciation event. The big brains of the hominid stocks from which we emerged made them extremely behaviorally flexible. Behavioral flexibility can be seen to contribute to morphological flexibility as Wills notes in the above quote. New environments select for new behavior that finds new niches that drive new morphological adjustments that give rise to new species that maintain new environments. Some of these species are simple phenotypic extensions of the ancestral species.

They are bigger, hairier or have longer necks. These new species are usually tracking simple fitness couplings as the fitness landscape itself fluctuates. Air gets colder, appendages get shorter (e.g. Neanderthals).

Other new species are products of a different, more rare and drastic process. These species' ancestors were cast adrift from their historical fitness couplings. Most of these species died off, but every once in a while they emerge from such bottlenecks with radical adaptations. Reptiles evolve feathers, warm blood and flight. Small tree mammals evolve wings and echolocation. In these speciation events, Darwinian processes that facilitate evolvability like bigger brains or Hox genes or exploratory processes with functional feedback tend to be selectively preserved and statistically end up playing larger roles in the radical phenotypes that emerge.

A good demonstration of this can be found in the average age of present day genera. The average age of lizard and amphibian genera is twenty million years old. The average age of mammal genera is six million years old and the average age of bird genera is four million years old (Wills, 1994). Part of this difference can be explained as the effect of bigger brains. Both mammals and birds have bigger brains (and therefore larger behavioral ranges) than amphibians. They are able to find and then adapt to a greater range of niches, which over time translates into more species being born quicker. Thus they have shorter average genera ages.

Another part of this difference can be explained by the unpredictable dimensionality of novel niches. Novel niches are novel phenotype-lifestyle couplings - novel phase spaces. A good example is mammals that can now locate food at night with sound. Bats have discovered a new way to couple to their fitness landscape. In fact, their fitness landscape itself has now changed radically due to this adaptation. The phase space of possible couplings for bats is radically different from that of loris or bush baby phase spaces, which presumably are similar to bat ancestor phase spaces. What bats did when they began to fly and echolocate was discover a vastly expanded phase space. The world contained multitudes of flying nocturnal insect environments and birds couldn't take full advantage of them.

In most cases of preadaptation the phase space discovered is tiny. Shorter appendages do not open a world of possibilities; they just allow you to keep warmer during the ice age. In exaptive cases like the discovery of echolocation, extreme adaptations can lead to the discovery of novel niches whose phase spaces are whole worlds of possibility. Exaptive cases can lead to massive radiations marked by additional morphological changes due to the effects of Deacon's relaxed selection. The lack of competition and a plethora of new environments drove bat evolution to an incredibly diverse distribution of bat species worldwide.

This brings us again to the issue of adaptive radiation. It seems that big brains can help animals find ways to live in novel environments and novel environments cause morphological changes that can ultimately result in speciation. In the case of birds the adaptive radiation and speciation was enormous, nothing else that size could fly. In the case of whales the radiation was less so, but still significant; fish cannot cooperate like mammals can (or apparently get as big as whales). Our hominin ancestors moved first onto the ground and later out into every land based ecosystem in the world and some special set of skills opened up those phase spaces for them. In most cases, changes in niches drive novelty. In the human case, those changes resulted in a unique kind of radiation.

Birds, bats and water mammals all changed morphologically when they radiated. Apparently, we did not. But is this true? Up until about one hundred thousand years ago there are a bewildering and growing group of hominin species. The archaic sapiens, most notably exemplified by the Neanderthals, seem to be just the tip of the iceberg. Throughout Europe, Asia and Africa anthropologists continue to find what appear to be distinct hominin subspecies. The picture we are beginning to form seems to indicate that whatever local form our cousins took they took it quickly. This conforms to the idea that these small local groups were sensitive to extinction yet exceptionally morphologically flexible. Like modern hospital viruses they were continually being wiped out, but always popping up with novel immunities subsequently.

When *Homo sapiens* eventually did pop up in East Africa between two hundred and one hundred thousand years ago they were coming out of a very complicated bottleneck event. We should not be surprised when groups emerge from evolutionary hotbeds with some very unique exaptations.[177] In our case, the exaptation was communal. Like each of the major transformations in the history of evolution, the human exaptation was social; it was cooperative. Instead of our individual phenotypes chasing certain external features like bugs at night, our group phenotypes began chasing external features like cooperative hunting, food sharing and communal childcare. Social organization, in all its technicolour, multimodal glory, was the primary feature of our adaptations. As a result, individual success was increasingly measured by social integration features, not individual morphological features like aerodynamics or hydrodynamics. The greatest constraint on human evolution was sociodynamics. When we changed we changed as a group, and individuals changed so as to sync with the group reactive face.

Humans haven't speciated like bats or birds or water mammals for two reasons. The first is obviously because they are such a new species. Two hundred thousand years is an evolutionary blink of an eye. The second reason is because the human exaptation, symbolic language, maps to whole reproductive groups, not to its individuals. In fact, language defines the group. All humans alive today are 'enlanguaged' and all those languages come from that original East African group(s). Non-language groups died off.

Most likely, what decided which of our individual ancestors reproduced and which didn't was social selection pressure within the group. A careful examination of the evidence should reveal that novel modern human traits are either cultural or, if they are individual, they are constrained by how individuals fit in to social structures not how they fit in to physical environments. The first humans died because either they didn't fit into the group or the group itself couldn't compete.

The human exaptation does not couple individually to its external environment like those other animal radiations. It couples to itself. It provides humans with a sense for adaptive reaction to features that exist, but not necessarily in their own immediate vicinity. Typically speciation is phenotypic change driven by immediate environmental conditions. Bats evolved

177 This kind of hotbed environment is also how the evolution of the first cells, eukaryotes and metazoa should be envisaged.

sonar to explore the dark, but the dark has different immediate conditions all over the world. So bats speciated to phenotypically fit each new set of 'dark' conditions. The Locative Mind allows humans to sense conditions that are not-here and not-now. It creates a group maintained reality whose features are distributed over time and space like the group is. This is the discovery of a new phase space. Within this phase space humans are not constrained by what surrounds them 'right now.' Humans didn't speciate because the conditions they adapted to were primarily social and secondarily physical.

*Homo sapiens* is a new species, but we are not like our other most recent, most successful cousins. In a certain sense we are more like the first protocells, the first eukaryotes and the first metazoans. The phase space we have opened up by coordinating our group operation is massive. It is so massive that at this point it seems unknowable. The world has only seen such a transformation a few times in its history. If we kill ourselves tomorrow we will just be one more example of failed radiation. "As biologist A. Meredith suggests, the pattern of sudden appearance, expansion, and then disappearance in the fossil record has much historical precedent and is ominous" (Margulis & Sagan, 1986, p. 228).

If we manage to stabilize our growth and find ecological balance we have a future. In this future we will not speciate physically until we once again experience conditions of low populations and massive death, conditions of selection. The other possibility is the discovery of new conditions, of unbridled radiation as in the colonization of space - conditions of relaxed selection.

Extremely novel behavior can lead to the discovery of massive phase spaces (i.e. novel niches). The evolution of birds is a great example of this. Birds evolved flight behaviors and this opened up a massive phase space of possible lifestyles to those first birds. Over time they have evolved into one of the most successful and highly radiated groups in the biosphere. Humans discovered symbolic cooperation. No other animal can do this so the possibilities language opens up for us are uncontested and still being explored.[178] However, it seems unlikely that we will speciate. Already the

[178] When life began almost four billion years ago just a few organic molecules discovered genetic reproduction and gave birth to the plethora of life we find now.

world is one big human gene pool. There are almost no isolated groups of humans left that could develop novel traits. What we will do is keep reproducing into open niches and keep populating the Earth. What drives this expansion is our ability to cooperate to find new technologies and ideas that enable us to capture, couple and transform our environment novelly. We will radiate and we will culturally differentiate and specialize, but it will not be our phenotypes that record this progress. It will be the ways in which we learn to coordinate group projects and group efforts. It will be our economics, academics, social media platforms and politics that will record our successes and failures, not our bodies.

# Conclusions: Human

The first proto-humans discovered a way to organize themselves into groups that were more than loosely connected individuals, each reacting to his or her own private adjacent possible. By stereotyping their communicative reactions to each other this group began to function in its environment like a unified organism. The group itself began to maintain an internal system of reference that enabled a repertoire of coordinated external interactions. It coordinated itself globally by potentiating the communication networks between members and more importantly, it maintained and reproduced this coordination by fixing this pre-potentiated network into the cultural phenotype over generations. Through enculturation (i.e. language acquisition plus iterated learning over generations) each new member was trained and embedded with the social information necessary to preserve the group's way of life.

Functionally, this is what happened to the first proto-cells, and to metazoa, only this time the information wasn't only in chemical information form (e.g. deoxyribose nucleic acid and other signal transduction media). It was in the behavior of exchanging symbolic information - symbolic language. The first humans were selected for this way to organize their groups because it enabled rich interaction with each other which further extended successful interaction with their environments. They were specialists in exploring new ways to sync to their environments, and they themselves were the most significant features of their own environments.

This particular path to reproductive success has a long and colourful history of producing some of the most important saltations in evolutionary history. Carbon molecules interact especially richly and presumably they led the phase space search for the kinds of autocatalytic interactions that gave birth to protocells. Carbon did this by contributing to the right conditions. It provided an interactive substrate for ubiquitous and varied communication from which regularities could accumulate. Eventually individual molecular regularities were selected to function together in reproductive groups like a unified organism. Those regularities were based on the cooperative and reproductive potentiation of autocatalytic DNA networks enabled by carbon connectivity.

Reproduction is highly canalized communication. Genetic reproduction is chemical communication with high information content. Interestingly enough this view of genetic systems is reflected in the words molecular biologists use. For example, Zweiger (2001) describes the chemical relationships in genetic reproduction in the following quote.

> "The way in which the molecules of life communicate has been likened to the way in which people communicate, so much so that linguistic terminology abounds in molecular biology. Nucleotides are known as *letters*, triplets that encode amino acids have been called *words*" (p. 143).

Gene sequences themselves are often referred to as 'sentences.'[179] This language analogy extends to the processes by which genetic systems reproduce themselves at the cellular level. The process of splitting and copying the DNA helix is called 'transcription' and the process of interpreting the transcribed pieces is called 'translation.' There are also 'proofreading' enzymes that correct DNA mutations. The study of genetic descent lines, i.e. cladistics, is often compared to the study of linguistic descent lines, i.e. philology (McFadden, 2000). There is something in how DNA networks interact and reproduce themselves that reminds us of language.

Language and genetic reproduction are both serial, digital, recursive and autopoietic codes that produce heritable product structures. Both also maintain symmetry by maintaining their product structures over generations. DNA codes for the transformation of cells and language codes for the transformation of cultures. Genetic symmetry unified disparate

[179] David Pearson, personal communication, November 26, 2012.

autocatalytic cycles into the bacterial cell and language symmetry unified disparate behavioral cycles into the socio-symbolic group.

This process of selection for intermittent communal unification spans the last four billion years - the entire history of life. In each case the unification was made real, in the sense that it became persistent, when communicative interaction was canalized into cycles of cooperation that were sufficiently synchronized to maintain that system past entropic death. That is, for this process to be successful, communication must be selected through cooperation and must, absolutely must, result in some new form of reproduction that is capable of maintaining the emergent group organization. This happened for eukaryotes, for animals and for social animals like ants and wolves. Ultimately it happened for humans too. History shows that when it does happen the new emergent organisms go on to populate and change the world. Their capacities to find and incorporate novel interactions in a world that never had them before drives a kind of compounding effect where more types of more interactions creates an environment where more maintainable novel relationships can become selected for persistence. This probably is linked to inclusion. Kauffman (2000) refers to a version of this idea as, "biospheres persistently increase the diversity of what can happen next. In effect … biospheres may maximize the average sustained growth of their own 'dimensionality'" (p. 4).

Interestingly enough, when you look at history, the tempo of these transformations is picking up. It took roughly two billion years for cooperative prokaryotes to discover mitotic reproduction, another one billion years for cooperative eukaryotes to discover meiotic reproduction and a half billion years later we discovered symbolic reproduction.[180] If you'll notice, each successive appearance of a new reproductive process supporting unified social systems occurs significantly quicker, roughly halving each time. This increasing rate of major emergences occurred despite massive extinction events like the oxygen holocaust and the Permian extinction. It would be interesting to see if the frequency of emergences in the universal symmetry domains had any statistical relationship to the frequency of the emergences

[180] This data is a rough estimate based on rough reports. Life began *about* 3.8 billion years ago. *About* 2 billion years ago eukaryotes got started. *About* 700 million years ago we have evidence for multicellular animals (Margulis & Sagan, 1986).

in the biological symmetry domains. For example, most of the interesting quantum and chemical emergences occurred a mere few hundred million years after the Big Bang, like most of life's interesting biochemistry occurred in the first few hundred thousand years of Earth evolution. In both cases subsequent evolution (e.g. generations of stars and generations of different cell groups) took much longer.

There appears to be some kind of ratchet effect occurring. A ratchet effect is an effect that moves forward and holds and then moves forward again. Despite intermittent halts, ratcheting processes don't lose ground. In this case what is ratcheting is the rate of emergent reproductive processes. Once bacteria radiated to cover the Earth it didn't matter how massive extinction events were (e.g. oxygen holocaust) they could never kill all bacteria. The bacteria that survived repopulated the Earth through binary fission. Likewise, once protists radiated to cover the Earth it became very hard to wipe them all out. No matter how great the extinction event, some protists survived to mitotically reproduce their level of complexity. Each reproductive process enables a greater range of complexity in the organisms it reproduces and thus maintains a greater spectrum of interaction in the biosphere. You can kill almost all of these organisms, but the few that survive maintain a level of organic complexity that can later re-radiate to fill the Earth.

Gould (1996) claimed that there is no evidence of organismic progress. His argument rested on the fact that species themselves do not necessarily get more complex. What he failed to consider was the possibility that processes like methods of reproduction can, and have, ratcheted up the tempo of complex organic emergence here on Earth.

Language is a candidate for the next reproductive ratchet process. Already humans have radiated to cover the Earth. What we haven't done yet is survive a major extinction event. One way or another, it is coming. If language enables us to maintain cooperative and reproductive viability through this event, then, and only then, will we join the other major transitions in the history of life. Then, and only then, will we start to be persistent members of this universe.

# Section 5:
# Mind

This section outlines how the human mind emerged from the socio-symbolic evolution of language.

## Mind Emergence

**FIGURE 22: HUMAN BRAIN AND NEURON**

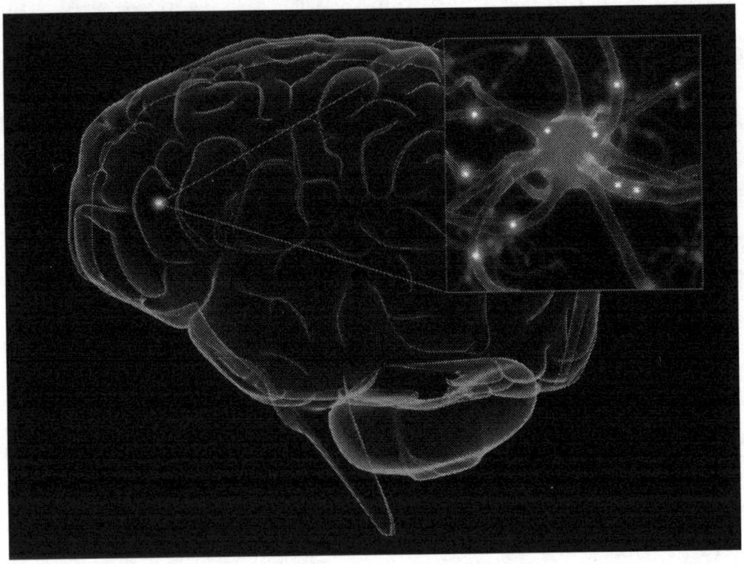

Source: Neurotic Physiology (2011)

The concept of mind is a minefield and stepping into a section titled 'mind' is tantamount to intellectual suicide. At the very least scientists and philosophers take their credibility into their own hands with such speculations. To minimize the mayhem caused by throwing the word 'mind' around it is important to emphasize that this section is simply an extension of the history being presented here. As such, mind is simply those interactions that make human behavior unique, our ability to interact with representational symbols and how that ability shapes our experience. Symbolic language capacity is mind. It is humanity's super-ordinate rule-set.

Historically, discussions of mind have assumed that human behavior is unique, but we can no longer afford to be so naïve. Traditionally, the idea of mind has included topics like consciousness, unconsciousness, reason, perception, memory, emotion, volition, self and imagination. To some degree or other, each of these qualities of mind has been shown to be present in animals. Thus, the line drawn here is not around the evolutionary history of consciousness in general, but between the kind of consciousness we experience and the kind that other animals experience. It is not perception in general, memory in general, volition in general or self in general that is spoken of here. When this investigation speaks of mind, it speaks of the specific concatenation of those qualities that constitute human cognition *as it differs from other animals*. Unless otherwise stated, the 'mind' referred to here is exclusively human.

For the most part, human cognition is the same as animal cognition. Our neurons look and act like theirs just as much as we look and act like them. Where we differ, is the human capacity to sense and interact with concepts that represent features that are neither here nor now. We can learn to sense and react to places that are not-here and events that are not-now by using each other to maintain a virtual map of not-here, not-now saliencies. Where animals find and maintain adaptive reactions by differentially preserving immediate physical interactions that happen to encourage adaptive reactions over lifetimes and generations, encultured humans can maintain adaptive reactions by preserving virtual interactions like myths and mathematics over conversations, generations, continents and cultural epochs.

Daniel Dennett (1996b) claims that most vertebrates are "Popperian creatures" (p. 101). Stuart Kauffman (2000) admires this distinction by describing Dennett's idea thus:

> Popperian creatures, in Dennett's fine phrase, have 'internal models' of their world and can 'run the internal model' with the clutch disengaged, rather than running the model in the real world. This allows us lucky Popperian creatures to allow our 'hypothesis to die in our stead' (p. 115).

What we as humans do is similar to what most vertebrates do. We create an internal model of the world and use it to maintain and reproduce ourselves.[181] But there is a difference between how we do this and how other animals do this. There is something about the "clutch" symbolic minds create that is qualitatively different from the clutch other vertebrates create.

This section will elaborate how the human mind is distinct from the animal mind. It will do so by showing how the human mind differentiated and specialized from the animal mind. In a sense, this investigation takes an immense history of interaction on the universal, life and human scales, and distills it down to one miniscule class of interaction, 'thinking like us.' To do this it needs to largely ignore general questions on general aspects of mind and focus on those aspects that generate and maintain the human phenomena. As such, this section is divided into two parts. The first part concentrates on how the kind of symbolic protolanguage previously proposed generated what we recognize as the modern human mind. The second part focuses on how that mind functions to shape the thoughts and experiences we consider to be exclusively human.

## Protolanguage to Mind

> *The whole mental shift of the Hominidae away from the rest of the primates is epitomized in the evolution of symbolic activity; and this, in turn, reaches its highest development in the uniquely human function of language. There is so much mentation involved in language that to think of it as a 'signaling system' or a social habit encouraged by survival value is worse than superficial, it is simplistic, not to say silly.*
>
> *Susanne K. Langer (1972, p. 345)*

[181] This fact that we both contain fathomable depths of symmetric pattern and are ourselves embedded in the fathomable depths of the self-same symmetric pattern is probably very close to what Leibniz (1962) meant when he claimed, "that every simple substance has relations which express all the others and that it is consequently a perpetual living mirror of the universe" (p. 263).

Susanne K. Langer's quote above is typical of social scientists.[182] What she is expressing is the disbelief that the way we make our livings has much to do with how our predecessors made their livings. This is exactly the kind of myopic chauvinism that has kept us from connecting our history with the history of life and of the universe. To be fair, only recently have we become aware of the almost unfathomable complexity involved in the evolution of each emergent class of systems in the universe. Only now are we beginning to see that the complexity we represent is built upon a foundation of many other levels of emergent complex relationships. Only by understanding how these other levels of complex relationships came to persist can we begin to understand how we came to persist. Most importantly, understanding this process will allow us to structure our future relationships so that they too will persist. Susanne K. Langer´s quote dimly reflects that age-old belief that we are special, that we are at the center of the universe and that God has a plan for us. Fortunately, we are not and he does not. Presumably Langer would have felt "silly" if she made the same comment about bats:

> The whole shift of [bats] away from the rest of the [arboreal mammalian insectivores] is epitomized in the evolution of [vocal/aural specialization]; and this, in turn, reaches its highest development in the uniquely [bat] function of [echolocation]. There is so much [vocal/aural interaction] involved in [echolocation] that to think of it as a 'signaling system' or a social habit encouraged by survival value is worse than superficial, it is simplistic, not to say silly.

How else could echolocation evolve if not as a "signaling system" encouraged by "survival value?" Could it be an accident? This kind of thinking is often referred to by evolutionary biologists as a 'Hopeful Monster.' It is the hope that some fortuitous confluence of genetic accidents (i.e. mutations) will

[182] Susanne K. Langer was an iconoclast and hero. She was not only one of the first female professional philosophers in America, but she also openly synthesized evolutionary science, aesthetics, linguistics and the philosophy of mind in one of the first true expressions of what we now think of as cognitive science. The negative impression portrayed by nit picking this quote is, in fact, homage to a great thinker who, as we all do, fell prey to a conceptual euphemism of her time, that we humans are special, and profoundly different from animals.

miraculously confabulate a new organic form all at once. 'You see, what happened was … one day a bush baby mother gave birth to a bat! Yeah, Yeah, kinda like the ape mother that gave birth to the first human!' This is unacceptable. We are not terribly unique.[183] When we look to explain our behavior (e.g. language, thought, mind) we cannot afford to just shrug and say 'Well, doesn't God play dice?' Extensive changes in functionality are worth understanding. They are structured by organisms like you and I moving into their adjacent possibles one step at a time. Bats eat, drink, play and make love. They experience the world like many other animals do, but with a little something extra. We too experience the world like other animals do, but with a little something extra.

At this point we are moving away from the evolution of language and towards how the evolution of language shapes our individual experiences. We are still moving from the general to the specific. The origin of the human personal experience, for the purpose of this investigation, is the origin of mind. This investigation argues that human personal experience is characterized by the subordination of animal sensory reactivity to socio-symbolic reactivity. Although we will refer back to the discussions of how animals generate adaptive behavior, the overall development will be focused on explaining those aspects of mind that are enlanguaged or encultured - the super-ordinate level of behavior.

Most of our mental experience is due to how our brains evolved to function over the last five hundred million years. However, in this investigation we are principally interested in how language has very recently changed that half billion year old brain. Theories that focus on non-symbolic mind phenomena (e.g. like the role of emotion in Antonio Damasio's somatic marker theory) will play a minimal role in the following chapters. This is not because they are not important in understanding ourselves. They are. But it is to the heart of our hopes, fears and desires that we are driving and any expression of them in human terms must ultimately be mediated by language.[184]

183 Every living organism, every species and every social group is unique. In this way we are completely mundane in our uniqueness. This quote refers to the fact that humans are just animals.

184 This may be overstating the case seeing as modern technology has made human expression, in the form of the visual arts, architectural arts, culinary arts, performance arts, etc. not only accessible but indefinitely maintainable. Symbolic thought so permeates these media,

## Group Minds

*"The kind of information that language was designed to carry was not about the physical world, but rather about the social world."*

*Robin Dunbar (2003, p. 220)*

*"Symbolic reference is at once a function of the whole web of referential relationships and of the whole network of users extended in space and time."*

*Terrence Deacon (1998, p. 452)*

We ended the last section with the first humans evolving a kind of group body map, or group sense halo, through symbolic cueing. This group map was a distributed system of adaptive reaction maintained by something like continuous locative updating. These adaptive reactions were like a 'sense for others' and depended on a virtual network of differentiated and specialized vocalizations that could cue adaptive configurations of differentiated and specialized cooperative action.

In essence we had the beginnings of a novel communicative network on a Darwinian evolutionary track. For a process to evolve by natural selection it needs to exhibit variation, heredity and differential survival. These early humans presented two new levels of variation for selection. The first was the way they interacted individually with their social and physical environments and the second was how they communicated, coordinated and reproduced this interaction socially on a group level. Those individuals who presented more cooperative behaviors linked to more distinct vocalizations were selected for cooperative behavior more often and left more progeny. Those groups that presented flexible coordinated behavior more often also survived differentially.

---

in the sense that they would be unrecognizable outside of the history of the symbolic experience, that none of these other human arts could have been realized without language.

Heredity here is also multilevel, but it starts with those who 'ping-back,' or respond, more often. This vocal feedback would have had to be consistent both in time and in form (remember, symmetry systems need to preserve some kind of structure through transformation). Stereotypical updates of locations and emotional states maintain an environment of regular information patterns. However, humans within this proto-symbolic environment would have also had to modify their vocalizations to differentiate this information and thus modify the reactive face they were broadcasting. Heredity is like-begets-like and the social stereotyping of broadcast information is the connection of like behaviors with like signals. Those distinct signals that incited distinct cooperative states were selected for. Similar to the language of cellular signal transduction, vocal signal-behavior links are also transformation-with-preservation-of-structure.

Differential survival is simply the preservation, by the social system, of those vocalizations that incite adaptive behavior. The buffalo that does not find two shoulders to stand between when the danger signal is broadcast, loses its children. Humans who failed to transform broadcasts into reactions well enough to behave cooperatively also died.

Two features of this account of the origin of language stand out. The first is that language requires minimal 'magical moments' to explain what has traditionally been explained using all kinds of intentional language and magical moments. Most theories of the origin of language depend on explanations like this one proposed by Dennett (1991) in his epochal book *Consciousness Explained*:

> Now it sometimes happened, we may speculate, that when one of these hominids was stymied on a project, it would 'ask for help,' and in particular, it would 'ask for information.' Sometimes the audience present would respond by 'communicating' something that had just the right effects on the inquirer, breaking it out of its rut, or causing it to 'see' a solution to its problem. For this practice to gain a foothold in a community, the askers would have to be able to reciprocate

> on occasion in the role of answerers. They would have to have the behavioral capacity to be provoked into making occasionally 'helpful' utterances when subjected to 'request' utterances of others (p. 195).

Dennett is presuming almost everything here. This description is one long magical moment (e.g. "now it sometimes happened, we may speculate"). It is also full of intentional language: "just the right effects ... causing [the ape] to 'see' a solution." Apes seek "help" and "information." "Askers" are "reciprocat[ing]." To be fair Dennett proposes this hypothetical event as a kind of mental exercise, not as a definitive explanation. Even so, it is a common practice in language origin theories to depend on such casual illustrations. Just so stories are very appealing. When challenged, theorists can always wave at their disclaimer and when they are not challenged they get credit for a good idea. Origin of mind and origin of language theories are both plagued with good magical moment ideas. Discerning investigators learn to be wary of words like 'innate,' 'induce,' 'appears,' 'spontaneous' and 'natural inclination.' They are classic red flags that mark magical moments.

What theorists are striving for is a naturalistic and fully scientific account of mind. This account must be demonstrably in accord with known scientific laws and that means it has to be to some degree testable and/or re-creatable. However, the thorough testing or measuring of systems that exist far beyond our own scale level (e.g. geological time, speciation, cosmological evolution, quantum scale) is beyond our current capacity. As a result, theories like Darwin's Theory of Natural Selection are extremely difficult to prove in the Popperian sense. In these cases we have come to accept a surrogate scientific method: if enough evidence can be produced to establish historical continuity and that continuity can be used to accurately predict and account for further phenomena, then the theory can be provisionally accepted as a work in progress. This is exactly the status of natural selection, continental drift, relativity and quantum mechanics. We accept all these theories and find evidence for them everywhere, but how exactly they work is still up for grabs.

A fully scientific account of mind may be impossible to prove right now. However, we can still build a functional theory from the history of emergence. The *Locative Theory of Mind* is just such a proposal. What made Darwin's Theory of Natural Selection credible was the logical continuity it provided for all biological phenomena. 'All animals are selected for how well they survive, check it out for yourself.' The greatest weaknesses of his theory were its discontinuities. How do complex features like eyes form? How come there are so few transitional fossils? What is the mechanism of heredity? These gaps, or magic moments, haunted Darwin till he died. They were his Achilles heel.

What would make a good theory of mind? The best theory of mind would show us how to predictably recreate mind in mindless animals (perhaps Chimpanzees) or how to create artificial intelligence. Failing that, the best theory of mind would be one supported by a massive amount of evidence of historical continuity and, in doing so; one that minimized the amount of magical moments. What makes the Locative Theory of Mind unique is two things. First of all it is supported by an over thirteen billion year body of evidence. More than any other theory of mind it proceeds logically up through universal and life processes to show how each step proposed makes sense within its natural context. Second of all, the Locative Theory of Mind minimizes the gaps between us and the rest of biology, chemistry, astronomy and quanta. In fact, it places our bodies, brains, minds and senses of self squarely in this universe in a way that no other theory has. It explains how the same process that made atoms and black holes, made us. The Locative Theory of Mind minimizes the use of magic moments to bridge the gulf between us and everything else.

The second feature that stands out is that the Locative Theory of Mind embodies language originally and principally in the group, not the individual (i.e. not in the brain). Language evolved as descent with heritable variation of human relationships (e.g. pragmatics), not neural relationships or genetic relationships. Even in the social theories of language origins the individual and/or his specific array of internal brain modules are traditionally chosen as the contexts within which the symbolic system came to exist. The problem with these theories is the lack of a credible platform on which the inception of a distributed network of symbol to symbol (i.e. self-referential) relationships can be not only generated but maintained.[185] Under what conditions can such a symbolic platform be regularly selected for?

There are millions of good uses for symbolic language, but humans didn't evolve language because they needed 'something like that.' They evolved language because along the way symbolic relationships had been selected for and then maintained by other systems of relationships. The best option for a platform that could support symbolic relationships from basic behavioral interactions on up to complex symbolic interactions is the social system itself. Language may not have evolved exactly as proposed, but if we consider humanity to be a major transition we should compare it to the other major transitions in history. In this case, the history of major emergences both in life and in the universe is unequivocal. Almost fourteen billion years of history supports a social emergence theory for the human mind.

185 For more on this see chapter "Language Origins."

# Socio-symbolic Emergence

## What was Language?

We are finally in a position to start the discussion about what language actually was. Too often theorists expect that modern language is identical to archaic language. It's easy to say that the ear-drum bones were originally jaw and skull bones, but the argument looks patently ridiculous if you do not have a fossil record to prove it. The danger of describing language origins is that it's easy to mistake ultimate function for original function. Robin Dunbar makes this mistake when he claims that gossip was the original motivating factor in the evolution of language. He makes this inference largely because psychological studies have shown that gossip is mostly what we use language for now. "You will discover, as we have in our research that around two thirds of their [people's] conversation is taken up with matters of import. Who is doing what with whom, and whether it's a good or a bad thing" (Dunbar, 1996, p. 4).

This is certainly true of myself living in the middle of Lima, Peru at the end of 2010, but I believe matters would change drastically if I found myself in an ice age climate with a few friends. I think discussions of how we could cooperate to deal with food, weather, tools, plans and nearby enemies/friendlies would become significantly adaptive in this case. The principle of allocation applies here.

> The principle of allocation presumes nothing about cause and effect, except that the effort put into feeding, or anti-predation, or reproduction tends to expand in evolution so as to fill the time made available to it. The expansion is halted by the dangers presented by the environment during the most difficult periods (Wilson E. O., 1975, p. 143).

The structure language took was selected for by how it enabled us to survive, "the most difficult periods." As previously discussed, "the most difficult periods" are those bottlenecks where exaptations are forged. However, after language originated it expanded to fill the free time it afforded us. Today humans no longer live under the imminent threat of extinction, but our days are just as long and we have to do something with them. Even now we typically dedicate forty hours a week to making a living by exchanging the information and things our communities need to function (e.g. the canonical job).

The question that opens our discussion of language here is not what language is, but what it was. What was it about the ancient proto-mammalian jawbone that made it evolve into the exquisitely intricate mammalian ear with its attendant capacity to interact with sound environments so intimately? What was it about proto-human pants, hoots, grunts, coos and trills that made them evolve into, 'history, economics, and literature'? The question is made doubly difficult because there is no fossil record in this case. For the moment we will have to depend on educated guesses that we can only hope will help inform our search for some kind of mosaic record of evidence.

Here is a candidate definition for the original human language: *The original human language was an emergent representational communication system that was maintained socially as an exploratory mechanism with functional feedback.*

Before unpacking this phrase, credit should be given where credit is due. This definition has been derived from two brilliant sources. The first half of it comes from Terrence Deacon's work where he imagines the mind as a, "representational system," kind of tacked on to a, "brain architecture mostly evolved in a world of concrete sensory and motor adaptations" (Deacon, 1998, p. 283). This concept has already been introduced in the "Language Origins" chapter at the end of the last section, but what is seminal here is that symbolic representation is *systemic* in nature.

The second half is from Gerhart and Kirschner's work on biological systems. They present a whole chapter of multi-level, multi-media organismic mechanisms that function to produce coordinated reactivity to the environment by over connecting components and then selecting for functional responses. If you frame shift this process onto human communicative networks you get a surprisingly similar outcome - coordination of function at the global level through selective connectivity at the individual level. This kind of process has already been simulated by scientists who model neural and cellular automata networks on computers. These scientists can train their nets to perform surprisingly complex operations simply by potentiating the connectivity codes of the system itself according to rules analogous to environmental conditions for selection.[186] This will be further discussed in the chapter on "Group Potentiation."

For now let us characterize symbolic language as it emerged for our ancestors. First of all, language was a representational system. Deacon (1998) argues compellingly for symbolic representations existing as parts of systems.

> It would be misleading, however, to suggest that those images [representations] are all there is to symbols, any more than the words on this page suffice in themselves to convey their meanings. They are merely neurological tokens. Like buoys indicating an otherwise invisible best course, they mark a specific associative path, by following which we reconstruct the implicit symbolic reference. The symbolic reference emerges from a pattern of virtual links between such tokens, which constitute a sort of parallel realm of associations to those that link these tokens to real sensorimotor experiences and possibilities. Thus, it does not make sense to think of symbols as located anywhere within the brain, because they are relationships between tokens, not the tokens themselves; and even though specific neural connections may underlie these relationships, the symbolic function is not even constituted by a specific association but by the virtual set of associations that are partially sampled in any one instance. Widely distributed neural systems must contribute in a coordinated fashion to create and interpret symbolic relationships (pp. 265-266).

[186] This is in reference to recent work done on cellular automata and artificial life in the area of complexity and computation (Cohen & Stewart, 1994; Lewin, 1992; Waldrop, 1992).

As a neuroscientist Deacon recognizes that there is no one-to-one mapping of symbols to neurons or even specific neural configurations. Nor is there a one-to-one mapping of symbols to the external 'things' they refer to. This came as a surprise to many theorists who expected that once we got the technology to look into brains we would find where certain ideas lived in the brain. What we found was, 'well that depends.' What does it depend on? Deacon's answer, and by proxy ours, is that it depends on other symbols. It is the virtual links between these symbols, the way they are connectively arranged in relationship to each other that give them meaning. Noam Chomsky calls the rule-set of how we all navigate this network of symbols "generative grammar." It is the rules that constrain how we transform one symbol into the next. Symbolic meaning is represented by how symbols preserve structure when they transform - symbolic symmetry.

Using the metaphor of maps, you could say that we are not speaking of physical maps, the kind of maps tacked onto walls or stuffed into glove boxes. We are speaking of the human ability to translate geographical saliencies into a series of meaningful squiggles, dots and lines, wherever we find them. This ability is a process, a symmetry system, not a thing. It's like 'mapping' in that it is a transformation process that converts massive amounts of 3D physical information into a simplified code that orients us. Symbols do not exist by themselves, they are just the components of a system of transformations and these components only exist insofar as they communally maintain relationships to other symbols. The waggle dance of bees that transform massive amounts of 3D physical information about food sources are also simplified codes that orient other bees. Buffalo are also social. They do not exist by themselves, they are just components of a social system of transformations and they exist individually only insofar as they maintain communal relationships to other buffalo. Lone buffalos do not survive.

To explain the origin of symbols you have to explain the origin of the symbol system. That system, or process, must be composed of components that somehow maintain regular interaction. This maintenance should come from the two sources that define a system: internal interaction and external interaction. It should come from regular component interaction (i.e. internal) and it should come from the novel interactions that system as a whole maintains with its environment (i.e. external).

Here is an example. Hydrogen and oxygen create $H_2O$ molecules when they interact under certain conditions. In this case, the covalent bonds of the atoms themselves maintain the molecular structure (i.e. internal interaction). However, when many of those $H_2O$ molecules interact they form groups we call 'water.' These groups have the well-documented feature of surface tension. Surface tension is the result of a mild ionic bonding between the $H_2O$ molecules themselves. The hydrogen and oxygen ends of the molecule have slightly polarized charges which result in a mild electromagnetic attraction between the molecules (McFadden, 2000). This cohesiveness gives water the external feature of surface tension which is responsible for the shapes of drops and the skating abilities of water strider bugs. In this case, the molecules themselves maintain the water structure, but so does how the group interacts with its light gas environment. Surface tension does not exist for individual $H_2O$ molecules, only for the group when it is in a light gas environment. Surface tension is an important feature that constrains where and in what forms those groups can exist. Surface tension is part of water's emergent reactive face. In the end, the thing we call water is maintained by how it interacts with its environment as well as how its components interact with each other.

Human interaction is considerably more complicated than $H_2O$ interaction, but both depend on internal systems of interaction (i.e. social cohesion) and external systems of interaction (i.e. the group reactive face) that give their groups emergent features as per their environments. It was the discovery of a symbolic means of communication that enabled human groups to organize and tune their reactive faces to their environments to a degree rarely seen in nature. The degree to which this symbolic substrate could be maintained over generations is the degree to which the cooperation it enabled became reproductively canalized (i.e. the degree to which symbolic language became a new reproductive system).

Thus human protolanguage was a behavioral system that constrained how group members adhered and reacted to each other, as well as how the group reacted to its environment. The original symbols were each other's reactive states as regularly expressed in vocalizations. The original symbolic system was the group members themselves, distributed throughout space and interacting over time. Modern language is a highly complex and modified descendent of this original system. It is still a distributed representational system, but, like all highly adaptive tricks, we have begun to move it into our individual internal systems to conserve it. Specifically, we have moved it into the part of our internal that was built to be plastic enough to capture as many good tricks as possible, our brains. For this reason the symbolic systems in our brains have taken the form of, "widely distributed neural systems [that] must contribute in a coordinated fashion to create and interpret symbolic relationships" (Deacon, 1998, p. 266).

So the first half of a definition for language is a review of the "Language Origins" chapter - language is a referential system. It is a system that is made of symbolic components. It is referential in that those symbolic components cause adaptive reactions in individuals to "real sensorimotor experiences and possibilities" (ibid) namely, to other's broadcasted experiences. Symbols originally referred to others. This symbolic net, or web became a distributed system because it was originally maintained by the physically distributed social group. Different vocalizations of different members represented nodes of distributed meaning over group time and group space. The second half of the definition of the original language is about how that system was maintained.

Language was originally maintained like all emergent systems are, by the differential survival of persistent interactions. Organisms that explore for a living, like us, make especially good use of exploratory mechanisms with functional feedback. If you'll remember, exploratory mechanisms are biological processes that organisms have evolved to enable environmental feedback to train themselves to become functional at a level of detail far exceeding that which they could differentiate and specialize themselves (e.g. genetically). They often work by creating a condition of over connection and then let environmental reactivity determine which connections will be preserved. Gerhart and Kirschner (1997) explain that this process, "is to generate without strict bias a large number of possible states and select the most appropriate one" (p. 147). The one that is selected is the one that functions best, the one that contributes to the most adaptive reactions, the one that enables survival and ultimately, reproduction.

Language is an exploratory mechanism with functional feedback. Constant pingback creates a state of over connectivity. This over connectivity is literally a way to explore the group sense halo. As the group interacted with its environment some of these connections are selected for their capacity to incite cooperative responses. As such, those vocal connections feedback into group functionality just as group functionality feeds back into the canalization of vocal connections. This is essentially a description of selective pressure for differentiation and specialization. The environment trains the group symbolic system to produce adaptive reactions that exceed individual capacities.

Without a pheromonal communicative substrate, ants were constrained to explore and react to their individual environments. With a pheromonal communication system, ants could chemically represent, within their group, where opportunity was in their environment and how they could arrange each other cooperatively to best take advantage of it. They used one single trick, one new communication system, to massively expand the phase space they could explore and the cooperative behavioral routines they could use to take advantage of what they found. Without a socio-symbolic system, humans were constrained to explore and react to their individual environments. With a socio-symbolic system, humans could symbolically represent where opportunity was in each other's environments. We used one single trick to massively expand the phase space we could explore and the cooperative behavioral routines we could use to take advantage of what we found.

When you ask what language is, you are asking not only what it looks like, but how it functions and, ultimately, how it came to function like that. This view of the emergent universe and emergent life proposes that the original symbolic language was a representational communication system that was socially maintained as an exploratory mechanism with functional feedback. To a great extent it has remained so. For this reason human groups or even societies should be seen as emerging units-of-selection at their own appropriate scale level and within their own appropriate interactive backgrounds. Within these social units of selection, language operates to maintain and reproduce the unit's organizational state, much like metabolic processes maintain and reproduce cells and the societies of cells called animals. It was the differential survival of the vocal representations of others that trained the language system to produce adaptive reactions at the group level.

Those who doubt the power and ubiquity of this kind of external environmental feedback to shape and maintain novel behaviors would do well to ask the water strider bug how it learned to stay afloat.

## The Old Poverty of Stimulus Argument

Language originated as, and remains, an exploratory mechanism with functional feedback. The representational system is group maintained because it functions to explore novel environments by reinforcing functional responses. In fact, the meaning of 'maintenance' has two time contexts. Metabolism can be seen as the maintenance of the internal state of the organism over a *lifetime* and reproduction can be seen as the maintenance of the external state of the organism as it represents a feature of the environment over *generations*. Clearly, metabolic maintenance can be compared to how symbolic representational systems maintain human groups (e.g. division of labour rules, gender roles, political, religious and economic institutions, etc.) or human individuals (e.g. the symbolic choices we live by). However, equally possible is a reproductive comparison. How is this socio-symbolic network reproduced in children over generations?

Children acquiring language can be seen in an exploratory-process-with-functional-feedback light as well. One of the first things children should learn (and perhaps the most important and thus the longest selected and the most deeply embedded in internal systems like genes) is to pingback. They should learn that vocal environments require vocal responses.[187] Some of those responses are adaptive. They produce positive reactions (affection/ excitation) or negative reactions (discipline/ inhibition). Others produce

[187] Children probably have an instinctual (i.e. genetically assimilated) tendency for vocal over connectivity. They have been selected to *include* a tendency to be highly vocal. This can be compared to Tomasello's (2009) experiments showing that infants have innate tendencies to cooperate. Both cooperative and vocal overconnectivity behaviors are so essential to language development that they have probably had over two million years to become assimilated, or 'included,' in the phenotype through preferential selection.

nothing. In this proposal the constant pingback is the over connectivity. At first, all sounds carry all meanings. The child babbles. The functional feedback is the potentiation (i.e. reinforcement) of certain vocal constraints by the environment (i.e. the mother, father, teacher, larynx, basal ganglia, etc.). The child does not need to create meaning from nothing. For him, everything vocal has meaning and cultural feedback (e.g. mom) selects the few thousand appropriate vocal modulations it will need to incite the appropriate cooperative responses it will need. The child learns to constrain its babble, preserving just the functional (e.g. socially persistent and recursive) elements and rules of the verbal spectrum. In this proposal the child doesn't learn distinct words as much as it constrains a continuous stream into the structure its environment syncs to, and it just so happens that this structure, for reasons we will discuss, is digital and serial. Perhaps we maintain and reproduce a digital and serial language not because the universe is made of discrete and ordered parts, but because rule-sets based on discrete and ordered interaction are the easiest and most efficacious cooperative codes to discover in design space.[188]

The canonical argument for universal grammar is that, "children produce complex sentences that they have never heard before, and acquire new lexical items on the basis of as few as one exposure in context" (Hurford, 2011, p. 475). How could they do this if they didn´t have a complex grammar system innately preprogrammed in their brains? This argument presents a false premise. Children don´t have to "produce" anything, not in the sense of production from nothing. They imitate all vocal sounds possible and then learn to constrain them to categories of positive results. They don´t produce the information. The information, or constraint pattern, exists in their environment and they just have to sync to it (note that this approach again presumes a massively interactive universe from which persistent interaction, in this case language, is preserved not created).

Metaphorically you could say that children don't produce the 'river' of language, they just learn to swim in it. Learning to swim requires certain physical, genetic and epigenetic predispositions, but actual swimming is

188 For more see the upcoming chapter "Language Simplification: serializing and digitizing the environment."

a matter of getting in the water. Children do not need explicit instruction in language, neither from a pre-evolved LAD module in their brains nor from their parents' genes. Language is made from children's brains interacting with parental environments. Their chemical and cellular substrates are preferentially preserved by feeding back positively with extended cultural constraint landscapes. Like most complex behavior, internal language information is very limitedly pre-determined because functional flexibility encourages processes of expansive growth from which external constraints select the specific suite of adaptive functionality that the organism embodies. As with all emergent phenomena, language requires an external environment from which it can be selected for persistence. This is essentially the problem with the Chomsky inspired nativist theories of language. They presume that language must be internally created from nothing when actually; it is externally selected from everything.

In Chomsky's defense, his original idea seems to have been misunderstood, simplified and vilified by many language theorists. Often, he seems to be set up as the mandatory straw man so that cleverer people can tell us how wrong he was and how right they are. However, even a cursory examination of Chomsky's original idea reveals he presented a thoughtful, even prescient, approach to a problem that had, up to that point, defied progress.

Chomsky (1976) approaches language like we approach any other biological phenomenon. "Why, then, should we not study the acquisition of a cognitive structure such as language more or less as we study some complex bodily organ?" (p. 10). This thought was radical in his time because philosophers and social scientists insisted that language was a learned, cultural trait, not a biological one. They felt justified in this presumption because, amongst other reasons, languages varied but organs didn't. The belief was that organs were the same in all humans, but language was specific to its culture. One was grown and the other was learned. Chomsky's initial claim to fame was unifying all languages under a 'generative grammar,' a symbolic rule-set. When he showed that syntax was a universal human trait, like organs are, it was natural to propose examining language as such. Given the subsequent popularity of sociobiology, evolutionary psychology, neurolinguistics, and all of the other cognitive sciences, Chomsky's initial idea has proven to be fantastically prescient.[189] In fact, when Christiansen and Kirby mention that it was Pinker's and Bloom's 1990 article that kicked off the biological interest in language origins, they miss a golden opportunity to focus on Chomsky's 1976 article as seminal.

So where did Chomsky go wrong in his analysis? He correctly identified the biological approach, but applied it incorrectly. He went wrong in two ways. First Chomsky (1976) thought language needed to be predetermined in the form of "advanced knowledge" (p. 11) and second he thought that language had to be internal. In 1976 the material reductionist paradigm ruled. Material reductionism has two relevant constraints in this context. The first is that all things are the result of historical lines of cause and effect. The second is that these historical causes can be exactly determined. In material reductionism, by determining the exact causes you can predict the exact effect. Neither Chaos nor Complexity Theories were seriously considered at this time.

These paradigmatic constraints led Chomsky to a simple solution. The rule-set that defined language in the form of a "Universal Grammar" must be a determinate locus of information that historically evolved. Some *thing* must have *caused* language. Since the environment was quintessentially diffuse and unhistorically unified, the language rule-set must be like a "complex bodily organ" that expressed innate pre-evolved information from within. This solution was particularly attractive because at the time behaviorism was the dominant paradigm and its claim that all behavior was learned from the environment was becoming increasingly questionable. Chomsky's Poverty of Stimulus argument was a reaction to the emerging scientific realization that *all* behaviorist explanations proposed "hopelessly underdetermined" environments.

The problem was that Chomsky put all of his eggs in the other basket. If language wasn't externally predetermined, it must be internally predetermined. In any case, the material reductionist paradigm of his epoch insured that it had to be predetermined somewhere. It had to be caused by some

189 Interestingly, both E. O. Wilson's *Sociobiology* (1975) and Chomsky's *Reflections on Language* (1976) came out at the same time.

physical process of development and since the brain was known to cause behavior, the brain was the logical place to locate the 'organ' that caused language. It was really quite a simple and powerful solution that conformed beautifully to the scientific trajectory of its time.

Things have changed. Over the last thirty five years Chaos Theory has drastically reduced deterministic hegemony while at the same time Complexity Theory has vastly expanded the simplistic view that behaviorism implied existed between systems and their environments. Over the years Chomsky has continued to focus on this internal locus of predetermined language knowledge, initially in the form of a Language Acquisition Device (LAD) and later in the form of genes. Unfortunately, the more we learn about the brain the less it looks like there is a LAD module. Like most other complex cognitive traits, language appears to be distributed all over the brain. Doubly unfortunately, the subsequent retreat to the fallback position of genes (e.g. it's all in the genes) is also looking bleak. As genes become recognized as less and less exclusively deterministic of any traits, much less complex ones like language, the argument that language is genetically predetermined starts to look increasingly like the God of the Gaps argument where language is presumed to exist wherever science can't find it.[190]

In the end, Chomsky's solution made something like the classic creationist fallacy. To this day creationists love to cite complex adaptions, like the eye, which they claim are far too complex to have evolved incrementally. What good is one part of an eye? Chomsky adopts something of this kind of logic when he makes the poverty of stimulus argument. Chomsky's phrase, 'Thus, it is clear that the language each person acquires is a rich and complex construction hopelessly underdetermined by the fragmentary evidence available" begins to look suspiciously like the creationist argument, 'Thus, it is clear that the *eye* each person acquires is a rich and complex construction hopelessly underdetermined by the fragmentary evidence available.' The solution that both groups propose for these problems is to invent a kind of

[190] The God of the Gaps argument is the argument that religion keeps changing its position according to what science can't explain. Science couldn't explain what made atoms a century ago, so God must have done it. We can't explain why the Big Bang happened today, so God must have done it. The God of the Gaps continually retreats in the face of scientific discovery. 'God' is thereby continually being restricted to the smaller and more distant *gaps* in our knowledge.

all-in-one solution. God must have created the eye. A LAD must have created language.

Chomsky vastly underestimated the symbolic environment that children grow up in. Of course, complex adaptations like seeing and speaking are partially internally predetermined. Every system is partially internally predetermined. Held & Hein (1963) put kittens into carts attached to other kittens so they never walked on their own. Although these cart kittens grew eyes they never learned to see properly because the development of functional seeing requires learning how to connect the motor systems to the visual systems. They never walked so they never developed a connected motor-visual system. Learning to see isn't genetic, it needs genes to present the material for selection (e.g. an eye connected to a brain), but without environmental potentiation (e.g. walking in the real world), 'seeing' never develops. Learning a language isn't genetic, it needs genes to help present the material for selection, but without socio-symbolic potentiation, 'thinking' never develops. Moreover, genes don't usually change permanently until the environment selects for them.[191] We have always needed genes, but only recently have those genes become selected for preserving linguistic predispositions.

Chomsky's poverty of stimulus fallacy is just a symptom of a much larger malady. The scientific community in general vastly underestimates the environments that select which structures can persist in this universe. Science's vast weight of legitimacy is exclusively and precariously balanced on the side of previous causes that can be isolated in experimental vacuums. In the case of language, the causes are presumed to be isolated in the brain. Environments are generally discounted. When environments are considered they are relegated to the mathematically over simplified and low dimensional generalizations that are typical of statistical analyses. They are rarely, if ever, seen to be the ubiquitous, simultaneous and continuous multi-layer levels of transformation of which the phenomena we wish to explain are made and within which these same phenomena survive.

191 Genes famously drift, but that drift is called 'drift' because it is disconnected from adaptive function.

Brilliantly Chomsky saw that biology had to be brought to bear on the problem of language. Unfortunately, like all of us, his solutions were constrained by the paradigm of his time. For him language had to be internally predetermined, just like an organ. We now know that organs like brains and eyes can never learn to see without incorporating information from their environments. For a functional eye, seeing incorporates the information around it. The extended physical environment it needs to learn to see is both external and preexistent.

Language is a sense like seeing. It is a modality that allows us to learn from what was around us in the past to extend what we can successfully do in the future. Of course, we need brains to learn language, but the language information is not in us, neither in a LAD organ nor in our genes, except to the extent that selection has been able to preserve certain predispositions for syncing to it into various internal systems over the last few million years. What determines the features of symbolic understanding is not genetic. Symbolic consciousness needs genes like molecules need quarks, but quarks do not adequately predefine the emergent behaviors of molecules. Chemistry is a different rule-set. Similarly, genes do not adequately predefine symbolic behaviors. Language is a different rule-set. Language is a social rule-set, a wave behavior.

There is a potentially helpful metaphor for trying to explain the mechanism of natural selection. The metaphor starts with the question of what Michelangelo's sculpture "David" is made of. A scientist asked this question would take a sample, runs some tests and then determine that David is made of marble from the Carrara region. An artist would disagree. An artist would say that toilets and columns are also made of Carrara marble, but they are obviously not "David." The difference is that toilets, columns and David all perform different functions in the human environment. They are preserved in different ways, at different levels and for different reasons - not for what they are internally made of, but for their external reactive face, for their 'functionality.' What makes David important, and thus preservable by us as a group, is not that he is made of marble. He performs a very different function for us. Seeing him inspires us. How does he inspire us? His power to transform inspires us.

David is not the block of marble that arrived in Michelangelo's workshop in 1500 A.D. David is what Michelangelo did to that block of marble. What Michelangelo did was transform that block of marble into a structure that inspires our belief in the power of man to perfect. Crucially, the value of David to humankind is not in the marble that makes him, it is in the marble that Michelangelo removed. David's value is in that block of marble's environment, namely one Michelangelo who eliminated most of that block to leave a structure that would persist in the human environment as long as human civilization persisted. Every piece of that marble block that didn't function to encourage the persistence of the resulting structure was removed, selected away. Information is selected from noise, organisms are selected from indiscriminate reproduction, entropy is selected from 'all possible states,' seeing is selected from billions of eye-brain-hand-feet interactions and language is selected from vocal overconnectivity.

Michelangelo didn't make the marble. Humans don't make language. We are born within a cultural stream of symbolic information and, with the help of all-the-reasons-our-ancestors-conserved to survive this stream, we learn to navigate it. The child's mind doesn't have to build a "structure-dependent rule" or a "structure-independent rule" because he doesn't have to make the sentence. Deep internal structure and wide external boundaries constrain the structure of his language. He doesn't have to choose or create anything. His genes preexisted as did his symbolic environment. He is born into his symbolic environment and grows into the constraint structures that persist within and around him. This symbolic environment is not static. It is moving through time like a stream. To a large extent we learn language like that block of marble learned to be David. Michelangelo was the stream that carved that block of Carrara marble into a beautiful, persistent and sublime form of human nature. Socio-symbolic language is the stream that carves our bodies, brains and genes into a beautiful, persistent and sublime form of human nature.

Christiansen and Kirby (2003) mention the constraints on language in the form of the inherent logics that give language its form. Philip Lieberman mentions that motor sequencing constraints extant in the basal ganglia also gave language a kind of preexistent 'order for free.' Obviously, both a child's genes and their cultural environment constrain their symbolic behavior. Chomsky's idea that the information in language had to be internally preformed and localized in the form of a language organ wasn't wrong it was just simplistic. In fact, it's not clear that he claimed otherwise. Like genetic information, symbolic information is maintained by a complex history of deeply conservative levels of internal pattern and selected by vastly creative future levels of external pattern.

Like universal biology, universal grammar could have evolved by many routes. It was constrained by an environment that only preserved certain interactions, but the interactions themselves need only be preservable (i.e. adaptive), not predetermined. As such, Michelangelo could have given David any number of different features, but he broke symmetries and gave David those that we see today. What makes David survive is not his 'marbleness,' nor his specific pose. Although marble is necessary, it is not sufficient to describe his success. David's success is in his humanity, in how he interacts with his human environment. What makes language survive is not its genes, nor its specific structure. Although genes and syntax are necessary, they are not sufficient to describe language's success. Language's success is in how it allows us to make and remake meaningful, creative and persistent future relationships with each other and the rest of our environment.

## Group Potentiation

Our ancestors taught themselves to talk by potentiating their group communicative network in a way *similar* to how neural nets are potentiated in the brain. When the infant brain develops, each neuron randomly connects to tens of thousands of other neurons. As the brain develops those connections that fire in sync with other connections are reinforced and preserved. Neurons that fire together, wire together.[192] Those connections that are out of sync die off. Out of sync with what? Successful neuronal

[192] My friend David Pearson has advised me that coordinated *inhibition* of firing is as informative to neural systems as coordinated firing. Thus, the phrase, "Neurons that fire together, wire together" is misleading because it implies that the information that defines brain activity is entirely made up of excitation reactions. More likely is that individual neurons fire chaotically and groups of neurons are characterized by how they inhibit individual firing to sync coordinated activity. I keep the phrase because it is handy and it captures the idea of synchronous interaction.

connections are synced to neighboring neural signals (i.e. other neurons in their environments). Globally, these neural networks are synced to external signals, be they from the pancreas, your mother or sun/moon diurnal patterns (e.g. other patterns in even more extended environments). Only those connections that maintain cycles of feedback survive. What is left is a trained brain, a brain that responds adaptively to what its eyes, hormones and others have told it. What trained it? Interaction with the brain's environment, your body (i.e. internal patterns) and your senses (i.e. external patterns), trained it.[193] Your genes help build a basic bauplan for a brain with massive neural over connectivity. That over connectivity samples all kinds of interactive experiences. Your brain experiences the world through its body and those experiences potentiate those neural connections that maintain adaptive relationships between all levels (e.g. body, brain, environment, etc.). Those experiences synchronize the brain. They shape it functionally. All neural connections that do not positively feedback with their environment die off. The next few chapters on language simplification will take a closer look at how this works with specific reference to Damasio's convergence divergence zones (CDZ) model of consciousness.

Humanity's predecessors were more or less communicatively connected. All they really needed was a simple behavioral adaptation for massive over connectivity. They got this in the form of developing a pingback locative sense for each other. Each person was then connected almost at random to every

---

193 Neuroscientists note that there is a hierarchy of conserved rhythms in the brain with the deepest, oldest parts of the brain (evolutionarily speaking) having the most highly conserved and canalized (genetically or epigenetically) oscillations. Examples would be hippocampal theta rhythms and homeostatic body rhythms like respiration. Mammalian cortices are the most recently evolved and external evolutionary layers of brain. They are also have the most chaotic and complex rhythms. In general, deep, old pattern is regular and simple while external, new pattern is highly chaotic and complex. Adaptive behavior comes from the selection for persistence of the interaction between the historical internal with the now and future external, be that the inner subcortical brain stem maintained by conservative historical rhythms syncing to the outer cortices or the outer cortices maintained by emergent synchronies with ever more distant and chaotic environmental patterns. It would be misleading to imply that there is no deep, internal, and preexisting pattern to neural interaction that is historically hardwired into brains. There is. It would be equally misleading to imply that this internal pattern was evolutionarily immutable. At some time in our evolutionary history these inner patterns were captured and canalized external relationships. Internal patterns come from a history of differential selection for emergent and persistent external synchronies.

other person. The signals in this case were not synaptic, but behavioral: vocal-aural.[194] As the group interacted with its environment, feedback, in the form of successful cooperative reactions, reinforced certain signals, canalized them. In this case those signals were stereotyped vocalizations not synaptic arrays. At first these vocalizations probably only modulated locative and emotional information, but by stereotyping this information to coordinate cooperative responses the group trained itself to react to its environment at a new level. Over time those vocalizations that communicated more specific information were reproduced preferentially and language began to maintain the differentiation and specialization of group adaptive reactions. What trained the group? The environment did. All the group added was a platform of massive communicative over-connectivity and selection did the rest.

Of course, there are significant differences between brains and human groups. There are also significant differences between ant foraging behavior, blood vessel formation and autoimmune systems. However, they were all selected for and maintained as one of nature's fabulous tricks - exploratory processes with functional feedback.

## Language Simplification: serializing and digitizing the environment

> *"Your only access to what is going on in your brain comes in a sequential 'format' that is strikingly reminiscent of the von Neumann architecture - although putting it that way is historically backwards."*
>
> *Daniel Dennett (1991, p. 215)*

Human language is serial and digital. It is serial in the sense that we can only understand or produce one stream of symbolic information at a time. It is digital in the sense that this stream of information is differentiated by distinct words. Words are symbols that primarily represent social references

[194] Vocal/aural interaction depends on the brain's synaptic substrate just as synapses depend on the neuron's molecular substrate. Vocal/aural interaction is just another extension in a long series of extensions of communicative systems into cooperative systems.

predate syntax (Christiansen & Kirby, 2003). If symbols predate grammar then the first thing we are looking for is the functional maintenance of symbols. Why would anyone maintain them? If symbols are detail poor now, they were extremely poor at the beginning. To be maintained they would have to provide information that the other senses couldn't. What was that information?

The senses provide iconic information about our immediate environments. We can see, touch, taste and smell bananas. Large brains like ours, and Pavlov's dogs, can learn to indexically extend that information to predict the near future. Both of these extensions are tied to our here-and-now state. Now, what if we learned to use others to relay information about their environments to us? This information would be tied to there-and-then conditions. The information source humans discovered and developed was what you need to know about what I'm up to. It was not-here information. But to convey not-here information one has to reduce it drastically. There is dead space between you and I, a communicative (and thus cooperative) boundary that separates your organism from mine - your behavior from mine. How do we cross this boundary? Luckily, this is a problem that nature has already solved. Like the transformation of information in signal transduction between soma cells, the symbolic signal is reduced (i.e. simplified) at the boundary (or space between differing systems) only to be re-elaborated by the receiving system. Language rediscovered digital communication to do what organisms like us have been doing for billions of years: sensory extension to include more cooperators in our cycles of maintenance and reproduction.

The digitizing of symbolic signals is not at all biologically unusual. When we discussed regulatory linkage in the "Life" section we saw that groups of cooperative cells communicate through various methods of signal transduction. More often than not, when one cell or cellular system in our bodies communicates with another, it must pass information through a boundary (often cellular or epithelial membranes) that mediates the transformation of information from one system to the other. This transformation typically involves a complex state change on one side of the boundary inciting a far simpler boundary interaction that incites a complex state change in the system on the other side, e.g. synapsis, transmembrane kinases, convergence divergence zones and phosphatases to name just a few.

This process of gross simplification at the boundary is very similar to what occurs between humans. The sender's complex, internal state of being is reduced to strings of single words that, when received, act to incite complex, internal states of being in the receiver. The boundary is the empty space (air) between the two systems and the signal is selected to be simple because only simplified bits are able to maintain structure (i.e. survive) across this dead space. Of course, the signal need not be digital, but digitization may be the easiest way for functional linkages to evolve. When different communicative media in the environment are sampled, digital solutions may be closer (in phase space) and/or more common, and thus sooner found and more often canalized. Signal transduction between internal biochemical systems often includes some kind of positive-negative charge or specific ligand or receptor interaction that modulates the transformation into distinct bits, or 'digital' interactions (e.g. neurotransmitters, ion gates). If distinct systems are selected for intimate communication, digitization may be a convergent feature of the kinds of communication media that are selected for in these cases. Digitization may be selected for as a functional constraint for sharing information cooperatively, regardless of whether the systems interacting are biochemical cells or vocal humans.

In fact, the digitization of communication is not at all unusual to engineering either. Claude Shannon revolutionized communication technology when he inspired us to convert analogue telephone systems to digital systems. The problem was that just amplifying the analogue signal also amplified the imperfections in the signal, the noise. Signals transmitted over long distances degraded. He revolutionized human communication by introducing digital coding where the signal was transformed into a digital code at the source, transmitted, and then decoded at the receiver. Shannon revolutionized human communication technology and ushered in the information age by rediscovering nature's own trick of digitizing communication to enable it to faithfully replicate itself. The device Bell labs engineered to decode the binary digits sent by analogue voices was called the 'regenerative repeater.' In a sense, humans also use each other like regenerative repeaters to communicate, cooperate and reproduce their descendants. Each of us transmits a vastly reduced chemical digital code, in the form of our DNA, to our progeny. On the other side, our children re-elaborate this code into a body built from what survived transmission and its environment. Like synapses, ion gates and humans, an emergent digital technology, in this case words, enabled us to reproduce cooperative behavior better.

The serial aspect of language seems particularly unglamorous. Language is a behavior and even complex behavior is serial. 'Complex' here means many distinct parts coordinated into successive and unified, but differentiated and recursive states, typically over long durations. As previously mentioned, behavior too has a bottleneck where the animal's response to a lot of simultaneous external stimuli must be collapsed to be processed and then responded to in a single series of coordinated actions. Only if this reactive face is serially coordinated in time and space can it function adaptively. Language is serial because it is constrained by the same process of coordination that constrains all adaptive reactions. Behavior in general is serial.[198] Any human experience mediated by language is likewise behaviorally constrained to distinct, unified, successive, differentiated and recursive states of being.

Together, the serial and digital medium of communication that maintained a group saliency map of symbolic adaptive opportunities resulted in a gross simplification of a parallel and continuous environment. Just as the first mitochondrial protists or multicellular metazoa must have groped about in their environments before they learned to maintain the new connective potentials of their cooperators to respond adaptively to novel saliencies in their environments, so did the first humans grope. New sense systems are weak. What the new human sense did poorly was see the environment. What it did for the first time was see it through another's eyes. What language gives up in depth of descriptive power, it gains in another mysterious power. No other sense can see beyond the here-and-now.

In the end, if cooperative communication links are differentially preserved, then both serial and digital solutions seem to be statistically probable. Thus, like many other communication systems, language was

198 See chapter "The Behavior Bottleneck" in the "Life" section.

selected to maintain reproductive capacity by serially and digitally limiting the noise to signal ratio and thereby extending the capacity of the system to grow, interact and maintain itself. Language is digital because it functioned to produce adaptive reactions in others. To do so it had to reduce massive amounts of here-and-now information to a differentiable vocal signal, one that can survive transport over a dead space boundary and be decoded by the receiver as a saliency to which it can respond in a timely fashion. It is serial because it is conveyed as a behavior and all behaviors are serial.

Language was a new and weak sense at first. To sense the environment through others, one had to read vocalizations and, at their best, serial-digital vocalizations are massively simplified representations of real environmental saliencies.

## Neural Digital Simplification: the CDZ framework

Antonio Damasio (1989, 2010) has a model for memory and consciousness called the convergence divergence zone framework (CDZ). His argument is essentially that neural connectivity patterns form coevally with sensory input patterns.[199] His model proposes that as we mature, the patterns of the world around the brain train its neural networks so that intelligence can be seen as the mapping of successful external interaction to internal connectivity. What is interesting is how he proposes external patterns are entrained in the brain. External pattern is transduced into neuro-chemical signals by multiple homeostatic and sensory portals (e.g. brainstem, eyeballs, eardrums, skin, nose, etc.). These diverse and simultaneous dispositions step into the brain by converging and diverging in specialized associative zones (CDZs) resulting in a series of global gestalt brain states. In this model signal is not fed forward continuously, but instead entire gestalt states of connectivity are incited and then collapsed in these CDZs through a kind of 'small world' neural architecture.[200]

[199] 'Sensory input' for Damasio is incoming information from both the body (i.e. interoception) and its environment (i.e. exteroception).

[200] Small world architecture is a mathematical graph that models networks of interactivity that support maximally flexible functional connectivity like genetic networks, internet networks, 'Kevin Bacon' social networks, and neural networks. For more see Strogatz (1998).

What he is proposing is that neural transformation is characterized by a process of digital simplification. In this process the details of the previous brain state is reduced (i.e. converged) in these CDZs, and then re-elaborated (i.e. diverged) in the next step. This could be described as a kind of global strobe effect where previous states of neural connectivity *inform* proximate states by constraining all *possible* states to a series of prepotentiated (i.e. memorized) *actual* states. This process of digital simplification disables the computational explosion implicit in multiplying the information passed from state to state.[201] By compressing global states into simplified states the brain is able to express unlimited information arrays within a limited network of neural connections. Digital symbolic language does much the same job for continuous experience. Like the CDZ framework, symbolic language allows us to express unlimited information within a limited system.[202]

This kind of digital simplification should be expected as a feature of systems that are potentiated by their environments. The massive amount of external information must first be compressed into signals simple enough for the brain to manage and then, if they are to produce useful behavior, those signals must avoid chaos and computational explosion when they incite further brain signaling. What Damasio is proposing is a way to understand how *all* complex brains generate intelligent behavior. Language is intelligent behavior and it should be no surprise that language too makes use of a process of digital simplification to produce complex adaptive behavior.

Our bodies are full of transductive processes that serve multiple functions. The molecule adenine serves in hundreds of different metabolic functions from its many jobs in genetic reproduction, as one of the four basic molecules that compose DNA, to the numerous interactions that require ATP (adenine triphosphate). Nature has a way of recycling materials and processes to create many different results. Redundancy, contingency and differentiation/specialization are just a few of the well-known biological process that depend on the recycling, or 'exaptation,' of old processes to produce novel functionality. Considering this, there is little difficulty imagining that digital

201 A. Damasio (personal communication, January 3, 2015).

202 Digitized systems can also dampen chaotic propagation (see chapter "System Interpenetration and Chaos Damping).

simplification may be a ubiquitous feature of complex biological functionality on many levels. Its reappearance in symbolic language is just one more level of functional expression.

## Physical and Symbolic Mapping

In the physical world, animals behave by stepping into their adjacent possibles. The physical world is represented to the animal as a location from which any number of adaptive reaction routines can be initiated. Only one routine can be initiated, but that routine is selected from the set of adjacent possible reactions. For animals this location can only be a specific place in space and time, the here-and-now.

How animals choose to interact with their environment is constrained by the set of behavioral routines they are capable of (i.e. the adaptive history and structure of their phenotype), the state of their bodies (i.e. internal state), the features within their sense halo (i.e. perceived external state). The phase space of the interaction of these three levels of organismic sync, combined with the physical environment, represent the entire set of transformations that an animal is capable of at any given time and place. Animals step into their futures by selecting behavioral routines that take them from one location in space and time to another - from one here-and-now to another. This model for behavior can be abstractly represented as something similar to exploring a fitness landscape. The landscape is defined by the three dimensions of space combined with the movement of the landscape (including the animal movement within it) as the time, or change, dimension. The animal travels this model by choosing from the constrained set of transformations available to it (i.e. the rule-set) at each here-and-now point. As such, each here-and-now point offers a unique set of opportunities, from which the animal selects one, to traverse a behavioral trail of linked moments in a physical world of real dangers and opportunities.

So what is the difference between human symbolic behavior and regular old human animal behavior? Not much. The two systems share certain features. For one, the structure of transformation in both systems is by way of serial lines of coordinated interaction. Behavioral routines and language are both serial phenomenon. For another, in neither system can the same thing exist in two places at once. Just like you cannot be in two places at once, you cannot express two concepts at once. If you want to transform the state you are in now to a desired state, you have to travel (i.e. transform), following a path of intermediate states (i.e. preserving certain structures). This will be more fully explored in the upcoming chapter "Women, Fire and Dangerous Things." On the other hand, there are also some important differences. Language does not share the exact same set of symmetry rules as other behaviors.

Like other behaviors, language is a symmetry system. It is transformation with preservation of structure. The ambient the rule-set of language exists in is the very specific ambient of human-to-human interaction. Conversely, the ambient the rule-set of other human behaviors exists in is the greater ambient of animal-to-environment interaction.[203] Most importantly, the symbolic rules humans use to effect symbolic transformations are not constrained by the physical rules animals use to effect physical transformations in time and space. That is, transformation from symbol to symbol is different than transformation from physical state to physical state, from place to place.

Concepts are virtual 'things' and they exist within a system of other concepts. All concepts are connected within the system by a rule-set that determines which transformations are allowed, namely syntax (and at a deeper level semantics). When humans think or talk they 'travel' from concept to concept using the transformations this rule-set maintains. Previously, Deacon referred to language as an, "associative trail," of, "virtual links." His model for language is very much like the fitness landscape previously described where humans express symbolic transformation as a movement from conceptual "buoy" to conceptual "buoy" - from moment to moment, from one state of connectivity to another, from one word to the next. Transformation in this symbolic landscape is characterized by two main differences from travel in a physical landscape.

The first difference is that the symbolic landscape that we travel is massively simplified in that all conceptual things (i.e. words) are tightly bounded (i.e. digitized). Unlike real things that are nested hierarchies of interaction

[203] Humans are animals. Human-to-human interaction is a highly constrained and distinct subset of human-to-environment interaction.

extending down to quantum levels, symbols have no interactive dimensionality outside the symbolic network. They have no deep structure outside of how they constrain human behavior. The second difference is that the adjacent possible for concepts is not constrained by the physical mechanics of movement. To *physically* travel from a valley bottom to a mountain top takes thousands of physical steps and, for our ancestors, days of effort. To *symbolically* travel from a valley bottom to a mountain top takes two steps, the transformation indicated by 'from' plus the one indicated by 'to.' The rules of transformation in time and space constrain travel in the physical world, but this constraint is largely absent in symbolic travel.[204]

The animal and proto-human mapping of their physical worlds was constrained by the mechanical rules of physical interaction. You and I are likewise constrained when we physically move (i.e. transform while preserving our structure). However, when our ancestors began to maintain a virtual system of representation between themselves, travel within this new system had different rules. Instead of mapping the myriad real and complex transformations of their physical world, this new system mapped the simplified and digitized saliencies of each other's worlds, of the social world. In this socio-symbolic world 'things' that existed very far away, or that existed a long time ago, could be immediately represented. They were present in the adjacent possible. Through language they could exist here-and-now at any moment by accessing (i.e. 'travelling') the group maintained system of symbolic associations.

Most likely, the representational process of maintaining a symbolic system started out as simple here-and-now representations for group reactions. This is the good old fashioned indexical cueing for coordinated social behavior that we started with. However, as our ancestors differentiated and specialized this system, and their mappings and reactions became more complex and extended, they would *not* have run into time or space constraints. The natural extension of this sense would interact, sync and reproduce, with use and with generations, to include more recognizable

204 To be more precise, physical constraints are so radically altered that they may appear to be absent. Symbolic language has physiological and environmental constraints (e.g. synapses or noise limits) but these constraints constrain very different kinds of interaction at very different levels and to very different degrees.

differences and more specialized reactions to those differences. Novel reactive faces like symbolic behavior sample their interactive backgrounds to find further emergent relationships. Over time, this natural process of sensory extension would include more and more distant, both in time and space, saliencies. These extended concepts would be maintained as long as the recognition of these extended representations remained adaptive - as long as they remained important to the group.[205] What this means is that symbolic communication is really just a specialized extension of good old fashioned animal indexical communication. The difference is that symbolic communication is not constrained to here-and-now interaction.[206]

Of course, sensory extension brings us directly to the subject of prediction. Since when has knowing the future not been adaptive?

## Predictive Extension and the Future

*"All brains are, in essence, anticipation machines."*

*Daniel Dennett (1991, p. 177)*

Like all emergent systems, the symbolic system was selected from and maintained by its environmental background. In this case this was the historical substrate of social behavior. Language is a metacommunication system that makes the group itself able to maintain and reproduce itself, adapting and transforming according to its ambient. Individuals exist within this social environment. Where individual behavioral systems create a behavioral trail of linked moments in a physical world of real dangers and opportunities, their

205 Perhaps a more exact way to say this is that concepts would be maintained as long as they didn't encourage decidedly negative symmetry behavior (i.e. destructive). Negative behavior could range from the relatively benign occupation of group mental space (thereby limiting innovation) to encouraging behavior that was socially destructive. Dennett is especially interested in the idea of language as a virus to be supported as long as it is not lethal. For more see (Dennett, 2007)

206 The incredible degree of success symbolic communication has afforded us has given symbolic language the appearance of being completely different from animal communication. It is not. It is a highly differentiated and specialized indexical communication system that has evolved without the native time and space constraints built into animal communication. For this reason many scientists argue that language is based on a built-in behavioral relationship with our environment (Bergen, 2013; Johnson, 1987; Lakoff, 1987; Noe, 2009).

socio-symbolic minds create an associative trail of linked representations in a virtual world of shared (i.e. social or imaginary) dangers and opportunities. Like all closely related symmetry systems the super-ordinate system does share features with the subordinate system. However, the rules by which the super-ordinate system interacts are not constrained by the rules of the subordinate system.

Starting about two billion years ago there were enough protists that they regularly interacted. For the subsequent one billion years protists were sampling different kinds of interaction with each other, they were exploring protist phase space. Some interactions were predatory and some were symbiotic (i.e. semi-regular). Most were otherwise (i.e. irregular). Most protist interactions resulted in no persistently reproducible relationships at all. However, eventually enough symbiotic relationships were sampled that a certain kind of maintainable one was discovered. This relationship was between the reactive faces of at least two protists. It was an external relationship, not internal (even though external relationships are maintained by internal relationships). The regular relationships that those protists maintained and reproduced were the first multicellular signal transduction systems.

Multicellular signal transduction is how multicellular cells 'talk' to each other in groups. Intra-cellular components like organelles, ribosomes and chromosomes also talk to each other, but in a different language. Signal transduction needs this intra-cellular cooperative substrate to exist. It needs each cell to maintain itself. However, multicellular signal transduction also expresses relationships that intra-cellular communication cannot. It maintains and reproduces relationships between organs and regulatory systems that single cells cannot hope to build much less maintain. Multicellular signal transduction extends the relational regularities each cell is able to make with all the others. Where protists typically have a sense halo of a few centimeters, multicellular organisms typically have meters, up to kilometers, of saliencies they can sense and react to. Multicellular signal transduction is an emergent symmetry rule-set that expands the dimensionality of what each individual cell can do to survive. Through cooperation, each cell in a multicellular organism is physically in sync with more of its environment, both its fellow cells and those external features they work together to sync.

Electromagnetism did the same thing for protons. It expanded the dimensionality of what protons could sync to. Gravity did the same thing for molecules. It structured extended nucleosynthetic environments, which in turn extended planetary environments for molecules to sample different types of autocatalytic relationships.[207] Eventually one of these relationships was the origin of life on Earth. In each case, the emergent systems depend on the maintenance of the subordinate systems. In each case, the emergent systems express a statistical step into the future as they represent a novel reactive face and an as yet unknown potential for further interaction.[208] For this reason superordinate systems include and depend on a history of subordinate systems, but they are not constrained by the rules of their subordinate systems. In this sense quantum, chemical, astronomic, prokaryotic, eukaryotic and multicellular systems are truly emergent. In this same sense, symbolic language is truly emergent.[209]

As animals, proto-humans navigated their physical ambients by using their behavioral systems. Part of this behavioral system is the building of feedback mechanisms that, over generations, extend the sense halo. This feedback creates a process where the earlier you can recognize coming saliencies (e.g. the four F's - feeding, fleeing, fighting and mating) and react adaptively, the more your progeny survive to populate the world. Because of this feedback process, most mobile animals, and certainly our ancestors, were in the business of predicting the future. As Daniel Dennett (1991) says, "the key to control is the ability to track or even anticipate the important features of the environment, so all brains are, in essence, anticipation machines" (p. 177).

The problem with predicting the future is how chaotic it is. No matter how extensive your sense halo is, or how many instinctual reactivity routines have been internalized (e.g. through learning, stereotypy or genetic assimilation), the environment is massively more complex. This complexity comes from the incredible amount of interaction that is happening around us, at all levels and time scales, all the time. This reality approaches a state of complete

207 Gravity maintained and reproduced suns, which maintained and reproduced new kinds of atoms, which maintained and reproduced new molecules.

208 Entropy is the clear precedent for statistical steps into the future.

209 For more see chapter "System Interpenetration and Dampening."

analogue transformation that surrounds us. It is analogue not because the natural divisions on the Planck scale or between fundamental forces do not exist, but because to the degree of sensitivity we are capable of at our size and lifespan scales, the ubiquity and continuity of transformation around us is, for our sensory and conceptual intents and purposes, undifferentiable. That is not to say that we do not recognize all kinds of opportunities that surround us, just that our senses are largely incapable of recognizing the tiny Planck or astronomic domain differences that are the only deep (in a metaphysical sense) divisions around us.

Once again, the problem with predicting the future is how chaotic it is. Every behavioral step we take into our futures is plagued with the incursion of chaotic interactions. A lion might appear, your brother's wife might appear irresistible, a butterfly in China, etc. When evolving feedback mechanisms that help us recognize saliencies and trigger adaptive reactions, there is a limit to how far ahead we can look. Such predictive mechanisms can only evolve to the extent that the ambient presents regularities. Things like the seasons are so regular that animals can internalize adaptive reactions to them over thousands of generations. Other things, like the invasion of novel competing species, are irregular and often cannot be incorporated into feedback mechanisms within a few generations. They are one-offs. They are broken symmetries. In fact, in an environment of continuous and ubiquitous interaction, the growth of the probability of one-offs interfering with future prediction mechanisms grows exponentially. Future prediction mechanisms suffer from both chaos and entropy as they step into the future. Once again, the environment is noisy. Ultimately, the problem with predicting the future is the massive and to some degree random interaction of everything around us.

Effectively, the degree of chaos in the ambient creates a limit to future prediction. This limit is often called a Lyapunov horizon. A Lyapunov horizon is the point beyond which effective prediction is impossible (Strogatz, 2003, p. 190). Equally, it is the point beyond which random interactions are so probable that nothing can be predicted. Animals have evolved brains and senses as future prediction mechanisms up to the point where information in their sensory halo can inform reactions that will be adaptive in their futures. As such, the future they are able to sense is limited by the noise in their ambients. Even their various noise dampening mechanisms fail over time and distance. Those mechanisms animals evolve to step adaptively into their futures (e.g. senses, memories, brains, instincts, alleles, etc.) are inevitably defeated by the exponential increase of chaotic interaction at each step. At some point the expense of maintaining senses for extending a more and more unpredictable future outweigh the benefits. The mechanism stops extending when the positive feedback process is drowned out by ambient noise. Thus, each and every animal's behavioral future, except to some degree ours, is limited by a Lyapunov horizon in their not too distant, real and physical future.

Interaction in the socio-symbolic ambient is different. The symbolic ambient is populated with digital virtual 'things,' not analogue physical 'things.' In this ambient, saliencies (e.g. words and/or concepts) are specific and determinate. It is a system of "meaning-buoys" that are differently potentiated (i.e. weighted with different connection strengths) so that interaction is limited to travelling from one determinate buoy to the next, one word to another. In this system there is far less noise. How much less is yet to be determined, but symbols do represent massively simplified real world structures or processes. Their interaction is a massively simplified phase space compared to the interactions possible as a result of the physical phase space of the three universal domains interacting around us all the time. Most importantly, only one transformation at a time can occur in a symbolic system. In comparison, think of the billions of physical transformations occurring simultaneously within and around you right now.

The symbolic system is maintained by internal, be they social or neural, interactions. These interactions are restricted to their own components (i.e. other symbols) and they are restricted to serial, digital expression. Like the metabolic system the symbolic system is buffered from messy, noisy, external, real world physical interactions. The super-ordinate socio-symbolic medium effectively separates symbolic behavior from the rest of the behavioral systems. Socio-symbolic information represented in our minds and cultures is only indirectly affected by the physical and biochemical information our bodies represent. This is not to say that one does not affect the other. They do, the internal is connected to the external. This is to say that chaos propagating from the physical environment is dampened by the translation of real things to virtual, symbolic things. It is the same principle expressed by all universal systems and explained in the chapter "System Interpenetration and Chaos Damping."

What this boils down to is that transformations effected in the physical world are sloppy affairs, with entropy propagating into them and chaos propagating out of them all the time, at all levels. Transformations effected in the symbolic world are highly bounded and determinate affairs. There are rules to how one word leads to another and uncertainty in the preceding word is left behind when the next word is introduced. Even though words embody indeterminacies on their own (i.e. vagueness), the rules by which they transform into the next word seal each set of indeterminacies behind the boundary of the next word. The transformation rules of the symbolic system itself constrains the propagation of chaos from word to word, from idea to idea. Each symbol is complete unto itself. Computer code was engineered to be digital (i.e. binary) for the same reason language was selected for it, to reduce uncertainty, to increase information.

Of course there must be a chorus of complaints at this simplification of a massively complex symmetry domain. For example, we should consider that each new word only partially restricts the preceding words to smaller and smaller meaning sets. Whether someone says the words, "arrested" or "shit" following the phrase, "That pig ..." largely constrains the possible meanings for the word, "pig." If he follows with "arrested" you assume he is speaking of a policeman. If he follows with, "shit" you assume he is speaking of a farm animal although you are still not certain. If he continues with, "all over the pen," then the meaning of "pig" becomes more and more determinate in the sense of representing a farm animal that snorts. This kind of symbolic back-propagation may be considered in analogy to system coupling. For example, the addition of an electron to an atom (i.e. coupling) further constrains the interactive potentials of that atom's other components - the relative arrangement of its other electrons, protons and neutrons. For now however, it should simply be noted that the socio-symbolic system emerged from the behavioral system and that symbolic transformation is simplified, chaos resistant and not subject to the mechanical constraints of physical transformation in time and space.

Although the socio-symbolic system originally evolved to maintain group level adaptive reactions, this was not all it was capable of. Jawbones evolved to help feeding, but that was not the only process they were capable of enabling. Once the socio-symbolic system existed, its novel reactive face created a new phase space, a new space of adjacent possible interactions. Such new phase spaces inevitably have boundaries. Sometimes these constraints are the barely adjusted ancestral boundaries, like a slightly more northern home range. Sometimes these constraints are more distant and unknown, like the size constraints on mammals (e.g. elephants on land and whales in water). In the emergent symbolic phase space, chaos does not inhibit extension. Another word can always follow the last. More importantly, the next word is not subject to growing uncertainty. It is just as determinate as the previous words. This means that extension in symbolic terms is infinitely possible. Linguists often refer to 'recursion' or 'discrete infinity' as unique and key features of the human language ability (Bickerton, 2003; Corballis, 2011). Partly because words are 'discrete' (read: 'digital'), they can be added to each other infinitely. When our ancestors began to explore where words could take them, they found no Lyapunov horizon, no limit to future predictions.

Through language we travel associative trails of virtual links that are highly uninhibited by chaos, time or space. Each concept or word is distinctly bounded and distinctly connected. The networks of association we create and maintain have rules; symbolic transformation must be digital and serial and it must conform to the rules of syntax and semantics. Because of these rules, the future was opened to us. The rules of language allowed us to start building multitudinous adaptive futures. The unconstrained extension of the symbolic system into the future (or the past, or Mars, or the center of the Earth) is *systemically* enabled.

A natural process of differentiation and specialization has led to the exploration and expansion of this symbolic phase space to its current dimensions. Since the beginning, language has created billions of adaptive futures. From this socio-symbolic interactive background humanity has been able to select for those ways of imagining our futures that have allowed them to survive and flourish. Together we have tuned our symbolic ability to find and maintain more and better adaptive futures for each other and for each generation.

One important reminder is that just because we can weave symbolic extensions into many futures, does not mean we can really tell the future. The real world around us is still permeated with chaos and its cousin entropy. Predictions are based on a fifty thousand year old history of preserved symbolic information as well as a half billion year history of what our senses and bodies tell us. Symbolic predictions may be theoretically infinite in reach, but the true transformations of the world are always shrouded in clouds of compounding probabilities and broken symmetries. Both emergence (i.e. positive symmetry) and entropy-chaos (i.e. negative symmetry) step into the future statistically. With language we can use symbolic information to narrow the immediate future into multitudinous more or less probable states, but our predictive powers still lose even probabilistic potency at distance. Ask any stockbroker, politician or newlywed.

## 'Inner Space': cooperative boundaries

> *"The most important feature of metazoan cells for complex multicellular development may have been their ability to seal off the outside world and generate extracellular space internal to the organism."*
>
> *John Gerhart and Marc Kirschner (1997, p. 18)*

> *Accordingly, I propose that a collection of autopoietic systems that, through the realization of their autopoiesis, interact with each other constituting and integrating a system that operates as the (or as a) medium in which they realize their autopoiesis, is indistinguishable from a natural social system.*
>
> *Humberto Maturana & Francisco Varela (1980, p. xxiv)*

Inclusion is part of what systems do. Systems are matter-energy arrangements ordered through space, transforming regularly (i.e. symmetry) to represent order through time (i.e. synchrony). When these systems are maintained and re-created due to certain persistent conditions, their ordered transformations create further ordered transformations because they present persistent and regular reactive faces to their environments. Some of these further novel interactions are more persistent than others, and the most persistent represent novel couplings, novel emerged systems. In this sense systems are inclusive. Because they are persistent - and because persistence requires reproduction - they start including more and more matter-energy within the boundary conditions that support them (see chapter "Inclusion").

Like abiotic systems, biotic systems have a long history of propagating symmetry. In fact, the most significant changes in life history have come from incidents of cooperative inclusion of one living system by another and the emergence of reproductive processes that maintain these cooperative inclusions (Smith & Szathmáry, 1995). As previously mentioned, inclusion involves, "the consistent extension of internal boundaries to include external regularities."

In the history of life each major cooperative union created a new and expanded 'inner space.' To function cooperatively (i.e. metabolically and reproductively) this new inner space needed to include new systems of communicative cooperation. For the first cells this was exemplified by the chemical systems of lipid boundaries and DNA transcription resulting in a new prokaryotic inner space maintained by a new cellular autocatalytic communication system, the chart of intermediary metabolism and binary fission. When Eukaryotes expanded the prokaryotic boundary to include proto-mitochondrial cells, that inner space of conserved interaction was further expanded. It now included not only the aforementioned communication systems, but also added the super-ordinate rule-set of multi ordinate simultaneous chromosomal transcription and an expanded set of signal transduction rules. This new communication system was built on the previous substrate systems, but it now functioned to maintain eukaryotic cells as a cooperative union through an expanded chart of intermediary metabolism and mitosis. The first eukaryotic cells experienced the world very similar to prokaryotic cells, but with a little something extra.

What is evident is that the process of life continues to include more and more external regularities within its internal boundaries and that this process involves the regular addition of more complicated meta-systems of communication that allow these enlarged, communal organisms to maintain themselves. Each major transition in life represents a jump in what can be called 'inner space.' This inner space is like an internal laboratory of conserved synchrony. It is the creation of an expanded internal environment where predictable (i.e. historically regular) interaction is necessary for maintenance (i.e. communication, cooperation and reproduction).

Inner space is a space of highly synchronous reactivity. It is systems of coupled oscillators interacting regularly with other systems of coupled oscillators. This space effectively extends what was the Lyapunov horizon of all of the cooperating (previously free-living) components. It does this by reducing the chaotic reactivity that previously limited the extension of prediction for each one. By including each other within an expanded inner space they create a new space of highly synchronous reactivity that effectively buffers their interaction from chaotic external interactions. Thus the internal becomes a chaos-buffering zone where experiments in synchronous reactivity can be fostered. In this case, the conserved core reactions maintain organizational stability while differentiation and specialization continues between internal systems. This results in novel interactions that preserve and make use of processes like contingency and regulatory linkage (see, "Internal Systems"). Under these conditions the inner space can be further compartmentalized, linked and regulated. It can explore novel communication and cooperation couplings in a safe, non-chaotic, symmetry-propagating environment. It maintains its own 'life' boundary conditions defining that delicate balance of self-organized criticality. The inner space is that environment of push-pull conservation and novelty generation which would be difficult to maintain in the noisy external environment.

Brett Calcott and Kim Sterelny (2011) discuss this kind of internalization of developmental cycles in the following quote:

> But environments do not just happen to organisms; organisms help make their own environment and those in which their offspring develop. Termites, for example, develop in a world built by and for termites. As a consequence, their developmental environment has been stabilized. Compared to their presocial ancestors, termite genes are expressed in a narrowed range of developmental environments, and hence the phenotypic effects of those genes are more predictable (p. 12).

Their point is that organisms have a history of creating new boundaries, like termite mounds, which in turn stabilize a new socially maintained internal environment (like termite nurseries in termite mounds) within which new developmental paths can be selected. In fact, Calcott and Sterelny go on to extend the idea to the evolution of the meiotic egg in the Cambrian explosion.

> We suspect that such increased parental control of developmental environments was crucial to the Metazoan radiation. In discussing inheritance, Maynard Smith and Szathmáry emphasize transitions in the kind and quantity of information transmitted across the generations. But for genes to have stable phenotypic effects, they must be inserted into a sufficiently structured and predicable developmental environment. It is no use just sending genes; the parental generation must build an environment in which those genes are used in the right way. As the developmental pathway becomes more complex, the gene-reading environment becomes as important as signal quality and fidelity. The egg is such a structured system. It is adapted both to function in an environment and to provide an initial set of triggers for gene expression (p. 12).

Please note that yet again it is the external reactive face, in these cases extended termite mounds and extended germ cell environments like eggs, that is instrumental in creating the conditions for internal novelty (e.g. specialized caste systems and specialized epigenetic gene expressivity). The prediction here is that all cases of true novelty (i.e. major transitions) have been and will be preceded by an innovation for an extended reactive face that will maintain an inner space whose conditions are conducive to selection for increasing degrees of cooperation. Many organisms like termites can benefit from the expansion of their cooperative ability to include more energy within their cycles of maintenance and reproduction, but a vanishingly few of them, perhaps like metazoans, manage to canalize this cooperation into novel reproductive modes. However, ultimately, this inner space is essential for the emergence of the novel reproductive mode and all of the structures it can generate and maintain.

Another way to express this inner space is as the space captured by the inclusion of novel saliencies in extended cycles of maintenance and reproduction. What was previously externally divided is now internally unified into an emergent social unity. This inner space is by definition larger and has more components than its component's inner spaces. This expanded inner space is also characterized by synchrony, by order through time. In a sense the creation of inner space is the growth of the propagative part of the spectrum of interaction. It expands the creative, push-pull part of the overall spectrum of interaction and thus expands the emergent social unity's capacity to sample and find new maintainable synchronies within that inner space.

As the quote at the beginning of the chapter indicates, multicellular animals created such an internal buffer zone of meta-exploration. This inner space is where somatic cell interaction and gene expression in embryological development are guided through safe routes to find new physical and informational expressions (i.e. new phenotypes that are eventually selected to become new species). The internal language these organisms developed was the extensive and multimodal platforms of signal transduction that couple all internal oscillatory systems. Thus the metazoan 'internal' is a novelty generating process that searches and exploits useful linkages within its own internal phase space. However, we must remember that this exploration is ultimately limited by how well its novelties facilitate successful external interaction (i.e. survival and reproduction of the system as a whole). In this way the organism explores its own organizational phase space internally and the product phenotypes are selected by successful external interactions. In a sense, by creating an environment for its components to safely explore each other over lifetimes, the phenotype explores its own environment over generations. At an even higher level, specialized human groups, like teams, corporations and governments are candidates for exploring the universe over our species lifespan and perhaps eventually radiating 'Earth life' into alien environments. To do this they would have to export mini-biospheres that reorganized, or 'included,' alien environments in mini human inner spaces. Again the matryoshka doll metaphor for scale free universal symmetry presents the same pattern at all levels.

Cells that are within animals have a sense (i.e. some way to consistently react adaptively) of the other cells in the animal because they are all interconnected in a network of communicative cooperation. Neurons and endocrine system cells, differentiated in time, space and communicative media, nevertheless react to each other regularly and adaptively to create globally coordinated actions. This is not so different to how language worked. It created an expanded 'social' inner space where global coordination could be facilitated by a novel communication system. This inner space is created and maintained by us, by our body's behavior, but it is not physical, it is informational. As a result, it can exist both in the group and within the individual brain simultaneously. Unlike physical systems, informational systems can exist in many places at the same time.

Originally we used others as representational landmarks to travel away from ourselves. They represented the adaptive extension of not-here information for us. Once the positive coupling to 'here' conditions was subordinated to 'not-here' conditions, it became possible to have positive feedback for interactions that occur outside our personal here-and-now envelopes. This creates the conditions for the growth of a system of exploratory process with functional feedback. Interactions occur at further distances and further times and the decisions they result in are selected to potentiate the symbolic connectivity arrays of successful groups. This feedback also affects within-group selection in the sense that individuals who are better able to expand their adjacent possibles using this new sense reproduce more than individuals who are less adroit. Eventually, selection through mediums of functional feedback, Baldwin effects and memory development internalized the symbolic sense in us in the same way all new senses have become internalized (i.e. those phenotypes that innately embodied those internal systems best suited to maintaining the symbolic sense survived differentially). Through this universal process of internalization, individuals began to maintain their own systems of symbolic landmarks internally. They opened a phase space for the propagation of a new symmetry. They began to be able to think symbolically.

What began as a symbolic system for group cooperation became canalized by moving into the individual. The best and most detailed context for how this process happened is given by Terrence Deacon in his book *The Symbolic Species*. For Deacon, language coevolved with humans. In this case language did most of the changing to fit our preexisting neurological and ontogenetic systems. *The Symbolic Species* is necessary reading for those who are curious.

Like in the first metazoa, the enlanguaged mind creates an internal buffer zone of meta-exploration. This inner space is where humans can explore the phase space of their symbolic adjacent possible safely without having to go anywhere or do anything risky. It is an additional level of interaction where we can "'run the internal model' with the clutch disengaged [symbolically]," to let our, "'hypothesis to die in our stead'" (Kauffman, 2000, p. 115). This internal symbolic system also functions as a symmetry propagating process that samples its symbolic environment in order ultimately to find and maintain synchronous linkages to its external environment. It is these external physical and socio-symbolic environments that then feedback and bound the language system itself. The physical and socio-symbolic environments both drive and bound the human imagination, but at different levels and through different mediums.

Beavers use their teeth to co-opt parts of their environment and maintain an inner space for themselves and their families. Eagles use their eyes to co-opt rustling foliage and grass hundreds of meters away to maintain an inner space for themselves and their families. Naked mole rats use their body odor and noses to co-opt nest mates and maintain an inner space for themselves and their family. We didn't use our teeth, eyes or noses as much as our ears and mouths to co-opt our social partners and maintain an inner space for ourselves and our families. Unlike beavers and eagles, we didn't co-opt inanimate saliencies. Like the mole rats, we co-opted each other. Unlike mole rats we had huge brains and moved through a harsh and fluctuating environment that required constant adaptive reaction. To survive we co-opted each other and by co-opting each other we opened up especially rich indices of remote interaction with partners who could reflect that rich behavior back at us. Thus we were able to discover the historically most powerful form of positive feedback, social cooperation, which in turn is opening up the unimaginably immense phase space for complex interaction that we are currently exploring. This feedback is so rich, it can sample and store so much positive symmetry, that we are a fast making everything close to us our partners. We started with each other and through them, we are co-opting everything around us, our tools, resources and technologies.

# Locative Mind Emergence

## The Locative Mind: where am I?

The term 'locative sense' refers to the power of language to give us a sense of other people's locations and reactive states. This sense evolved like other vocal/aural sensory extensions mammals recently discovered - namely sonar and echolocation. This chapter will introduce the concept of the 'Locative Mind.' The Locative Mind is different from the locative sense in the same way that minds are different from senses.

Senses are key processes that animals use to extend their capacity to interact successfully (i.e. capture/couple) with their environments. The physical senses are the ability to represent options for adaptive reaction through a process of mapping external physical saliencies. The symbolic sense, language, is the ability to represent options for adaptive reaction through a process of mapping external socio-symbolic saliencies. Both of them seek to extend adaptive reactivity to saliencies ever more distant in time and/or space. Both function to extend adaptive futures.

'Minds' are ill defined. In this investigation 'mind' refers to the process that results in the human experience of enlanguaged consciousness. What we want to get at here is that part of the process of generating human consciousness that is unlike the way animals generate consciousness. The key difference between animal and human consciousness is the way we interact as a group to generate, maintain and reproduce a virtual map of the human environment - a virtual map of what has been important to us. When this investigation speaks of the mind it is referring to this over fifty thousand year old process of communal symbolic system generation.

The word 'locative' refers to how we, as individuals, orient and navigate ourselves in this socio-symbolic ambient adaptively. The Locative Mind is that part of the consciousness generation process that locates us within our socio-symbolic ambient as opposed to that part of the consciousness generation process that locates us within our here-and-now physical ambient.

Physical senses generate what could be called a map of physical saliencies, a state of physical awareness. They represent to the brain the disposition of those features of the immediate environment that have historically triggered adaptive behavior.[210] The locative sense is a sense for reacting to virtual saliencies, a state of socio-symbolic awareness. Once the group began maintaining a system of symbolic interaction, individuals within the group could check out what was salient to others by calling. Like a bat checks its ambient for physical saliencies, individuals could check the social network for conceptual saliencies. When the group was talking it was adaptive to pay attention and react adaptively. Paying attention meant to some degree inhibiting the other senses and elevating the locative sense ('Shh! I hear something!'). In this way individuals could follow and react to what was salient to the group by following what virtual representations were being communicated.

Animals like us react to their environments, be they internal or external. When the internal environment represents 'hunger,' the 'find food' routine is triggered. That routine either completes its cycle or another trigger is represented, let us say 'predator nearby,' and a different routine is initiated. In this way animals orient and navigate their ambients, moving from one physical trigger to the next, from one routine to the next.

Humans operate on this same system. We react to our physical environments using our senses. The difference is that we have evolved another sense. Some may prefer not to call it a 'sense,' it could also be referred to as an extra source for triggering adaptive behavior. This sense is the locative sense and it allows us to represent saliencies that are important to the group. These saliencies are internal to the group. When one individual represents 'hunger,' the 'find food' routine is triggered within the group. In this case hunters coordinate their activities to gather, trap or hunt food before they may become hungry themselves. They may even learn to gather

210 The lack of certain features in an environment that usually has them can be as informative as their presence. In the same sense, the inhibition of neural firing can be as informative to brain pattern formation as their firing.

enough food to survive long winters. They use each other to extend their potential for future adaptive reactivity. If someone identifies a group of nervous hippos nearby he communicates this symbolically and a group defensive routine is initiated. In this way the first human groups oriented and navigated their ambients, moving from one symbolic trigger to the next, from one social routine to the next.

Orientation and navigation in a socio-symbolic environment was modeled on orientation and navigation in a physical environment.[211] The difference was that the group actively maintained the socio-symbolic environment while the physical environment was maintained by larger scale systems, biosphere systems. This is an example of the building of a group boundary defined by the maintenance of symbolic connectivity to maintain a social inner space. Individuals within this social inner space were selected for how good their choices to operate on either symbolic triggers or physical triggers turned out to be.

When reacting to the locative sense, humans move from symbol to symbol, from idea to idea. These symbols are located (i.e. differentiated) within the symbolic system by how they are connected to other symbols. Just as paw prints in river mud lead to 'predators nearby' behaviors, the symbols 'paw prints' lead to the symbols 'predators nearby.' Only in the latter case the behavior is speech not action. Two children by a fire can enact the symbolic behavior without ever having seen 'paw prints' or 'predators.' They can do this because their brains have been trained to react to the representation of symbols as well as physical things. In this way humans who are tuned to the locative sense at any one moment can travel from symbol to symbol. Each symbol they encounter presents a new array of further options for the next symbol.

In the phrase, "Hello, my name is Camelo," each word added probabilistically decreases the chance that the next word can be any number of options. Likewise, after seeing paw prints, the hunter may smell musk or hear movement that additionally probabilistically decreases the chance that the salient animal can be any number of options. Some symbol combinations are heavily weighted to specify the next symbol. In fact, some symbols may be so strongly correlated with others that they come as unitary bundles,

---

[211] Once again see chapter "Women, Fire and Dangerous Things."

like idioms. Other symbols may have many distributed connections that are weighted more by what symbols context them, as in the case of the "pig" that either "shit" or "arrested."

If you smell something delicious you close your eyes and breathe in deeply. Once you identify the direction you start moving in that direction and stop intermittently to check your sense of smell. Eventually, you close in on the pie on the windowsill and reward yourself with a slice. In this case a physical sense allowed you to represent options for adaptive reaction through physical saliency mapping. The symbolic sense allows us to represent options for adaptive reaction through symbolic saliency mapping. You stop and listen to your mother when she says, "If you want a slice of pie I left it on the kitchen window sill, it'll be cool in about fifteen minutes." In both cases we move from trigger to trigger, from link to link, to zero in on desired effects. Just like you learn to trust your chemical sense of smell you also learn to trust your symbolic sense of your mother. We use all of our senses to find adaptive futures.

Again, when you smell something delicious you often close your eyes and inhale deeply through your nose. By closing your eyes you inhibit your sense of sight to heighten your sense of smell. Presumably, by inhibiting other senses you are able to devote more neural resources to exploring the one sense channel you have identified as adaptive at the moment. In this way we all tune our senses to changing saliencies in our environments. Part of navigating the symbolic system is inhibiting the physical senses and focusing on reacting to the locative sense.[212]

The Locative Mind is a system of distributed representations. This system can be likened to a network of interconnected nodes (i.e. symbols) that represent highly simplified human saliencies (i.e. concepts important to coordinated human interaction). Using this system requires that the user first inhibit reactivity to immediate conditions. The person suppresses the here-and-now environment of sensual reactivity so that he or she can focus on symbolic reactivity (e.g. 'close your eyes and think'). In the symbolic world, as in the physical world, all things are connected differentially. However, unlike

212 Obviously, this has a lot to do with forming the selective conditions that differentiated and enlarged the human forebrain where attentional control is concentrated. For more see Deacon (1998).

the physical world, all things in the symbolic world are meaningful. Nothing represented in the symbolic world is maintained unless it was adaptive and subsequently maintained as part of the adaptive capacity of the network itself.[213]

Travel (i.e. symbolic transformation) within the Locative Mind is similar to travel in the physical world in that it is serial and to a limited degree digital (e.g. trigger to trigger). It is different in that, unlike the physical senses, the extension of the locative sense is not constrained by increasing chaos or the physical here-and-now conditions. Everything within the system is not only meaningful, but has a finite and constrained set of connective transformations available to it. Transformation within the Locative Mind is largely unconstrained by the mechanical laws of motion. Laws of pragmatics, neural architecture, grammar and recursive serial-digital structure are principle constraints on transformation within the Locative Mind. That is to say that the Locative Mind is constrained by a different rule-set than the one that constrains the physical body.

The Locative Mind is a symbolic state of awareness. It orients each of us within the much larger socio-symbolic network. Imagine a network of symbolic concepts something like a city map. Each concept (i.e. symbolic location) in the network is differentially connected to other concepts (i.e. locations) by associative pathways. One way or another any place is accessible from any other place, usually by a number of different routes - but only by one route at a time. All the 'landmark' concepts in this network represent features of the human experience that have been socially canalized as important or meaningful to coordinated human behavior. Like many city landmarks, they have been built by human experience. When we concentrate on the locative sense we navigate this map. We locate ourselves at some beginning concept and begin to travel to a desired concept.[214] We travel

[213] There is also the possibility that symbols can be viral, that they can reproduce within the system without giving any selective benefit to the host. For more see Dawkins (2008) and Dennett (2007). The problem with viral symbols is that by definition they don't contribute to the maintenance and reproduction of the system. Therefore, they must be accounted for by various unsustainable mechanisms. The system may always have room for a certain amount of noncontributing viral ideas, but these must always change because they are not necessary for persistence.

[214] A 'desired concept' need not be explicit. It may by unconsciously or emotionally motivated as in the instinctual 'seek' behaviors most vertebrates have. In this context, ancient

towards this destination by transforming one concept into another (saying/thinking one word/phrase after another). As we travel other opportunities for different destinations become apparent. Perhaps someone mentions a different possibility. Perhaps you change your own mind. Perhaps you get 'lost.' The Locative Mind is an internal, personal version of the external socio-symbolic network. When we think and/or speak we locate ourselves within this associative map and by using it with others we help maintain and potentiate our community's socio-symbolic network.

In much the same way that the inner space of the body is the principle environment that preserves the cells and regulatory systems of the body, the inner space of the socio-symbolic network preserves the individuals and institutions of human groups.

This metaphor of the Locative Mind process will be more fully developed in the following chapters. What is important at this time is to begin to see thought and speech as ways to dissociate from immediate sensory behavior and to engage in socio-sensory behavior. Both thought and speech are behaviors that can access what has been preserved socially as meaningful to human adaptive interaction. One of the things that makes these behaviors such powerful sources for the extension of adaptive futures is their ability to access this network of conserved group meaning (i.e. history of meaningful human saliencies) through any symbolic concept. Like the DNA double helix, the way the socio-symbolic network structures information is as a conservatory of what has worked before. Unlike DNA it can be accessed from any concept that is presented or occurs to you at any time.

The Locative Mind locates us within a personally potentiated version of the socio-symbolic network. It motivates how we transform (or travel) this concept-map by locating its symbolic-self at the transformation locus of simplified digital symbol/concepts and then transforming these concepts according to the rules of the potentiated socio-symbolic network. Like all behavior, symbolic behavior is serial. It moves from concept to concept using neurally mediated behavior routines.[215]

---

and subconscious processes trigger a kind of 'wander' behavioral routine where the desired goal is unknown but no less 'desired' in a motivational sense. For more see Panksepp (2010).

215 This will be explained in the chapter "Women, Fire and Dangerous Things."

Additionally, the Locative Mind maintains and organizes a personal narrative history of our interactions to the degree that we need to differentiate ourselves from others and specialize our interactions within the socio-symbolic network over life spans. On a group level, our Locative Minds identify and locate our symbolic selves within the socio-symbolic network itself. Ultimately, the Locative Mind is responsible for the maintenance of a symbolic 'self.' The following chapters will explain how the Locative Mind became selected for persistence. For now, the Locative Mind needs to be personalized to start building an understanding of how we create and maintain our unique awareness of 'self.'

## Personalizing the Locative Mind

Up until now language and the Locative Mind have been discussed in the context of the group. This is because evolution rarely, if ever, works in giant leaps of complexity. In the punctuated equilibrium model species may seem to disappear or appear rapidly in geological time, but new species and new exaptations do not appear overnight. The building of complex adaptations like language needs to be supported by reasonable intermediate steps. These intermediate states need to be adaptive in and of themselves.

In defending science from creationists biologists have been fond of pointing out that the eye needn't have magically appeared fully developed. A spot of light sensitive skin is better than nothing. A slightly concave spot of light sensitive skin is even better, and so on. What language theories have largely lacked is a credible model for showing the steps language would have had to take to become what it is, an entirely new self-organized system of communication, cooperation and reproduction. Although the model proposed here is largely unfleshed out and speculative it is consistent with the universal pattern of emergence. Some body, and more importantly some brain, didn't create language. Like all other major emergent systems, language began as a group adaptation. It was a distributed system of communication that fed back into itself as it coordinated the group and eventually selected those groups that could maintain this hard won and valuable organization - those groups whose individuals could maintain and reproduce it within a social environment.

However, if the goal is to shed light on the way we experience reality, on our hopes, fears and desires, then what began as a group adaptation must be reduced to an individual one. Each of us experiences a reality very different from our proto-human ancestors. We not only talk so as to behave appropriately with others, but we also think to ourselves. Somehow, language had to have become internalized.

Stephen Jay Gould (1996) argued convincingly that we are not in the age of mammals - despite what popular science books may imply. To his mind we are still in the age of bacteria. His point was that just because a few mammals finally emerged does not change the fact that bacteria are the first, longest, most durable, most distributed and most prevalent life forms to exist on Earth. Regardless of whether he is right,[216] bacteria have always existed in the history of life. They are the basic life system from which all further systems descended and upon which all further systems depend. Like protons are the substrate for the multiple levels of chemical and astronomic order in the universe, bacterial cells (i.e. prokaryotes) are the substrate for the multiple levels of biological order in our biosphere.

Our bodies are also organized at these multiple levels. They operate on organal, cellular, molecular and quantum levels of constrained release of energy. If any of these levels failed to be maintained, all animals would die. The transformation of language from the social to the individual level is similarly dependent on substrate systems. Principally, language is a group adaptation. It evolved as such and the group continues to be the substrate level upon which language is maintained. However, this system is so adaptive that we have incorporated it into ourselves. Given the context for language origins proposed here, it is important to characterize how this could have happened.

Language originated externally to the individual yet internally to the group. The symbols that constituted the system, as well as the rule-set by which these symbols transformed into other symbols, were principally maintained by the group. Like groups composed of many cooperating cells (e.g. animals), the human group is defined by those individuals that maintain

[216] I believe there is good reason to say we are in the age of Metazoa because even though bacteria may outlive, out reproduce and outweigh us, Metazoa now produce more biomass and consume more energy than bacteria does in 1/8 of the time bacteria has had to radiate. Metazoa are the super-ordinate rule-set for biological transformation on this Earth now.

and reproduce the symbolic system (i.e. the cycles of cooperation that coordinate their maintenance as a whole). The most obvious candidate for the original medium of group communication was some combination of speech and gesture. Over time, language evolution has differentiated and specialized the media of system maintenance including, but not limited to: gossip, storytelling, religion, politics, pictographs, phonetic alphabet, writing, math, science, and most recently texting and Facebook. However, as Deacon stresses, the evolution occurring within these interdependent systems is actually multilevel co-evolution.

As the group adapts to the environment, individuals within it adapt to each other. As the symbolic group maintains and reproduces itself, it divides and radiates out into the environment, birthing new language groups. Both internally and externally, this kind of radiation is fueled by processes of differentiation and specialization. This differentiation and specialization is a result of novel interaction along many scale levels of constraint and organization - many levels of internal to external. Negative feedback from group-to-environment interaction caused some groups, and some ways of communicating, to go extinct. Individual genetic and epigenetic differences in neocortical connectivity allowed some individuals to focus their symbolic sense longer than others, causing differentiation in individual capacities for language. These individuals consistently out reproduced others within their group, affecting the species phenotype, population genetics and cultural differentiation.

Much the same happened in multicellular groups. The massively complex and multilevel systems of signal transduction that coordinate cell cooperation within animals was not achieved because cells wanted to cooperate. They were achieved because complex external interaction (i.e. persistence of the functional whole) selected for differentiated and specialized internal communication mechanisms (i.e. complex internal interactions). This differentiation and specialization occurred at the intracellular level with the same DNA differentially activated within different soma cells. It also differentiated inter-cellular, organal, skeletal, muscular, nervous, endocrine, sensory, instinctual and social behavior systems. The multilevel co-evolution of symbolic systems should be expected to parallel the multilevel co-evolution of multicellular signal transduction systems with the caveat that language has had far less time to interact, differentiate and specialize.[217]

This process of feedback and co-evolution at many levels results in both the human phenotype and the language system altering themselves to better accommodate each other. Part of this alteration is the internalization of aspects of the socio-symbolic system itself. The voice box is lowered as a result of selection for more variety of sounds. The basal ganglia are co-opted to help organize syntax. The neo-cortex is expanded as a result of selection for more memory and attentional control. Another part of this internalization is the alteration of language itself to sync with the human brain.

There was strong selection for coordinated symbolic interaction. The result of this co-evolution was the internalization of parts of the symbolic system itself. As a system that coordinated group interaction, symbolizing was highly successful in producing adaptive reactions. This was largely due to the fact that symbolizing enabled extended future telling (e.g. creation of not-here, not-now adaptive futures) under almost any circumstances. As each member participated in more and more not-here and not-now interactions these extended-time and extended-space opportunities became part of their behavioral adjacent possible. When they engaged in communicating ideas like 'what would Suzy do if I gave her flowers' positive feedback became a very real force for engaging the group network for your own benefit, 'she'll want to make a baby with me.' This kind of positive feedback would work on many levels, but the upshot is that any natural propensities one has for engaging the future telling powers of the group network for one's own benefit would be rewarded. People that symbolized better, procreated more. It is not coincidental that most hunter-gatherer 'head men' are both prolific and eloquent speakers. They also typically have a disproportionate number of offspring.[218]

---

217 Multicellular differentiation and specialization has been developing for over five hundred million years (even longer if you take into consideration other examples of bacterial and protist multicellularity). Socio-symbolic differentiation and specialization has probably been developing for no more than two million years.

218 The source for this paraphrased information is currently unknown to the author. Any help would be appreciated.

In natural history, each case of the internalization of good interactive tricks is unique, some land animals maintained vascular respiratory systems, other's tracheal respiratory systems (e.g. insects). What is inevitable is that if the trick is good and the conditions that maintain it remain stable, the populations to which the trick is an adjacent possibility start to over represent those phenotypes that best take advantage of it. In doing so, the average phenotypes change to incorporate those tricks. The most obvious choice for the physiological system that most rapidly and fully incorporated the language trick is that most plastic and impressionable of human organs - the brain. This idea needs more unpacking, but for now we should, and do, expect the brain to carry the lion's share of language adaptations.

Socio-symbolic language became internalized as our brains evolved ways to plug into it, couple to it and eventually capture it. Our ancestors were living, cooperating, competing and dying in two environments. The one was the physical environment they had pre-evolved senses and behavioral routines for. The other was a newly established group symbolic environment in which the rewards were immense, but only accessible to the degree to which you could engage and eventually internalize the symmetry of the socio-symbolic system itself.

## The Origin of 'Self': who am I?

On the subjects of 'language' and 'self' Terrence Deacon (1998) says:

> Language functions as a sort of shared code for translating certain essential attributes of memories and images between individuals who otherwise have entirely idiosyncratic experiences. This is possible because symbolic reference strips away any necessary link to the personal experiences and musings that ultimately support it. The dissociation allows individuals to supply their own indexical and iconic mnemonic in order to reground these tokens in new iconic and indexical representations during the process of interpretation ... This is crucial for the development of self-consciousness, and for the

> sort of detachment from immediate arousal and compulsion that allows for self-control. Self-representation, in the context of representations of alternative pasts and futures, could not be attained without a means for symbolic representation. It is this representation of self that is held accountable in social agreements that becomes engaged in the experience of empathy, and that is the source for rational, reflective intentions. According to Vygotsky, this sense of self emerges slowly as children mature. It becomes progressively more facile in perspective shifting and at the same time consolidates greater control over the other aspects of self that derive from non-social sources, such as the experience of pain and effort, the arousal of basic drives, or the physical boundaries of control over events (p. 452).

In many ways the Locative Theory of Mind is a variation of what Deacon has already said. Similarities include what he refers to as the symbolic stripping away of, "personal experiences and musings," which have been previously referred to as 'language simplification.' It is the reduction of the massive amount of sensory information to stripped down, digital, serial and communicable representations. The "dissociation" he refers to is the result of symbolic communication that enables individuals to dissociate from their own experience and imagine the experiences of others. They can fill in these stripped down representations from others by supplying, "their own indexical and iconic mnemonic in order to reground these tokens in new iconic and indexical representations during the process of interpretation." Deacon also notes the necessity for the inhibition of here-and-now reactivity, but calls it the, "detachment from immediate arousal and compulsion that allows for self-control." Where we diverge is in our treatment of the origin of the human sense of self. Deacon simply presumes the emergence of a sense of self as a result of, "symbolic representation." He then moves on to mentioning the social context of the self and ultimately lays the expletive burden on Vygotsky with the vague and dismissive phrase, "this sense of self emerges slowly as children mature."

There are probably two reasons for this cursory treatment. First, Deacon has his plate full with his primary theme, brains and language. In this kind of work, the dangers of getting sidetracked in the fascinating tendrils of your work are very real. He rightly chooses to stay focused. The second reason is that Deacon presumably has no obvious platform on which to build a theory for the emergence of a sense of self. There is no physical evidence of where, when, why or how the brain magically manifests a self, and the brain is the boundary beyond which, as a neuroscientist, he is loath to theorize. Deacon's presumption is that internal 'symbolic representation' in individuals somehow caused a 'sense of self,' but what if a 'sense of self' is a constraint or side effect of being embedded in a socio-symbolic network? Once a socially maintained symbolic network is predicated as the environment that shaped human mind evolution then there is a new platform from which to work. This new 'social' platform leads to a unique search for the origins of our 'selves.'

So what would engaging the language system have been like for our ancestors? First of all there is an entire physical world of sensory information to which she already reacts. In addition to this, there is starting to be another world of symbolic information to which she can react to. In the physical world her experience is grounded in the cycles her body is experiencing and the sensory information of what is around her. She is absolutely limited to these here-and-now sources of experience.[219]

In the symbolic world she is not so limited. She can use her symbolic sense to read other people's localities, other people's ambients. She senses their ambient through their symbolic broadcasts. Like all senses, she uses this information to additionally excite or inhibit her own behavior. When she does this the transformation locus that is constraining her behavior is not her own bodily here-and-now, it is located elsewhere. The act of interpreting remote behavior in others requires that she locates her reactive face elsewhere. She can now put herself in other places through interpreting other's broadcasts.

The interpretation process is a process of 'reading,' or 'making sense of,' another person's point of view. This other point of view is a different location

[219] This here-and-now experience includes those behavioral routines which have evolved to recognize present saliencies and generate future reactions. In this way, they can see the future, but their futures are limited to the saliencies that they recognize in the here-and-now.

in time and space. The interpreter senses the interpretee's interactive location through their broadcast. "This is possible because symbolic reference strips away any necessary link to the personal experiences and musings that ultimately support it. The dissociation allows individuals to supply their own indexical and iconic mnemonic in order to reground these tokens in new iconic and indexical representations during the process of interpretation." The locative sense can read foreign ambients by interpreting object-environment reactivity as if it were its own body-environment reactivity. Where a bat co-opts the sound properties of its physical environment, she co-opts the sound properties of her social environment.

Broadcasted sounds are now symbols and they represent environmental saliencies that interact with each other in much the same way that she interacts with her physical surroundings. In this way words become 'imaginary' representational locations of interaction that react to each other like her body reacts to saliencies at its own location. Words represent saliencies in her extended social environment. They represent localities of interaction. With words 'she' can temporarily be anything, anywhere. She has motive force to interact with things in her environment; with language, the symbolic things (i.e. the nouns) she represents have motive force to interact (i.e. to 'verb') with other symbolic things (i.e. other nouns). The river can now 'take' her child away. She can imagine the river 'taking' things the same way she 'takes' things. Through language she can also imagine herself as someone else, even someone dead, or not yet born, or even someone who never did - or will - exist, a god perhaps.

This locative power of imagination results in a personal displacement.[220] On the one hand she has a body in which she interacts with the physical world. On the other hand she has a nascent mind in which she can also temporarily represent interaction at other localities in a very different socio-symbolic world. Symbols represent remote behavior. In her Locative Mind, through language, she can experience any interaction for which she has words. In a sense, she can temporarily leave her embodied experience behind and travel in other bodies, other ambients. She can tell stories from other points-of-view. But this new kind of remote behavior creates a problem; 'How do I

220 See chapter "Locative Sensing" to review the locative power of imagination.

differentiate between the body and experience I always am and the other points-of-view and experiences I can temporarily engage in?' The language system operates on principles of differentiation. What interacts symbolically is what we have different names for. So the problem then becomes 'What do I call myself when I am away from myself?'

It's curious because normally we think of a sense of self as a unity, 'this is the one and only person I am.' And in one very important way we are right. All behavior is unified - it is coordinated to produce serial adaptive reactions. It is the behavioral Von Neumann bottleneck mentioned in the 'Life' section. All animals experience behavior as serial coordinated reactions to changing immediate conditions. In this sense, all animals have always had a unified sense of self. What happened when humans began engaging the socio-symbolic system was the first splitting of this unified experience. It was a 'differentiation' that we ended up specializing. When our ancestors used symbols to represent remote behavior they learned to locate agency somewhere other than their own here-and-now. This ability to sense remote localities was initially selected for persistence because it allowed them to represent cooperative opportunities with other humans, but it represented these opportunities by giving the represented object agency. The result was that all represented objects become determined by how they interact. Apples grow, like we do, they ripen, like we do and they fall, like we sometimes do. We gave all objects agency when we learned to sense each other as objects with agency. By symbolically locating agency elsewhere we split our own unified sense of agency. Are we reacting to our own here-and-now or are we reacting to the story we are telling, to Suzy Q's there-and-then? Where and when 'I am' now depends on what behavior we are engaged in - eating or thinking?

This put our ancestors in a strange position. Now they were regularly represented within the socio-symbolic environment as an object of agency. From *other's* eyes *you* become visible. When our ancestors began using each other's senses to coordinate group reactions they were developing a sense for each other. This sense operated on symbols that allowed individuals to begin to see the world through each other's eyes. An unavoidable result of this is that you begin to run into representations of yourself as seen through other's eyes. You are an important part of their symbolic environment. For the first time you are forced to imagine yourself.

The uniquely human sense of self is what happens when we travel away from our embodied-self using the socio-symbolic network. When we do this we need a name for the place we start from and always return to. This place will be called the 'embodied-self.' Biological systems typically transform by differentiating and specializing. Language is no exception. It grows by canalizing special names for features that are important to generating adaptive reactions. In the case of the self, an essential feature for anyone engaging a group maintained symbolic network is a symbol for herself. As the symbolic interactions she is capable of become more complex she must travel more. The more she travels, the more she locates herself within a network in which other's images of herself are crucial to coordinated social behavior.

What is curious is that the sense of self is not a unity. It is a differentiation, a division. It is the difference between the normal sensory experience that all animals are constantly bound to and the symbolic sensory experience we are capable of. Who are you when you are reading *Treasure Island*, or listening to a friend's personal account? Where are you when your friend explains to you how Suzy Q must have felt when you broke her heart? You are certainly always physically right there, but you can also temporarily be symbolically in other places, other times too. If you manage to put yourself convincingly in Suzy Q's place last night, you will hopefully learn from it. In just such a way *we* coordinate our reactive power to interact successfully in our socio-symbolic environment. In fact, it is because I am right here in front of 'you' that I am 'Camelo.' Because people like you have trained my brain to see myself through your eyes, through our eyes, through the system's eyes. Only within this system that I embody and operate can I escape that animal consciousness where everything is my reaction to everything immediately within and around me.

As organisms we have had a sense of self for almost four billion years. As humans we have had a different sense, a second sense of self for perhaps fifty thousand years. This second self is a social self and for it to be a shared adaptive saliency of course we have to give it a name. That is what we do, give names. The only problem is, you cannot call everybody 'I.' We each need differentiable names within this system.

Our ancestors made good use of symbolic systems. As a result though, they found themselves socially maintaining language avatars for themselves. These avatars were the representations of each other within the symbolic system. They were functional necessities for inciting coordinated interaction within the group. The system preserved them when they occurred because they enabled coordinated social behavior. The result of the maintenance of this systemic coordination was an environment that selected for and preserved the idea of selves. In a sense, *'I'* exist for *'them.'* Before there was a socio-symbolic world there was no need to differentiate the self. There was only one self, trapped in this here and this now.

Antonio Damasio (1999) has a different view of the role language plays in the creation of self. He sets up his argument by noting that, "when I was in medical school and in neurology training, I remember asking some of the wisest people around me how we produced the conscious mind. Curiously, I always got the same answer: language did it" (p. 107). He goes on to vigorously deny that language can produce consciousness:

> If language operates for the self and for consciousness in the same way that it operates for everything else, that is, by symbolizing in words and sentences what exists first in a nonverbal form, then there must be a nonverbal self and a nonverbal knowing for which the words 'I' or 'me' or the phrase 'I know' are the appropriate translations, in any language. I believe it is legitimate to take the phrase 'I know' and deduce from it the presence of a nonverbal image of knowing centered on a self that precedes and motivates that verbal phrase. The idea that self and consciousness would emerge *after* language and would be a direct construction of language, is not likely to be correct (p. 108).

Damasio's argument is that the self is based on what he calls "core consciousness" and language is a phenomenon of the "extended consciousness." For him, the "extended consciousness" emerges from a preexistent "core consciousness" and thus a sense of self must precede language.

This view is both brilliantly insightful and ultimately misleading. His "core consciousness" is analogous to what has been called 'animal' or here-and-now consciousness in this investigation. It potentiates adaptive reaction from the here-and-now embodied-self by positive feedback (a version of which he refers to as "somatic markers"). All animals, including us, operate on this system of consciousness. It is the animal, or embodied, sense of self previously referred to, and the "nonverbal self" Damasio refers to.

However, what Damasio fails to realize is that the unique sense of self humans feel *is* a result of language. Brilliantly, Damasio realizes that consciousness is a preexisting state, a prerequisite for any sense of self. It is true that a sense of 'I know' must precede language. All animals coordinate their reactions from some localized state of physical orientation that could be called 'I know where I am and what I should do now.'[221] However, Damasio's subsequent attempt to show how the human sense of self is an extension of this animal consciousness, ultimately creating, or 'motivating,' language, is misleading. It does not follow that language is a result of this physical, here-and-now "I know" locus of motivation, quite the opposite. It was through the external maintenance of a second locus of motivation that the human sense of self was created.

Orientation and adaptive reaction within the socio-symbolic ambient creates a second sense of self - a second locus of transformation. This second sense of self exists in an environment different from the physical here-and-now. It exists in a socio-symbolic environment. The embodied-self does not constrain interaction in this environment and thus transformation can be incited at a different locus of motivation. This second locus of motivation is the point at which transformation is occurring in the symbolic environment. It is the thing (noun) that is acting (verbing) within the symbolic environment. It is what is falling, taking, or hunting. How does that represented thing act? It

[221] The distinction being drawn here is infamously problematic for the simple reason that using words to represent an unsymbolic state of mind is a fool's errand. Top philosophers have tried and failed to establish the distinction. The best expression I am familiar with is the distinction between awareness and consciousness where awareness indicates the ability to successfully orient and act within an environment and consciousness indicates the ability to do the same symbolically. In this sense awareness is non-symbolic consciousness. Animals do not ever think 'I know.' However, the only way for us to imagine their here-and-now form of awareness is through symbolic phrases like 'I know where I am and what I should do now.'

acts as you do. It reacts as you do. It does things. It has goals. 'It' also transforms. 'It' also preserves a structure. This second locus of action, this remote point of virtual transformation is the 'location' of the symbolic-self.

Damasio is right, 'self' and 'consciousness' did emerge before language, but the human version of 'self-consciousness' emerged after, and because of, language. The human sense of self emerged within and because of its socio-symbolic environment.

The human, "sense of self" is a social reflection. It is only because we are able to exist within a socially mediated network of representations that we can see ourselves as other than what we directly experience in our here-and-now. Our feeling of being more than our embodied selves is maintained by constant social reminders to see ourselves as actors within a socially maintained web of representations. This web of concepts is vast. We can imagine our 'selves' wrestling with monsters on Mars or kissing Cleopatra two thousand years ago. Without this mirror of socio-symbolic interaction human self-consciousness is impossible.[222]

Together, the embodied-self and the symbolic-self refer to our new and very human sense of self. They refer to our unique ability to displace our embodied selves by behaviorally engaging the symbolic network. This displacement differentiates where and when your embodied-self is as opposed to where and when your symbolic-self can be. Differentiation within the symbolic system is marked by a new name. What confuses us is the presumption that our sense of self is a unified, indivisible, undifferentiable whole. This misconception stems from the inescapable truth that physical interaction has only one context, the here-and-now of the embodied-self.

Ultimately, all behavior has always been unified as serially coordinated reactivity triggered by immediate conditions. However, we have failed to recognize the other ambient we interact in, the socio-symbolic system. In this environment our symbolic selves, our socially maintained avatars, interact to produce coordinated social behavior. Only because of this fundamental division do we need a name for ourselves. In this case we have more than one way to say it; I, me, myself or Wayne. All of these express the division

[222] Interestingly, real mirrors have long been the intuitive tool used for testing 'self' consciousness. Most recently these tests have included apes, dolphins and elephants; all of whom it turns out can learn to recognize themselves in mirrors.

of the embodied-self from the symbolic-self. Each of these discriminations differentiates me from my animal cousins, my presymbolic ancestors and from 'you.' All of these symbolic discriminations together are part of my very unique, and very human, 'sense of self.'

## "Consciousness Explained"

> *Human consciousness is just about the last surviving mystery. ... What, then, is the mystery? What could be more obvious or certain to each of us then that he or she is the conscious subject of experience, an enjoyer of perceptions and sensations, a sufferer of pain, an entertainer of ideas, and a conscious deliberator? That seems undeniable, but what in the world can consciousness itself be? How can living physical bodies in the physical world produce such a phenomena. That is the mystery.*
>
> *Daniel Dennett (1991, p. 21)*

Perhaps a more apt title for Dennett's seminal book, *Consciousness Explained* would be *Human Consciousness Explained* or *Self Consciousness Explained.* Dennett largely focuses on demystifying our human experience and spends very little time with primate, animal or cellular states-of-being. Dennett's book is about why we have the feeling that there is a "central meaner" inside us and how that feeling is different from what is actually going on in our brains. He is interested in how our state-of-being is neurally created. The, "last surviving mystery," is why do we have that feeling and, for him, the solution is because of how the brain actually works. And, in a limited sense, he is absolutely right.

Fortunately, there is more to it than what happens in our brains. Our senses of self didn't burst forth from an ancestor's mutated brain like Athena bursting out of Zeus' forehead. True, the "central meaner" feeling is a product of how our brains operate now, but then the real question of human consciousness becomes, 'Why did our brains come to operate so?' The answer to this question remains unknown, but the Locative Theory of Mind suggests that it would have something to do with the serial and digital constraints on symbolic representation. Simply put, we have a central meaner feeling because symbolic meaning for us can only be represented in digital (i.e. simplified and unitary) symbols, serially represented.

The Locative Mind only exists to maintain the transformation locus of one concept to another. Like all other behavior, our brains may be generating it by coordinating continuous, parallel brain/body systems, but the resulting behavior can only be expressed through the Von Neumann Bottleneck of behavior. The Locative Theory of Mind suggests that we have a central meaner feeling because meaning can only be expressed from one locus of symbolic transformation at a time. As such, the symbolic-self can only make 'centralized' meanings. Just as our embodied selves are trapped in one single here-and-now at a time, so is our symbolic-self trapped in one single symbolic transformation at a time. The logic of language constrains the making of meaning to one locus of transformation at a time. The rule-set of language is constrained by the rule-set of behavior. Language is a behavior and behavior is serial. Thus the rule-set of behavior itself constrains us to being central meaners.[223]

But the feeling of being central meaners is just one feature of human consciousness. In a greater sense, what Dennett would like to explain is how our brains manifest the experience we feel. He wants to explain consciousness. How do we do it? It seems obvious that there had to have been external conditions that remained constant enough, over long enough time, to differentially select for a substantial rewiring of the brain. As usual, the best way to start imagining this is to consider how organisms have done this kind of thing before.

Imagine that you are a bacterium. Even though you exist in a social context, primarily you interact individually (with yourself and with features of your environment), you grow individually (by synchronizing the features

[223] Interestingly, this explanation for why we feel like we are central meaners implies that each of us has two loci for generating meaning. The first is the aforementioned locus of symbolic transformation and the second is the embodied locus of physical transformation. Perhaps meaning for us is so difficult to understand because until now it has been understood as a unitary phenomenon when it is really an entanglement between how our embodied and symbolic selves are constantly creating different kinds of competing adaptive futures. Should I eat the delicious chocolate or look sexy tomorrow?

you interact with) and you divide individually (by duplication, the most synchronous process of all). Your cycle of existence includes you and your personal ambient. One dramatic day you are swallowed by a larger bacterium. But you do not die. Inside the large bacterium you start to interact with your new environment. Inside yourself you have the same metabolism, but now it must capture and couple to features of the large bacterium's metabolism. In this, a tenuous accord is struck and you are still able to grow and divide. As you divide some of your offspring are included in the larger bacterium's divisions and your symbiotic descendants do well too. Rather quickly, your descendants begin to change. The way they used to communicate with their environment must be changed to now include new ways of communicating with their new 'social' environment. This new accord, this new language of interaction, initiates a cascade of adjustments in both host and symbiont. The result is a system that emerges that is capable not only of novel internal interactions (i.e. differentiation and specialization), but even more importantly; it is capable of entirely novel external interactions. These novel external interactions ultimately allow the new unified cells (i.e. protists) to radiate all over the world. They change the surface of the Earth. Now, you interact together, you grow together and you reproduce together. Your cycle of existence is now socially bound to other cells. Together you are co-creating a communal adaptive future.

This is the story of how each cell in your body became what it´s ancestors were one billion years ago, free-living protists (i.e. eukaryotes). This happened. Now consider the human adaptation. Imagine you are something like a clever chimpanzee. Even though you exist in a social context, primarily you interact individually, you grow individually and you reproduce individually. Your cycle of existence includes you and your personal ambient. One dramatic day the most important feature of your environment becomes your troop mates. Inside the troop social dynamic you have the same sensory, instinctual and behavioral tools available to capture and couple to your environment, but the environment has now changed. Now, what is important to react to, is others. After a tenuous period of adjustment your descendants begin to be selected for their abilities to communicate adaptively inside this new social environment. This new accord, this new language of interaction, initiates a cascade of adjustments in both host (i.e. group) and symbiont (i.e. individual). The result is a system that emerges that is capable not only of novel internal interactions (e.g. mental and cultural) but, even more importantly, it is capable of entirely novel external interactions ("The same thing we do every night Pinky, try to take over the world!"). Humanity starts to change the surface of the Earth. We interact together, we grow together, and we reproduce our culture together. Our cycle of existence is now socially bound to each other. Together we are co-creating a communal adaptive future.

This story could be more easily compared to the origin of multicellularity where groups of cells had to create new languages (i.e. signal transduction) to come together cooperatively and reproductively within epithelial boundaries. Less easily compared are abiotic systems where electrons or asteroids are caught up, coupled and eventually released from certain symmetry products like atoms or planets. An electron captured by an atom communicates within that atom in a different 'language' than it communicated with other universal systems before capture. The bump and grind of it is that 'language' is a word we can use for the symmetry rule-set that additionally binds us and ultimately defines us as novel communal systems. Symbolic language is not the first symmetry rule-set to emerge and certainly won't be the last.

What does this mean in the context of explaining consciousness? Somehow, in some way we need to start considering how we operate within the system we ourselves have co-created. How does becoming part of a socio-symbolic network shape our personal experiences? Like mitochondria (the descendants of the cell swallowed in the first example) our body and our state-of-being has undergone significant alteration to adapt us to this new socio-symbolic environment.[224] Any kind of explanation of consciousness that does not take into consideration the processes that shaped it (and continue to shape and bound it) is partial at best.

Dennett probably gives the best model around for how the modern sense of consciousness is generated within the brain. His Multiple Drafts Theory is eminently plausible. However, he fails to show us how we emerged

[224] Obviously, mitochondria and soma/germa cells have undergone far more extensive physiological alterations simply because of the amount time they have been coevolving.

from what came before. He fails because like almost all cognitive scientists, he fails to flesh out the history of the systems upon which our brains, bodies and minds are built - the context within which, together, they make sense. Any theory that promotes the internal emergence of a major natural novelty has missed the boat. The jawbone didn't become ear bones because that's how jawbones and brains roll. The jawbone became the ear bones because better hearing of salient *external* features was adaptive. Brains didn't become minds because that's what brains do. Brains didn't, in fact couldn't, invent language or thought unless you give them historical conditions (i.e. environment) in which the rewiring is selected for over a reasonable amount of time. Once again, true novelty comes from external interaction.

To explain consciousness you need to see it in the context of what life has always done, what systems have always done. By coming together under a new system of communication humans were able to coordinate themselves in ways that no other animal can. However, the adaptations canalized to maintain and reproduce this new system created a new environment internal to the group, a symbolic environment. Within this new socio-symbolic environment adaptive reaction required the differentiation of the embodied-self that you continually experienced and the symbolic-self that sometimes motivated socio-symbolic behavior - the symbolic-self to which other's constantly referred. Unlike the embodied-self, the symbolic-self could be invoked externally by others, and imagined under any time or space conditions, it could be anywhere, anywhen - it was different.

The socio-symbolic network was a group communicative network evolved to adapt to group needs by selecting for those individuals predisposed to plug into it. Selection for predisposition created a new level of physiological selection. It rewired the brain to access the symbolic network by doing what all adaptations do, regularize it internally. 'Stereotypy': the internal canalization of good tricks by any and all means possible. By internalizing parts of the system logics of the socio-symbolic network, we became permanently capable of accessing those parts of the network our brains could sustain. For example, much of the brain evolution since the advent of the symbolic network has likely been to inhibit here-and-now reactivity in order to extend symbolic capacity to produce adaptive futures (e.g. prefrontal cortex rewiring). Another big part of this brain adaption would have been be the creation of a memory system that supports the kind of connectivity symbolic networks have (e.g. neocortical rewiring - more on this shortly). As inhibition and memory systems evolved, more of the socio-symbolic network could be internalized and the symbolic-self could travel more extensively (e.g. thinking). More time away made people more aware of the difference between the symbolic-self that existed within the group and could be anywhere at any time, and the embodied-self to which you always returned and from which you always acted in the here-and-now. With the evolution of a symbolic memory capacity those intermittent and transitory moments of symbolic-self reactivity could start to become permanently stored and accessed, slowing coalescing into a relatively consistently maintained narrative life history - a 'narrative self.'

Imagine this; every noun you learn is connected to other nouns by different grammatical relationships (verbs, articles, prepositions, etc.). Each of us learn language by building a map of these relative connections from our socio-symbolic environment. Then we use this network by traveling in it to navigate real life experience and thus our socio-symbolic landscape grows and evolves as we strengthen some and lose other connections to make sense of our personal experiences. Now the more you use a certain noun or concept the more you explore its connectivity within the network. Concepts like 'Mithras' are rarely used and thus relatively under-connected or under-defined. Most people cannot explain what Mithras is. Now consider your own name. Your name is one of the most common and highly connected, or over-defined, concepts in your entire life. Your narrative self is your memory of that bundle of historically potentiated connections that define how your name, and other symbolic self-referents like pronouns, has interacted with other concepts over the course of your life. The narrative self is your memory of how those concepts that refer to yourself fit into the symbolic network that you build over a lifetime. It is the story told by the grab bag of myriad socio-symbolic connections that have described yourself to yourself.

The narrative self is not how you fit into your social group. It is how your self-created socio-symbolic avatar fits into your self-created concept-map.

Who you believe you are and who you are, are two different things. The narrative self will be revisited in upcoming chapters.

Our sense of consciousness, our state-of-being, is likely to be produced very much like Dennett claims. In that way, it is a product of the brain. However, the brain is so because it has adapted to a larger context, a socio-symbolic context. Our beliefs, hopes and desires stem from the interplay of our animal substrate of here-and-now behavior with our added superstrate of socio-symbolic behavior. It was the interaction of individuals within the socio-symbolic network that enabled our uniquely human sense of self to emerge. This sense of self was born of the integration of the physical and symbolic contexts. It was the interplay of the two that ultimately defined our unique state-of-being - the conscious human mind.

## "Women, Fire and Dangerous Things": animating concepts

> *"But if the categories originally only translate social states, does it not follow that they can be applied to the rest of nature only as metaphors?"*
>
> *Emile Durkheim (1915, p. 18)*

An essential part of the argument offered here is that language started as an external interaction in the form of a socio-symbolic network. In this case, a burden lays on the Locative Theory of Mind to present evidence that the symbolic communication that originally functioned on a group level, later became internally represented in individual brains. Without some evidence that the brain captured and eventually canalized aspects of this symbolic system, that the direction of causality was from external to internal, the argument for the Locative Theory of Mind is less compelling. So before moving on to the co-evolution of a memory system that enabled the internalization of parts of the socio-symbolic network it would be helpful to get a more nuts and bolts idea of how the internal experience of language (i.e. thought) functions in regards to external experience (i.e. behavior).

George Lakoff is one of the preeminent language theorists[225] of the 20th century and he calls conceptual interaction, "cognitive semantics." What follows is based on his influential book, *Women, Fire, and Dangerous Things*.

The title of *Women, Fire, and Dangerous Things* refers to a fundamental element of Lakoff's cognitive semantics, namely that, "experience is structured in a significant way prior to, and independent of, any concepts." By this he means that there is a "basic-level physical experience" that provides the foundation for, "basic-level concepts" (Lakoff, 1987, pp. 270-271). Lakoff attributes the fact that physical experience has a pre-conceptual structure to research done by Mark Johnson (Johnson, 1987). What they are both saying is that the internal human experience of thinking and talking is organized by certain external experiences. Specifically, symbolic behavior is structured by physical behavior.

For Lakoff (1987), symbolic concepts are based on two kinds of physical interaction. The first are what Lakoff calls, "basic-level categories in the physical domain" (p. 270). These are physical structures that we recognize and react stereotypically to without the aid of symbols. A few good examples are in the title of his book, "women," "fire," and, "dangerous things." Each of these phenomena is recognizably structured and stereotypically reacted to throughout the human species. What differs in different languages (and thus different minds) is how categories of these basic concepts are connected in relation to each other. In Dyirbal, an Australian aboriginal language, "women, fire, and dangerous things" are all part of the same linguistic category. You and I should find this strange. For us, those things are rarely associated with each other. What this shows is that different languages build different categories with different concepts that are differentially connected, or associated.

According to Lakoff (1987, pp. 12-57), various scientists support the idea that basic-level categories are constructed by a pre-conceptual level of

[225] Technically, Lakoff is a linguist but like so many great theorists his interests and impact cannot be measured within the tightly defined boundaries of just one discipline.

recognition. At this level, "THE CATEGORIES OF MIND FIT THE CATEGORIES OF THE WORLD" (p. 32, his capitals). What he means by this is that we typically recognize basic interactive things first. Then, we create subordinate and super-ordinate associations. Lakoff (1987) says:

> The picture Brown gives is that categorization, for a child, begins 'at the level of distinctive action,' the level of flowers and cats and dimes, and then proceeds upward to super-ordinate categories (like plant and animal) and downward to subordinate categories (like jonquil and Siamese) by 'achievements of the imagination.' 'For these latter categories there seem to be no characterizing actions' (Brown R., 1965, p. 321). This 'first level' categorization was seen by Brown as having the following converging properties:
>
> - It is the level of distinctive actions
> - It is the level which is learned earliest and at which things are first named
> - It is the level at which names are shortest and used most frequently
> - It is a natural level of categorization, as opposed to a level created by 'achievements of the imagination' (p. 32).

These basic-level categories are the foundation upon which language is built. Curiously, this foundation is created by "distinctive action." That is, how we interact with a thing determines how we conceptualize it. Do you pick it, pet it or trade it: flowers, cats or dimes? Our categories are created by the things we most commonly interact with. Additionally, the structure of those associations is the structure of that category. Obviously, Dyirbal people 'distinctively interact' with women, fire and dangerous things. For them, women, fire and dangerous things all signify a similar class of interaction. Feel free to conjecture what that might be.

The second class of physical interaction we base our concepts on is what Lakoff calls, "Kinesthetic image schemas." Where basic level categories can be likened to 'things' (i.e. nouns), kinesthetic image schemas can be likened to transformations or actions (i.e. verbs). His primary example of image schemas is the 'container schema':

> The CONTAINER schema defines the most basic distinction between IN and OUT. We understand our own bodies as containers - perhaps the most basic things we do are ingest and excrete, take air into our lungs and breathe it out. (…)
> The CONTAINER Schema
> - Bodily experience: As Johnson points out, we experience our bodies both as containers and as things in containers (e.g. rooms) constantly.
> - Structural elements: INTERIOR, BOUNDARY, EXTERIOR
> - Basic Logic: Like most image schemas, its internal structure is arranged so as to yield a basic 'logic.' Everything is either inside a container or out of it …
> - Sample metaphors: The visual field is understood as a container, e.g., things *come into* and *go out of sight*. Personal relationships are also understood in terms of containers: one can be *trapped in a marriage and get out of it* (Lakoff, 1987, p. 272).

Image schemas are based on the 'distinctive interaction' of universal systems. As previously mentioned, fundamental to systems are the distinctions of 'internal,' 'external' and 'boundaries.' These distinctions are embodied in image schemas as well. They have "structural elements" whose relations define the whole. As Lakoff says, "The 'basic logic' of image schemas is due to their configurations as gestalts - as structured wholes which are more than mere collections of parts. Their basic logic is a consequence of their configurations." In this case, Lakoff's 'gestalt' is what this investigation calls a 'system' defined by symmetry. The rule-set of image schema symmetry is what he calls its "basic logic." The rule-set of container schema symmetry is, "Everything is either inside a container or out of it" Like electrical charge or electron 'spin,' many interactions are structured by transformation with preservation of just two states; positive - negative, left - right, and in - out (in order: charge, spin and container). Image schemas then, are embodied logics of physical interaction. They are composed of physical interactive processes that our bodies have canalized (either 'memorized' or otherwise 'incorporated').[226]

Lakoff notes a series of other basic image schemas our bodies use to organize experience including the PART-WHOLE schema, the LINK schema and the CENTER-PERIPHERY schema. Each of these is conjugal to both Symmetry Theory and its significance to theories of mind, but due to time constraints we will focus on the image schema primarily responsible for motivating the metaphor that will be used to describe the symmetry of language, the SOURCE-PATH-GOAL schema:

The SOURCE-PATH-GOAL Schema

- Bodily experience: Every time we move anywhere there is a place we start from, a place we wind up at, a sequence of contiguous locations connecting the starting and ending points, and a direction. We will use the term 'destination' as opposed to 'goal' when we are referring to a specifically *spatial* ending point.
- Structural elements: A SOURCE (starting point), a DESTINATION (end point), a PATH (a sequence of contiguous locations connecting the source and the destination), and a DIRECTION (toward the destination).
- Basic logic: If you go from a source to a destination along a path, then you must pass through each intermediate point on the path; moreover, the further along the path you are, the more time has passed since starting.
- Metaphors: Purposes are understood in terms of destinations, and achieving a purpose is understood as passing along a path from a starting point to an endpoint. Thus, one may *go a long way toward* achieving one's purposes, or one may get *sidetracked*, or find something getting *in one's way*. Complex events in general are also understood in terms of a source-path-goal schema; complex events

[226] This subject of canalized, or 'incorporated,' relationships could probably be interestingly compared to the growing interest in body maps (Blakeslee, 2007) and embodied cognition (Shapiro, 2011) and embodied intelligence (Pfeifer & Bongard, 2007).

> have initial states (source), a sequence of intermediate stages (path), and a final state (destination) (Lakoff, 1987, p. 275).

Lakoff was one of the first linguists to realize that concepts (i.e. symbols) and how they transform (i.e. language) are fundamentally metaphorical. Both concepts and the relationships they have to each other are structured by physical things (especially our bodies) and the way they relate to features of their environment. In a sense, physical interaction is the substrate from which symbolic interaction emerges. This mapping of concrete interaction to conceptual interaction makes language and thought metaphorical by nature. Language interaction virtually represents physical interaction. The transformational logic that drives conceptual relationships is metaphorical association. The relationships symbolic things can have with each other are metaphorically derived from, and constrained by, physical relationships. Ultimately, symbolic interaction is derived from physical interaction. We call this symmetry rule-set 'metaphor.'

The source-path-goal schema is the fundamental structure preserved in transforming (i.e. mapping) concrete physical interaction to abstract symbolic interaction. What this means is that both speech and thought arrangements are, *"motivated by the structure of our experience"* (Lakoff, 1987, p. 276). Language is just another behavior that identifies goals and travels to satisfy them. In a sense, both intentionality and agency seem to stem from billions of years of selection for added behavioral substrates. The last substrate we added was socio-symbolic.

Before we had a symbolic sense, our ancestors behaved by sensing and associating reactive routines to certain patterns of sensory information. Their behavioral routines had very proximate goals that contributed to either the maintenance or reproduction of themselves or their descendants. The symbolic sense became selected for and maintained because it could additionally coordinate group behavior. A side effect of this coordination however was the extension of spatial and temporal sensing (i.e. prediction) to an unprecedented degree. Symbolic language did this by using others to filter and relay ambient information. In doing so, the fundamental systems logic of behavior (e.g. source-path-goal) became the systems logic of language (e.g. source-path-goal). If you'll notice, this is exactly the metaphor that has been used to describe the symbolic transformations that language preserves. It is the 'travel' metaphor. Presently, a way to envision this travel metaphor in language symmetry will be proposed. Lakoff's recognition of the fundamental importance of source-path-goal image schemas as the organizational substrate for semantic relationships will become central to this proposition.

It is important to note that animal here-and-now experience does not make image schemas explicit (i.e. conceptual or symbolic). Animal experience is implicit. The animal goes over there not because it holds a specific image of a goal in its 'mind' all the way over there, but because its neural/endocrine system has primed it to respond to certain sensory cues that in the past (either learned through its own experience or by its ancestors and is now stereotyped as instinct) have fed back positively in the form of opportunities to maintain and reproduce itself. This is not symbolic behavior (although it is often anthropomorphically expressed as such). This is pre-symbolic behavior. The symbolic behavior we experience, and wish to explain here, is the shared, willful, complex and coordinated evaluation and execution of goals that we symbolically represent in mind and that causes us to modify and evaluate reactivity over extended periods of time and space. We want to explain behavior like, 'I want to build a house that looks like Michael Tracy designed it over the next three years.' Or, 'I want to be a psychiatrist when I grow up.' It is notable that human symbolic behavior is like all other behavior in the sense that it seeks to extend adaptability from the here-and-now. What makes it unique is the degree to which this new locative sense can extend and maintain adaptive futures beyond here-and-now conditions.

Where Lakoff is misleading is in his invitation to imagine pre-conceptual experience as divided into two categories: basic-level categories and image schemas. This division essentially differentiates things (i.e. basic-level categories) from processes (i.e. image schemas). While animal consciousness can probably recognize something like 'things' and 'stereotypical relationships' in the patterns that trigger their behavioral routines, it's not at all clear that animals naturally differentiate the two so explicitly. Explicit recognition of a goal as a 'goal' is different than implicit reaction trained to capture or couple to a salient feature of the environment. What it means to be human is related to having explicit and extended goals and choosing from many imagined futures to behave so as to maximize the probability of achieving those adaptive goals. Once again we come to the extraordinary power of imagination and the extension of possible futures that language enables.

Lakoff's characterization of natural reality as containing separate pre-conceptual 'things' and 'processes' is probably a result not a cause of language. Structurally, language divides things (i.e. subjects, or 'structures') from processes (i.e. predicates, or 'transformations'). Language made goals explicit 'things' to be 'achieved.' Before language, our ancestors would not have differentiated, 'bear coming *out* of the cave' (to use the CONTAINER image schema), from just 'bear' (to use the basic-level category). Both patterns would have incited the same behavioral response sets - RUN!

Lakoff is right, our language is based on physical experience. However, in animals that physical experience may be less differentiated than he thinks. Either experiences are so differentiated that each instance is unique or they are differentiated by the behavioral responses they incite. In either case the division he proposes becomes moot. Animals cannot understand the world in terms of nouns and verbs.

What Lakoff shows us is that the metaphorical substrate of language is constructed and maintained like all other behavior, by internalizing external interaction. Initially this metaphorical substrate was a social behavioral network, but through time we individually adapted to this social network by learning it, internalizing it, embodying it. This process of embodiment included the co-option of many of our brain's and body's ways of making sense in this world - of behaving in this world. In this sense, our language maps to our behavior. However, like most large exaptations, this complex process included many of Gould's 'sloppiness' and 'quirky design' hallmarks.

In a sense, language high jacked its adjacent sensory system (e.g. vocal/aural brains) to help us socialize. In a similar sense our brains high jacked their adjacent chewing systems (e.g. jawbones/skulls) to help us hear. Our brains can now organize behaviors sensorially or they can switch over to use the underlying neural substrate to organize behaviors symbolically. From this point of view it does not seem at all surprising that Lieberman found evidence of language performance co-opting the basal ganglia to organize complex speech performances much like it organizes complex physical performances.

The way we interact with features of our environment, our physical behavior, is the foundation for how our concepts interact with each other, our symbolic behavior. Exactly how this happened can be debated, but the mapping of physical logics to symbolic logics indicates that language was originally an external interaction. If language was originally a group adaptation for flexible, coordinated group interaction then the interactions it motivated became the environment to which individual brains had to adapt. That symbolic environment directly mapped physical interactions because successful physical behavior is what maintained and reproduced the original symbolic network. Only through time were our brains (and bodies and genomes) able to do what they do best, begin to capture and reproduce these external patterns.

As Terrence Deacon (1998) suggests, this co-evolutionary process probably occurred over the last two million years. He mentions that not only did our brains get significantly bigger, but in a few areas they are wired significantly different than our ape cousins. This kind of change takes time. It is likely that much of the last two million years we have been communicating symbolically in groups. However, the origin of the personal mind, as a result of successfully internalizing this socio-symbolic network (i.e. 'thinking'), may be a much more recent development.

There is an additional source of evidence supporting the hypothesis that language originated outside the brain and only later became internalized. Early in the twentieth century Lev Vygotsky noted that children learn "social speech" first, beginning at around the age of one. This kind of language functions to communicate with others. At around the age of three the child starts to talk to itself in what could be seen as a primitive version of 'externalized thinking.' At around the age of seven this externalized thinking disappears along with the appearance of the ability to think internally (Vygotsky, 1934). Evidentially, the ontogeny of language is from external social interaction to internal personal interaction. For each of us language functions first as the social utility of 'speaking' and only later does it develop the additional personal utility of 'thinking.' Although Haeckel's famous claim that 'ontogeny recapitulates phylogeny' is not true, it is undeniable that fetus development displays morphological constraint path vestiges from its phylogenetic history. Previously, we saw that metabolic systems also retain the vestige constraint paths of their autocatalytic precursors. The fact that children grasp language first as an external social behavior, not as an internal brain behavior, is not in itself conclusive. It is, however, further evidence that symbolic language had external, social origins.

## Memory: doors to the future

> *To space, time, and causality in biology, Morowitz adds 'memory.' Biology, he claims, is the bridge between physics and history ... All life has chemical memory that cannot be dated by direct measurement. The metabolic memory of modern cells most likely even predates the most ancient rocks.*
>
> *Lynn Margulis (1998, p. 77)*

> *The most primitive sort of conceptual memory is enough to give the human world something which the ambients of other creatures do not have - a time dimension, which extends backward into the past as well as forward to a future. Animal acts are all forward-directed; the influence of the past upon them is physiological, and though it enters constantly and systematically into subsequent motivations of reflex movements and of fears, desires and expectations, it is those immediacies that are felt, not any image or idea of the bygone events which 'conditioned' them. Just as animal life is lived 'here,' 'there' and en route from one place to the other, but not in a geometric space, so its 'time' is a present always heading into a future, but not a homogeneous temporal dimension in which earlier and later events are ranged.*

> *Dimensions are conceptual principles, which require symbolic presentation to let them emerge as spontaneous elementary abstractions. This is probably one of the greatest steps dividing man from the rest of the animal kingdom, although it is also one of the earliest.*
>
> *Susanne K. Langer (1972, p. 337)*

All systems have memory. This memory is the historical relationships that maintain and define them. Systems *are* regular historical relationships. It is the maintenance of those historical relationships, one could call this 'memory,' that maintains them in their present and future environments. Even the fundamental forces are just historical regularities. 'What has worked before' is what maintains systems.

Like nucleosynthetic astronomic systems, organisms changed the way universal systems accessed 'what has worked before.' Gravity doesn't seem to operationally exist in the quantum world. However, given certain quantum and chemical substrate conditions, gravity does emerge as an organizing force for the maintenance and re-creation of astronomic systems. Solar systems maintain themselves through the regular re-creation of gravity fields within galactic environments. They are gravitationally constrained levels of autopoietic galactic organization.

Organisms maintain themselves through reproducing an internalized environment within which the interactive background is constrained to a level of organization that the whole system can be regularly re-created. Organisms access 'what has worked before' through DNA interacting within a highly differentiated and specialized cellular environment. The degree to which cells internalize the conditions-that-have-worked before is perhaps life's greatest novelty. In a sense, the first cells reinvented memory when they canalized DNA structure. They internalized this chemical memory in the cooperative behavior regularly reproduced within their emergent boundaries. These cells were selected for their autocatalytic ability to re-create the chemical information of 'what has worked before' in an entirely new way. Like gravity revolutionized the structure of the universe, biology organized and accessed system memory in a way that revolutionized the structure of the Earth's surface.

As highly complex groups of cells, humans are part of this historical arc of memory evolution. One of the most significant adaptations the emergent symbolic brain had to make was in the way it too reorganized memory. Unfortunately, human memory is a complex subject that can only be cursorily examined here.[227] As such, memory will be examined here in its functional role within the context of the Locative Theory of Mind. This means that memory will not be dealt with as it exists internally (i.e. as per neuroscience), but as it exists to facilitate external interaction (i.e. as per behavior or prediction).

In the behavioral context, memory is the name for a certain class of tricks our brains and bodies have evolved to facilitate the capture or coupling of salient patterns. Like 'the senses,' memory is a general name for the many processes our brains (and bodies) use to couple external patterns to internal patterns.[228] Senses and memory are both part of the same suite of brain processes that help animals constrain steps into their adjacent possibles thereby contributing to the re-production of those futures that include their own persistence. Memories are constraints that were canalized by previous sensory experiences.

Memory processes in the brain are highly distributed and extend from temporary changes at the synapse level up to permanent potentiation changes throughout the brain (e.g. protein fixing of axon and dendrite arrays). Over generations, selection for different kinds of memory result in the genetic (or epigenetic) fixing of new types of brains that better access what has worked for that descent line. A good example is the protist behavioral proteins that evolved into animal synapse proteins as demonstrated by Emes & Grant (2012) and Grant (2008).[229] Together, all these different levels

[227] Human memory is so complex because of its historical levels. Cognitive memory depends on neuron memory. Neuron memory depends on molecular memory. Molecular memory depends on atomic memory and atomic memory depends on quantum memory.

[228] The broad application of the concept of 'memory' could be interpreted to refer to all bodily features evolved to maintain and reproduce the organism. However, in this instance 'memory' will be constrained to the neural context. It would be interesting to investigate memory in this broader context in the future.

[229] For more see, "Adaptive Response and Moving into the Adjacent Possible."

of memory contribute to recognizing patterns and producing adaptive reactions. Memory uses information from the past to create adaptive futures. In animals like us, memory processes are especially evident in the brain. For the rest of this chapter 'memory' will refer strictly to brain, or neural, processes unless otherwise indicated.

The problem that memory processes face is how to organize themselves to inform adaptive reaction. Many animals are hardwired to instinctually respond. This is stereotypy, genetic assimilation, Baldwin effect, internalization, inclusion etc. It is the fixing, through preferential selection, of adaptive reactions into the cycles of maintenance and reproduction of the physical system itself as it samples novel adjacent possibles over time. The problem in these cases is that external salient features need to remain stable over many generations for the phenotype to find and canalize some kind of coupling mechanism that can start the positive feedback process and eventually hardwire these good tricks. Trial and error takes time and consistent conditions.

However, organisms like us were selected to react to fluctuating environments. We often use memory processes to: 1) recognize conditions that trigger certain adaptive reaction routines or; 2) coordinate and complexify those adaptive reaction routines themselves. Very loosely, these two categories fit into the creation of long term memories that the organism uses to cue the behavioral routines and short term memories that the organism uses to coordinate and canalize these routines effectively. Both of these types of memory overlap to capture or couple salient patterns.

The challenge for memory is how to filter massive amounts of sensory information for regularities and then train the brain to recognize those regularities, and then trigger responses in a timely and adaptive way. As Damasio's CDZ framework proposes, memory for animals like us is the potentiation of the brain to react adaptively to its environment. This potentiation occurs within the organism's lifespan as well as over generations (e.g. behavioral learning vs. selection, ontogeny and phylogeny). Over lifespans brains reinforce those memory associations that prove adaptive. Over generations memory processes have evolved by the preferential preservation of those kinds of brains and bodies that potentiate (i.e. that can be trained) more adaptively. What is interesting is that up until now this potentiation of the brain over lifespans has always been locally reactive. Saliencies must physically present themselves before memory processes can learn to associate them with adaptive reactions.

Principally, memory systems are trained by immediate feedback. This can be thought of as a kind of 'remember this pattern, it is important' process, and often these patterns indicate the proximity of the famous four F's. Symbolic communication creates a new context for this feedback. In symbolic interaction, concepts of gods no one has ever seen, or places no one has never been to can be represented in immediate conversation, they are present in the immediate environment to be reacted to. These concepts are available for positive or negative feedback potentiation even though what they represent is not physically present. When they are represented, personal or social reactions can mark them as memorable (leading to either positive or negative reactions) or not memorable (leading to no reactions, no transformation with preservation of structure). The memory system that emerges from this kind of selective potentiation is functionally different from the reactive memory of animals.

Memories that can be expressed symbolically are often referred to as declarative memories. An example of a declarative memory is, 'I know that Napoleon was short.' I never met Napoleon, but my socio-symbolic network (e.g. high school history class) has preserved this information and trained my brain to preserve it. Declarative memories are different from non-declarative memories (languageless memories) in two ways. The first difference is how they are accessed; the second is how they are potentiated (i.e. preserve structure when they transform). Functionally, non-declarative memories are those animal-like memory processes we have just discussed. They are accessed by here-and-now environmental triggers. The environment presents patterns that are recognized/remembered and the appropriate response is triggered. These memories are accessed by physical here-and-now saliencies in the environment.[230] Whether it's catching a fly with your

[230] There was a study done on cats where the neural inhibition of their body movement when they were sleeping was blocked (the inhibition was blocked). These cats 'played' while they were sleeping. They went through 'mouse catching' actions. This would seem to indicate that they were acting on memories, but since they were sleeping they were not actually

tongue or playing the viola, animals and humans have trained their brains to 'know how' to interact with their immediate environments adaptively.

On the other hand, declarative memories are strictly human. They are accessed by symbolic triggers. The symbolic environment that triggers them is socially maintained in the form of culture and individual thought.[231] External symbolic transformations like conversation, reading, or storytelling can all trigger declarative memories. Likewise, internal symbolic transformations like 'thinking to yourself' or dreaming can also trigger declarative memories. Declarative memories can be both preserved and accessed from a network of symbolic saliencies that not only are distributed throughout space (as per the distribution of symbolic society members), but are also distributed throughout time (as per the distribution of symbolic concepts that have been selected for persistence over the last fifty thousand years - at least). Only humans have trained their brains to 'know that' concepts that represent neither here-nor-now saliencies can be used to additionally organize behavior.

The creation of declarative memories is similar to non-declarative memories because both can be potentiated in the here-and-now and both are to some degree internally canalized in the form of instincts.[232] For now we will ignore the ultimate instinctual process and focus on the initial potentiation process.

People learn to avoid touching fire the same way they learn to speak and think symbolically. They achieve both by remembering the positive or negative feedback they receive when they interact with fire or concepts. In the one case, the pain of touching fire trains yours brain to recognize it and in the other, your mother or father train your brain to recognize "HOT!" things are not to be touched. What is different are the environments that

---

sensing anything in the normal way. Presumably memories can be accessed through dreams (e.g. not-here, not-now interactions) but since it is principally waking interaction that is preferentially selected, dream memories will not be considered in this investigation. For more see (Strogatz, 2003, p. 86)

231 The word "thought" here means language guided thought, the kind of thought associated with 'talking to yourself' or 'mental reasoning.'

232 This is not to say that declarative memory content, like specific facts, are, or can become, instinctual. This is to say that our faculty for making and accessing declarative memories in general has become to some degree internally preserved.

present the patterns to be recognized. The physical environment typically transforms very predictably, over hand-eye saccades, steps, and breaths - over a consistent landscape of daily bodily functions, inertia, textures, moon-cycles, smells and seasons. What is recognized by the non-declarative memory processes has an inviolable temporal and spatial continuity to it - that of our physiological coupling to our physical surroundings. It is this inviolable physical continuity of our environment that constrains and structures the source-path-goal rule-set of adaptive behavior. The symbolic environment has a very different temporal and spatial continuity. The symbolic environment is constrained by the rules of how concepts interact, not how physical entities interact. Physical landscapes and symbolic landscapes are constrained by different rule-sets.

The logic of how one concept incites another is very differently constrained than the logic of how one physical change incites another. As previously mentioned, concepts can be represented minimally and they can transform very quickly, crossing mountaintops and generations in the time it takes to utter prepositions. The logic that threads symbolic transformation is syntactic and semantic, not physical. As a result, the memories potentiated in declarative processes are accessed by conceptual triggers that map relationships that are often unconnected to physical environmental constraints. What this creates is an additional memory process, a symbolic or declarative memory system, one that accesses patterns outside the here-and-now environment. This symbolic memory process constrains behavior by learning to react adaptively to a symbolic landscape that transforms by preserving a different kind of structure.

For one thing, adaptive behavior in the symbolic landscape is accessed differently. In the physical landscape adaptive behavior is limited to choices available from a physically adjacent possible. Access to memories that can potentiate adaptive behavior is restricted to one point of access, the here-and-now. The socio-symbolic system allows as many points of access to past experiential information as there are concepts remembered. In the socio-symbolic system the adjacent possible is tremendously de-constrained. In a way, remembering the concepts themselves is the limit to how many access points one has to past information. This past information comes in the form of how each concept point is differentially connected to others in our individual concept-maps and, of course, how these symbolic transformations constrain our behavior. The trained associations they make is the meaning we make. We call training these associations, 'learning a language.'

Organisms are always transforming themselves both in time and space. One of the main ways they do this is by reproducing themselves. This investigation proposes that language is a new form of reproduction and it supports this proposal by comparing historic transformations. In this vein, the declarative memory system is a crucial component of language as a novel cooperative and reproductive symmetry domain. Preserving concepts outside the here-and-now context enables multiple points of access to past information. Ultimately, socially maintaining a network that accesses past experience through symbolic associations enables access to extended future and past sensing for each member. This makes language an exceptionally valuable form of behavior for sampling vastly extended adaptive futures. Through language we can sample, or 'sense,' interactions unimaginable to other animals. However, it would be suspicious if similar sorts of functional complexifications hadn't happened to other reproductive systems in the history of life. Most likely, they have.

Binary fission occurs by dividing a single loop of DNA. Transcription starts at one locus on the loop and continues around the loop until it reaches the beginning locus at which point the two loops separate and the prokaryotic cell divides with one DNA loop in each half. This reproductive system accesses historical information (e.g. bio-chemical memory) serially, from one locus of transformation. To a degree, the simplicity of this serial transformation limits the complexity of the prokaryotic cells it produces. This limited complexity is reflected in their limited ability to interact in complex ways with their environments. Simple cells interact simply (compared to complex cells).

After almost two billion years of interaction, prokaryotic binary fission finally evolved into eukaryotic mitosis. Mitosis is a far more complex process of transformation. In mitosis there are multiple simultaneous replication loci on multiple chromosomal loops.[233] This process represents a massive increase

[233] Basically, mitosis not only adds more DNA in the form of multiple chromosomes in the nucleus, but it can also reuse the same codons and genes on one chromosome to do different tasks by changing the context in which it is activated. By contrast, binary fission is a

in computational complexity. In essence the parallel recursivity enabled by this new reproductive process opened up a whole new developmental phase space for eukaryotic cells. Each section of DNA can now be accessed from multiple transcription points to contribute simultaneously to multiple developmental tasks. Essentially, the same feature that some researchers claim is special about language evolution is true for genetic evolution - increased recursivity.[234] The result of opening this new phase space of recursivity through mitosis was larger, longer living and more complex eukaryotic cells. This increase in complexity can be seen as the expanded morphological potential of eukaryotes to develop complex ways to maintain interaction with their environments. Eukaryotes represent an extension of sensory and behavioral inclusion within an expanded phase space of interaction. They were able to include more within their cycles of reproduction and maintenance then prokaryotes. For exactly this reason eukaryotic cells revolutionized the surface of the Earth two billion years ago.

Like binary fission, non-declarative memory systems are also constrained to process information from a single locus, the here-and-now. As a result, the relationships they enable animals to have with their environments are relatively simple (compared to human relationships). Information accessed through the non-declarative memory systems is constrained to run through the here-and-now, the Von Neumann bottleneck. The immediate conditions of the environment essentially constrain what can be meaningful to the animal.

Conversely, declarative memory systems have access to as many relationships as concepts they maintain. Any concept can lead to whole worlds of possible relationships. Any concept can be either socially or individually called up at any time or under any conditions. Declarative memory systems can access, organize and potentiate information to a degree that represents an immense increase in computational complexity over non-declarative systems. In a similar way, mitosis can access, organize and potentiate the historical information in DNA to a degree that represents an immense

one-way trip from the beginning of a single chromosome to the end. There is little internal recursivity (i.e. looping sections of codons) in binary fission compared to mitosis.

234 Some people who have focused on language recursivity are Corballis (2011); Hauser, Chomsky, & Fitch (2002); and Studdert-Kennedy & Goldstein (2003).

increase in computational complexity over binary fission. Mitotic cells are far more complex (in the sense of more and different parts) than binary fission cells. In both genetic and memory situations, the new system emerged from, and depends on the maintenance of the previous systems. In both cases they emerged in cooperative networks (e.g. mitochondrial symbiosis and human social symbiosis). In both cases they were preserved because they enabled expansion of the abilities to interact in complex ways with their environments. They opened new phase spaces by including more within their cycles of maintenance and reproduction.

Mitosis was an emergent cooperative medium that revolutionized the dimensionality of the eukaryotic adjacent possible. It tremendously de-constrained the eukaryotic adjacent possible. Eukaryotes were able to maintain a new reproductive language that extended their ability to sense, capture and couple distant temporal and spatial adaptive futures. Typically eukaryotes have longer life cycles and larger sense halos. The proposal here is that mitosis isn't the only emergent communicative medium to have radically altered how their emergent organisms discovered, maintained and reproduced extended adaptive futures. Two billion years ago eukaryotic cells revolutionized the surface of the Earth. Humanity is currently revolutionizing the surface of the Earth.

Ultimately, the problem with memory is not the obvious value of storing information from past experience. The problem is access. How does the organism control getting back to the salient features of past experience in order to inform current behavior? Through one road (e.g. here-an-now reactivity) or many (e.g. reactivity to as many concepts as can be mentally and socially maintained)? There are probably many ways to canalize here-and-now memory systems, as per the diversity of non-declarative memory systems. Motor skills, priming, learning habits, classical conditioning, habituation and sensitization are all good examples of non-declarative memory systems. However, since animal brains have only one structure for indexing[235] and accessing past experiences, animals are constrained to match present stimuli to past stimuli and potentiate their responses accordingly.

235 "Indexing" in this context means a structure or system logic which organizes memories so that they can be adaptively accessed.

This type of animal memory is functionally very different from human memory because it needs to be incited by something right here, right now. These iconic and indexical memory systems can only access adaptive futures through immediate environmental conditions.[236] They are eternally bound to a much lower dimensional here-and-now awareness.

The human memory system has an additional way of indexing and accessing past experience and it functions by massive simplification. We essentially use each other to compress complex phenomena into simple symbols. The socially pre-potentiated language system is the virtual structure humans use to travel away from their individual compulsions to react to immediate environmental stimuli. The structure of this language system allowed for a direct connection to past experience by bypassing immediate reactivity and using the new internal, socio-symbolic representational environment to access past experience at *many* or *any* concept loci (for more on the increased efficiency of socio-symbolic communication see Appendix 2).

Basically, the timeless and spaceless structure of socio-symbolic systems allows feedback to 'now' decisions from any concept point. This new route to the past (or future) opens up the possibility for new social and neurological systems to evolve that support: 1) selection for a new kind of brain organization that indexes and categorizes past experience by connectively potentiating and maintaining highly simplified concept points and 2) the use of this new kind of memory to produce extended adaptive futures by accessing vastly expanded libraries of symbolically indexed past experiences. Over time, patterns of accessing this information through concept points can be sampled, potentiated, reorganized and maintained through functional feedback by personal symbolic interaction (e.g. thinking) or socio-symbolic interaction (e.g. talking, writing, computing). Over lifetimes and generations, over individuals and groups, our sense of the future is systematically extended by

[236] As mentioned, iconic and indexical memory systems can only extend survival by making proximate predictions based on immediate (here-and-now) stimuli. For an interesting discussion on the human expansion of accessible memory see Donald (2001, pp. 157-164).

symbolic memory. For the first time in the history of animal life our sense of the future is largely unconstrained by here-and-now conditions.[237]

Animal memory opens one door to the future; symbolic memory opens many. How we choose to use this expanded sensorium is different question.

## The Locative Mind as a Concept-Map

Within our phylum (i.e. Chordata) decision-making is principally maintained and reproduced by two metabolic systems working together (i.e. synced) - the endocrine system and the nervous system. The endocrine system is hormonal and, "the function of hormones is to 'prime' the animal ... Hormones affect the intensity of its drives, or ... the level of its motivational states" (Wilson E. O., 1975, p. 153). Much of this priming can be thought of as what we commonly call 'moods.' Whether we are hungry or annoyed can significantly mediate how we react to events in our daily lives. In a way, the hormonal state induced by the endocrine system is the principal environment (i.e. neuro-chemical background) within which the neural nets in our brains interact to constrain behavioral decisions.

As discussed, our brains are potentiated networks of neurons that have learned to sync to external patterns in such a way as to organize internal interactions to maintain adaptive external interaction - behaviors. This investigation proposes that the original socio-symbolic system was potentiated in an analogous way. The neurons in the brain have to learn to fire together to wire together and, through socio-symbolic communication, so did coordinated proto-human groups (for more see chapter "Group Potentiation").

The picture of mind that emerges from all this interaction is one of multilevel synchrony of related processes to produce what feels to each

[237] It should be noted that each major transition in life has been accompanied by a jump in functionality. Each transition has expanded the constraints of here-and-now interactivity. Autocatalytic molecules are constrained to ionic and covalent bonding at specific loci. Prokaryotes expanded this to a multifunctional cell wall. Eukaryotes expanded the multifunctional cell wall to an even higher multifunctional cilia and sensory boundary. The story continues with the human symbolic jump of functionality fitting smoothly into the end of a long history of expanded adjacent possibles. Thus, although it is true that the human symbolic sense of the future is largely unconstrained, this is a unique development only within the history of multicellular animals, not within the history of all life.

of us like a unified experience. Both Dennett and Deacon have elegantly shown us how intricately interwoven these various levels can be by building theories that draw from cellular, organal, social and evolutionary levels of transformation. However, to get to some kind of functional understanding of what our minds are we are once again going to have to build some kind metaphor that fits with how we metaphorically understand reality. Lakoff would be proud.

The metaphor for the individual Locative Mind to be offered here is of 'concept-maps.' The idea of concept-maps is based on Deacon's idea of mental, "buoys indicating an otherwise invisible best course, ... a specific associative path" (Deacon, 1998, p. 265). These buoys are symbolized 'concepts,' or 'concept groups'[238], and we travel from concept to concept along trained associative paths. Each concept represents a, "virtual set of associations" (ibid) that can be imagined as pathways connecting it to other concepts.

The best analogy for this kind of system is the brain itself. Neurons are interconnected by up to tens of thousands of connections. These connections are pre-potentiated because they are the result of an internal genetic/epigenetic predisposition for massive overconnectivity that has been culled to maintain only those cycles that sync to adaptive external signals. But this process does not stop. The strength of each connection is not only pre-potentiated, it is also adjusted with each current use by how it syncs to extra-neural patterns (other neurons firing). So each neuron is connected by a real array of biochemical associations. This would appear to indicate a distinction between neural-maps, which are connected by 'real' biochemical associations, and concept-maps, which are connected by 'virtual' associations.

Presumably, the virtual system of concept-maps is maintained by the real neural system in the brain, but like all emergent systems there is an added level of constrained release of energy, a super-ordinate rule-set. This added level of constraints is trained into the brain by the socio-symbolic

[238] Concepts can also be seen as word groups. 'Mary had a little lamb' may, in some cases, be remembered or used as a functional unity. Word groups can also be nested meanings, as in high dimensional words like 'constitutional' which have alternative sets of associated meanings depending on their contexts. 'Constitutional' means different things in scientific, philosophical or political contexts. 'Concepts' include symbolic meaning as expressed in either individual words or in functionally unified word groups.

network. Our personal concept-maps are trained and maintained by our cultural concept-map, our socio-symbolic network. In this sense, our brains are synced to the extra-personal patterns of our cultures, or 'encultured.' This external pattern of interaction is maintained on a different communicative, cooperative and reproductive substrate than an individual's neural electro-chemical pattern. That is, the rule-set that maintains the socio-symbolic map is behavioral whereas the rule-set that maintains neural maps is physiological. Neural structure is organized to capture, couple and canalize external patterns. Brains learn to sync us to our environments. They do this by mapping different kinds of saliencies like, barometric pressure, seasonal temperatures, predation behaviors and conspecific behaviors. Much of the recent massive expansion of the human brain is due to the specialization of humans in syncing to conspecific behavior. Specifically the human neural specialization is in the development of a new symbolic sense that allows us to socially create and maintain a virtual map of adaptive behavioral relationships that each of us can use to extend our own awareness of adjacent adaptive opportunities and in doing so extend our group's social awareness of adjacent adaptive opportunities. The concept-map then, is the communal store of information that each of us learns to sync to that couples our social behavioral rule-set to our personal neural physiology rule-set.

So, the concept-map syncs inter-neuron to inter-personal relationships. It maps external social information to internal mental information, albeit with some of Gould's exaptive sloppiness and quirky design hallmarks. This is a process of transduction between two very real and very different systems and thus technically should not be called virtual. The sense in which symbolic-to-neural transduction is not real is the same sense in which quantum-to-chemical transduction (e.g. the quantum-classical boundary) is not real. [239] However, because the socio-symbolic ambient has not been recognized as constitutive for thought there is a need for a word to represent that ghost

239 In the computer context one cannot argue that virtual is 'not real' or 'different from real' either. All virtual interactions in computers are a result of interaction between your biological organization and its electro-mechanical organization, transduction between two very different systems. These interactions are real. The patterns they effect on this world are real. They are simply not understood well enough within the reductionist paradigm to be called 'real.' In fact, virtual transduction of this kind is a good example of positive symmetry, of systems selectively preserving emergent synchronies as they statistically step into their futures.

in the machine that syncs neural interaction to symbolic interaction. In this investigation this word will be 'virtual.' For now, we will continue to use the word virtual not because the interaction is not real, but because we still do not know exactly how this transductive process formed and constrains our 'minds.'

At a metaphorical level it may be easier to understand the concept-map as networks of symbolic loci with pathways connecting them, like a trail system in a forest or a road system in a huge city, or even a vascular system that carries blood to every cell in your body. In these systems the size of the pathway represents how often it is used or its 'connection strength.' Take the Freudian practice of free association as an example. The therapist gives the patient a word or short phrase and the patient names concepts they associate with that word. What the therapist is hoping to discover is how that patient's brain has been trained to connect certain concepts to others, how he personally structures meaning. Does the word 'Father' lead to associations with 'jerk,' 'shut up' and 'belting' concepts or to 'long talks by the fire' and 'knock-knock joke' concepts? Each of us is continually adjusting these webs of association by constantly adjusting the connections (i.e. the transformations that preserve associative structure) in our symbolic environments. Our declarative memory systems maintain a virtual map of these symbolic associations that is both externally trained (e.g. socially) and internally reinforced (e.g. used regularly). This map is based upon the socio-symbolic map we each inherit from our cultural environment, but it is potentiated by our individual histories of interaction.

Originally, the socio-symbolic network would have only maintained and reproduced rather simple concepts like 'good,' or 'you.' As sampling and selection continually extended cooperative interaction through both time and space, this system would have differentiated and specialized much like the soma cells in the first cooperative multicellular organisms of the Cambrian explosion did.[240] One result would have been individual brains 'learning'

[240] This process can also be found in the transformation of proto-synaptic proteins in protists to eventually produce neural synapses in animals. Research done by Emes and Grant has identified proteins in single celled protists that produce 'behavior.' These same proteins differentiated and specialized into neurons and brains in multicellular animals. For more see Emes & Grant (2012) and Grant (2008).

(and eventually 'including' or 'evolving') capture and coupling mechanisms that enabled further systemic differentiation and specialization of adaptive concept-connectivity. Another result of this co-evolution would have been increased efficiency (e.g. cycle time, flexibility and/or fidelity) of interaction and a complexification of rules for interaction (e.g. syntax) that could support the extension of ever faster, more efficient and/or more complex behavioral routines. Unfortunately, there is no opportunity here to go into these processes in detail. Perhaps most distressing is the lack of time to compare the phylogeny of cellular signal transduction with that of symbolic language.[241]

The concept-map is a personal network of symbolic associations. It is derived from the socio-symbolic network by each of us learning how to orient and navigate the symbolic information in our cultural environments. What the concept-map does is transduce virtual (i.e. behavioral) relationships into neural relationships and use that store of neurally potentiated information to orient and navigate the cultural environment. A concept-map is as indispensable to our symbolic self as a physical environment is to our embodied self. A concept-map is essentially the symbolic environment that the symbolic self learns to locate itself within. In a sense the symbolic self is defined by where it locates itself within the concept-map. The symbolic self is a serial and digital locus of symbolic transformation that is defined by how its concept-map constrains how it can step into its adjacent possible. Our personal feeling of agency and intentionality comes from how we locate ourselves within the concept-map and how the transformational rule-set of our concept-maps constrains how we can act to transform that location.

Presumably children don't learn to become *self*-aware until they have learned to create and maintain a personal symbolic environment (i.e. a concept-map) within which 'being' anywhere makes sense. Awareness can be thought of as a location from which the world is known to you. It is your current disposition relative to that network of known saliencies to which you have inherited or built up knowledge in the form of adaptive reactions to the world. It is the unique place from which the adjacent possible is

[241] It is probable that our current level of understanding of how both multicellular and human systems developed is so poor that this kind of inquiry would lead to grossly speculative conjectures.

oriented and accessed. It is the location from which any transformation can be localized and the structure of that transformation can be preserved. The world only begins to make symbolic sense when we develop the capability to locate ourselves in a concept-map and use its previously stored information to maintain and extend our adaptive futures. Life only begins to have symbolic meaning when we know where we are, how we got here and how we can use our knowledge of how we got here to navigate ourselves to a better place. Before that, we are just babes in the woods, existing, but without any definitional points of reference. To create meaningful lives we learn to orient and navigate this vast network of pre-potentiated socio-symbolic memories by 'downloading' them into our plastic brains so we can operate remotely, so we can think for ourselves. We learn to locate our 'selves' within a larger socio-symbolic network.

## What is It Like to Think?

> *"The relation of thought to word is not a thing but a process, a continual movement back and forth from thought to word and from word to thought."*
>
> *Lev Vygotsky (1986, p. 218)*

We all know what it is like to think, but as neuropsychologist Chris Firth points out we are actually only consciously aware of about ten percent of what our brains are doing (Frith, 2009). He describes this as the brain letting us know what it is doing on a 'need to know' basis. The result is that the part of our experience we have conscious access to is a small part of what is going on. In this sense it might be interesting to spend a moment trying to characterize what is actually going on during that ten percent when we feel like we are thinking.

Symbolic thinking is the result of training the brain to internalize socio-symbolic concepts and then using our embodied kinesthetic image schemas to transform these symbolic concepts. As mentioned, our huge brains have probably had over two million years to experiment with different ways to internalize socio-symbolic associations. Extension of memory and prediction powers is what brains have been selected for since before they were brains (Emes & Grant, 2012; Grant, 2008) and symbolic communication opens up vast vistas of not-here and not-now prediction potentials. The modern human brain is the result of a short period of selection for differentiated brains that specialize in modeling and preserving the kind of socio-symbolic interactions that produce ever more extended adaptive futures.

The transduction of external symbolic interaction (e.g. talking) into internal symbolic interaction (e.g. thinking) required finding a way to map complex behavioral patterns to neural patterns. For these neural patterns to model symbolic interaction they needed a rule-set to constrain and motivate symbolic transformation. Where interaction with a wide world of physical saliencies could constrain and motivate individuals to cooperate symbolically, what could motivate just those internal symbols to interact cooperatively? The answer is simple, the internal rule-set for symbolic cooperation was modeled on the external rule-set. To get somewhere, to transform adaptively, you need according to Lakoff (1987), "A SOURCE (starting point), a DESTINATION (end point), a PATH (a sequence of contiguous locations connecting the source and the destination), and a DIRECTION (toward the destination)" (p. 275). Thought is internally modeled behavior. Behavior is serial and based on the source-path-goal schema. Thinking is a system of symbolic symmetry that is motivated and constrained by the behavioral bottleneck as expressed in the source-path-goal rule-set.

What this means is that we should be able to describe how we think as analogous to how we travel. Importantly, this comparison is not fatuous. The behavioral and symbolic systems are synced by symbolic transduction. 'Thinking' is transformation within the symbolic environment. 'Travelling' is transformation within the physical environment. When travelling we change locations. When we change locations different opportunities for adaptive behavior are presented to us in the form of women, fire and dangerous things. These saliencies are recognized when they enter our sense halos and our embodied brains cause us to react to them as they have been selected and trained to. A Bayesian feedback process of adjusting neural error weightings according to the success of our reactions then helps train the brain to interact more and more adaptively over time, change, interaction, lifetimes.

When we think symbolically we change concepts (i.e. move through symmetry transformations from one concept to the next). When we change concepts the new concept presents different opportunities for further association. This array of further associations, this adjacent possible, is like the 'symbolic halo' of that new concept. Changing concepts, like changing locations, presents new arrays of potential associative behavior. Words like, 'women,' 'fire' and 'dangerous things' each lead to different associative pathways with other concepts. The success of these different symbolic pathways is established by the same Bayesian process that trains physical behavior. The positive or negative results of the chosen transformation creates an error weighting that helps train the brain to symbolically interact more and more adaptively over time. For exactly this reason legal systems do not hold adolescent behavior to the same standard of culpability as adult behavior. Adolescent brains are still insufficiently trained to make consistently rational behavior (for more on rational vs. instinctual behavior, see Appendix 3).

Our brains have been potentiating adaptive behavior for hundreds of millions of years - since before our ancestors even had brains (Emes & Grant, 2012; Grant, 2008). Thinking symbolically is just one more behavior to be adaptively potentiated. Ultimately, thinking is like moving around except that what is selected for and entrained in your brain is adaptive socio-symbolic behavior not adaptive physical behavior.

When I try to picture this idea of 'thinking' I remember walking out to the barn to milk our cow in pitch-dark predawn. To see the path through the bush I usually had a flashlight. The beam of light it cast illuminated the trees and bushes closest to me with bright colours and sharp definition. As the features of the forest got farther from me they lost colour and definition and eventually they faded out into the surrounding darkness of the Canadian bush. Thinking symbolically is like moving within a forest of associated concepts but, only being able to focus on those associations that are closest to you. You may have a specific goal (e.g. the milking stall) or you may have a vague goal (e.g. 'get away from here' or 'I wonder what's over there?') but in either case you focus on those features that surround you to try and determine those patterns that will best lead you to your goal. Most thinking is like going to the milking stall every morning; it's habitual. Minimal choices need to be made seeing as you are just re-traveling well used associative pathways.[242]

Making new associative pathways is much harder. Novel ideas are novel ways of associating symbolic concepts. This is much more like Alexander Mackenzie charting the unknown territory of Western Canada. All he could see was what immediately surrounded him. One river branch ran to the north and another to the south. From such a fork he couldn't know which one would ultimately lead west. Sometimes we follow associative pathways that lead us away from our goal and we have to backtrack. We all make mistakes. Other times we choose associations that lead us right to our goal. The histories of science and politics are full of ridiculous mistakes as well as stupendous successes.

Once a novel idea like Symmetry Theory has been associated with another novel idea like the Locative Mind it takes a certain amount of rehearsal and fleshing out of the details to form a navigable map. If you look at a map of Mackenzie's voyage from his day (Figure 23) and compare it to a modern map (Figure 24) you will see considerable improvement over time. Extending your own personal concept-map is much like this. The more time students spend rehearsing mathematics, for example, the more refined their mathematical map becomes. With practice they eventually come to know the mathematics landscape well. The more you explore the west, the better your western map becomes.

[242] See Appendix 3 for more on instinctual and habitual symbolic behavior.

**FIGURE 23: 1793 MAP OF WESTERN CANADA (THE MACKENZIE TRAIL)**

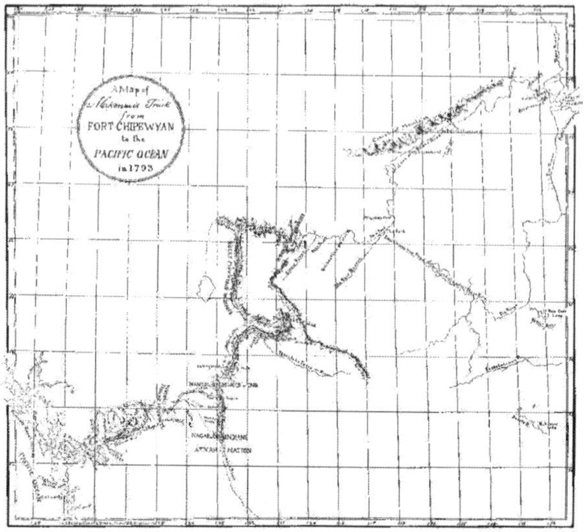


Source: Gutenberg (2011)

**FIGURE 24: MODERN MAP OF WESTERN CANADA**

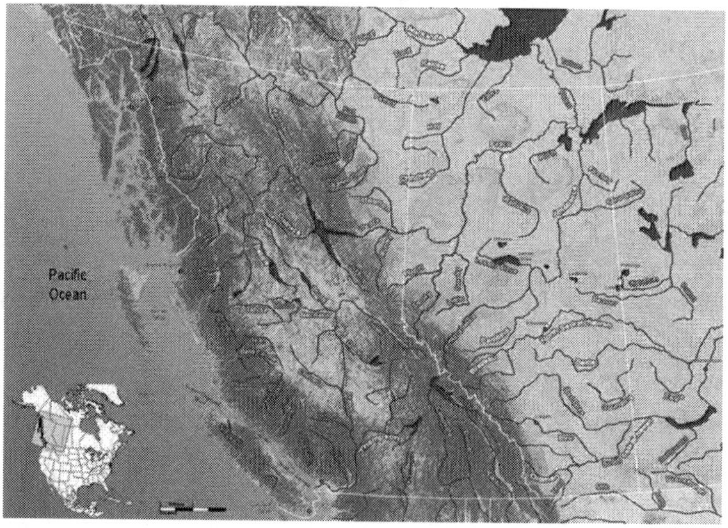


Source: Wikimedia Commons (2006)

Over lifetimes each of us learns to create and navigate concept-maps. These concept-maps are all the associations we were taught or learned. The rules for creating our own map are largely culturally maintained by the socio-symbolic system, but we also differentially maintain the weightings these associations have through our personal experience and use. What we read, write, say, watch, hear, think about etc. all determine how we adaptively interpret and react to our cultural environments. In the end, the ways each of our symbolic selves is constrained by our personal concept-maps is what is unique about our lives. Loss of this map through death or debility is loss of the personal symbolic environment that defined and potentiated our symbolic-self. This is why memory diseases like Alzheimer's are so tragic, the loved one's personality changes. Ultimately, the self you think of symbolically (i.e. the narrative self) is derived from your personal concept-map of symbolic associations - from the history of how you found, practiced and made those associations that made your life meaningful.

When we think, we are travelling our concept-map, moving from one concept the next. The concepts we choose temporarily 'light up' their proximately associated symbols and as we choose to associate with further concepts they light up further conceptual pathways, further associative choices. Like moving through the forest with a flashlight we can illuminate many different choices from any one position, but we can only choose one pathway. How we choose that pathway is constrained by how we (e.g. you, me and our ancestors) have chosen previous pathways. As we learn and get older our ability to recognize those patterns that lead to desired goals become more finely tuned. If we become experts at some symbolic task then our level of pattern recognition becomes almost automatic.

The language system creates an alternative environment from which to select for adaptive behavior routines. This environment is created externally by the socio-symbolic network and internally maintained in our brains as potentiated networks of symbolic associations, or memories. In this memory system each symbol is a potential access point to the information contained within the system as a whole. Conversations that start out about who is sleeping with whom can end in dense plasma thermodynamics or in waffle toppings. All concepts are connected, but each of us connects them differently and to different degrees of specificity. Spiders build webs that both define and constrain their experience. All spider webs are characteristically species specific, but each web is unique. All concept-maps are characteristically culture specific, but each personal web of meaning we build is unique.

When we think, we are transforming concepts along the lines of how those kinds of concepts have been previously transformed. How we have been taught they transform. How we remember they transform. When we think, we are simultaneously accessing and recreating the personal concept-map we have constructed over a lifetime of symbolic interaction. When we interact symbolically with others we are simultaneously accessing and recreating the socio-symbolic network as a whole - the very same network our own personal brains continually access, re-create and maintain.

## Discourses on Mind

### Personality: hopes, desires and fears

Like fingerprints, we all have hopes, desires and fears. Like fingerprints, the specific features of each are unique. Each of our particular concatenations of hopes, desires and fears define us, they identify us as unique in a sea of similar patterns. These collections of stories about ourselves are like our behavioral fingerprints. At best, for the enlightened and the honest, they serve to define how we remember our past actions and how we should mediate our future actions. More often this behavioral fingerprint is what folks tell other folks so they will be thought well of, so they will fit in.

Fit in to what?

So they will fit in to other people´s patterns of expectations. So they will sync to other people's reactive faces. So they will fit in to their socio-symbolic ambient - their culture.

The concept of the 'narrative self' is common. Both Dennett (1991) and Damasio (1999) refer to it and both identify it as something like the self-maintained 'story of ourselves.'[243] As such it is both a feature of our declarative memory systems and a feature of our group's socio-symbolic memory. The memories associated with the narrative self are by definition declarative and thus can only have two sources: our own symbolizing brain or the symbolizing brains of others.

From childhood we are taught to see ourselves within the socio-symbolic ambient. We are given a name not to remind ourselves that we exist (how could we forget?) but to remind ourselves that those around us will constantly refer to that feature of the local symbolic ambient called

243 Damasio (1999) often uses the phrase "autobiographical self," but he is referring to the same phenomenon. See chapter "Consciousness Explained" for more on narrative selves.

'Camelo.' Who 'Camelo' is to myself is a collection of symbolic associations in my mind that were partially created by myself thinking about myself and partially by others referring to me. When others refer to me (think of your mother tickling your chin, "good boy, aren't you a gooood boy") they are potentiating those associations that contextualize 'myself.' They are connecting my concept of myself to other concepts in my concept-map and these connections locate me within my web of meaning. They are co-creating my image of myself. For some reason I have always thought of myself as a 'good boy.'

Dennett refers to this symbolic avatar as a 'center of narrative gravity' presumably because, like an aggregating planet, your symbolic avatar gains mass and definition as time passes. If you are not rehearsing events the way you want to be seen in relation to them, then others are rehearsing events the way they want you to be seen in relation to their own personal contexts. Somewhere in the middle 'Camelo,' as a shared locus of personal and social narrative gravity, is constantly being added to, defined and transformed.

Maintaining a narrative self is a social imperative. Autistic individuals do not function socially because they do not create a narrative self. They do not have hopes, desires or fears in the same sense we do. Typically they neither have children nor write books (which is to say they do not physically or symbolically reproduce). Part of the maintenance of your narrative self is the creation and maintenance of a behavioral fingerprint that others can recognize.

'Camelo is a great guy but he's never on time.' This tells others that I am a good cooperator except in the contexts where being on time is crucial. If I am honest with myself I realize that others think that I am never on time and that my idea of who I am includes a history of not being timely. If I am not honest then I deny the fact. Part of who we are is balancing what others say about us with what we choose to personally rehearse. Often the degree to which we overemphasize our own self-created version of ourselves is the degree to which others refer to us as delusional, egocentric or self-involved. For many people, their self-professed hopes, desires and fears have little to do with their actual actions.

Part of creating a narrative self 'reactive face' is maintaining a set of hopes, desires and fears that distinguish us socially. Primally, each of us has the same hopes, desires and fears. We all hope to live, flourish and reproduce. Additionally, we desire those saliencies that we believe will help us realize our hopes and we fear everything that interferes with these processes. Individually, these features are a common denominator throughout the species. Socially, on the other hand, our hopes, desires and fears distinguish us. They identify our narrative selves within a sea of other narrative selves. Like ants or bees that use pheromonal or other biophysical signatures to differentiate each other's cooperative utility, we use our personalities (defined by concepts like "hopes, desires and fears") to differentiate cooperative utility to each other. How else would we predict each other's actions to cooperate successfully if we didn't attribute different action potentials to each other? Since the beginning, language has functioned to potentiate social networks.

Like all systems we are differentially selected to include each other in reproducible cycles of maintenance. Like neurons, humans function within networks according to how they have learned to react to each other. Part of training these reactive potentials is canalizing a system by which you can present a reliable reactive face to others in the system and reliably represent the reactive faces of those you are connected to. Those that do not sync to maintainable and reproducible cultural cycles end up under bridges and in psych wards.

Once again we can see parallel processes in different systems. 'Action potential' in the neural context is electro-chemical. In the social context, 'action potential' is behavioral. Both exist in an exploratory-process-with-functional-feedback context and both serve to sync internal connectivity with behavior that is externally adaptive. Both also exist to potentiate a synchronized social entity whose emergent functionality is greater than the combined functions of their components (e.g. the brain and the group).

As previously declared, it is to the heart of our hopes, fears and desires that we are driving and any expression of them in human terms must ultimately be mediated by language. Many of you will be disappointed to hear the truth. Your hopes, desires and fears are versions of my own, of all of ours. They exist to help our phenotypes maintain and reproduce themselves. What you really expected was an understanding of why your specific hopes, as they differ from mine, came to be. How language and experience have mediated those primal hopes, desires and fears for you is a unique story. It is a personal story.

Your personality developed like your epithelial cells when they differentiated and specialized to make the gripping pattern unique to your fingertips. The deep socio-symbolic pattern of your culture (think DNA) is the substrate for the development of your personality (think fingerprint). You start with certain language based cultural constraints (imagine the constrained release of energy that DNA embodies) and then you interact socially with your environment (like DNA interacts with its cellular cytoplasm and neighboring cells). Like cells and all natural systems, your own personal history is full of broken symmetries that make its trajectory unique. Each of us samples the human environment differently. However, your creation of a self, with specific hopes, desires and fears, is limited to the kinds of constrained release of energy that shapes human experience. Some of those constraints are physiological (e.g. DNA). Some of them are cultural (socio-symbolic). Some are internal; some are external. In the end your particular mix of hopes, desires and fears is like your body, it is a unique expression of combined levels of historical interaction. In the same way that your body develops by syncing to its physical environment (diet, altitude, local bacteria, etc.), your personality develops by syncing to its social environment.

When we think of our fears we do not think of those fears that we all have (i.e. looming objects, snakes, darkness, death, pain and disease). We think of those fears that distinguish us from others. I fear my work not being accepted. I fear inane crowds. The degree to which these fears affect my socially perceived actions, combined with the degree to which they affect my personally perceived actions is part of what constrains my narrative self. Good friends do not ask me to go to inane social events. What is uninteresting about our hopes, desires and fears is that yours are fundamentally the same as mine. Thus, when we mention our hopes, desires and fears to others we are engaging in defining ourselves in our social context. How are we different? How are we interesting? How can we be useful to each other?

When we think about, rehearse and eventually canalize our own personal arrays of hopes, desires and fears we are also engaged in defining ourselves in our social context. We are creating, with the help of those around us, the way we will interact with others. We are creating our own social action potentials by defining ourselves functionally within the socio-symbolic system. Ants socially co-create workers, queens, nurses and warriors. Humans socially co-create hunters, gatherers, nurses, social workers and pizza delivery men. Undoubtedly, as we get to know our brains better we will find that brains socially co-create many different families of neurons specializing in many different functions on many different levels to the tune of many different rhythms.

None of us want to think of ourselves as cogs in a machine. But neither do we want to be willfully delusional about the universe we are part of. Are we the individuals we have been taught we are or are we all part of a greater organism, a humanity animal? More importantly, since we do have the power of choice and an extended sense of our history and future, how should we choose to see ourselves in the future? Should our hopes, desires and fears be individual or collective? When should we act on which ones and to what degree?

Western cultures champion the individual. God has been dead for almost two centuries and with him went any sense of collective right or wrong. In God's place have sprung up the existentialists who endorse meaning for meanings sake.[244] In the modern existentialist world you are the only one who can decide what is right and what is wrong. You elect your government and it will survive according to your approval. In this environment intelligent and insightful movies and books are those that caste 'meaning' and 'life' as personal, ephemeral, chaotic and unknowable. More than often they leave you bewildered (e.g. thank you Woody Allen). But what could be less ephemeral than life? Your DNA, the chemical information in you, has survived for over three billion years. The symbolic information 'in' you has survived for

---

[244] The term 'existentialist' is also poorly defined. Many existentialist thinkers would not self-apply the term. In the context of this investigation 'existentialism' refers to the fragmentation of philosophical thought characteristic of 20$^{th}$ century philosophy (Structuralism, Post-structuralism, Modernism, Post-modernism and epistemologically based arguments that can be loosely characterized as semantic - e.g. "Oh, that's just semantics!"). Existentialism here refers to philosophies that are fundamentally based on ideas of relativism - e.g. "God is dead ... we cannot know anything objectively for sure ... therefore everything is measured by its utility relative to individuals - subjectively."

over fifty thousand years. What could be less unknowable than language? The language of quantum mechanics has enabled atoms to reproduce and persist for thirteen billion years. Our language has enabled us to reproduce and persist throughout the biosphere for the last two million years. This is not chaos. This is order of the highest order, maintaining and reproducing itself to the logic of selection for persistence. Is it even possible that we are *just* individual wills cast adrift in a big wide universe? Is there really no logic, no order, no unifying pattern to being human?

Sue Savage-Rumbaugh (1997) echoes the current physical reductionist opinion of the source of our hopes, desires and fears when she says:

> The brain is a physical system whose operation is governed solely by the laws of chemistry and physics. What does this mean? It means that all thoughts and hopes and dreams and feelings are produced by chemical reactions going on in one's head (a sobering thought) (p. 76).

This thought would be very sobering if it were true. It would mean our hopes are entirely chemical. The current disconnect between science and the humanities is largely maintained by the responses that most of us have when we are told that our hopes are chemical - namely, "baloney."

Essentially the current scientific paradigm is telling us that an hourglass is a bottle neck or music is a CD. Yes, telling time with an hourglass requires a constriction point. Yes, CDs have music recorded on them. However, telling time requires someone who not only knows what 'telling time' is and wants to tell time, but who also creates the glass container, fills it with the right material and flips it regularly. A CD needs musicians to create the music, engineers to record the music and a public who pays to listen to it on CD players they both constructed and bought. To say that our hopes and desires are chemical brain events is to say that telling time is a bottleneck or that music is a CD. It is the epitome of necessary, but not sufficient.

Hourglasses need bottlenecks to tell time. Humans need brains to think. For that matter, organisms need DNA to reproduce. In none of these cases, however, can the part be said to be the whole. DNA needs cells that collect and pass them the right molecules. Hourglasses need humans to build and flip them and human brains need human bodies that interact adaptively with human social environments to produce the kinds of hopes, desires and fears you want explained - the kind that aren't just chemical.

Until we see ourselves for what we are, our choices will be flawed. Our hopes, desires and fears are not just chemical. They are not just personal. They are also social. If we allow our personal (or god help us our chemical) hopes, desires and fears to hold exclusive right to what is true, then we are denying how nature has organized itself, how it has organized us. Your mind is the product of billions of years of propagating symmetry. You should be proud of it. But you should also realize that 'your' mind is just a part of 'our' mind, that what is right for you should also be considered within the context of what is right for us.

Even more importantly, what's right for us humans should also be right for our environment. It is the current environment, and no other, that both maintains and reproduces us.

## Beyond the Brain: the socio-symbolic mind

Symbolic language is what differentiates the human mind from the animal mind. Like other animals, our bodies 'speak' quantum, chemical, astronomic, binary fission[245], mitotic, meiotic and instinctual languages. Each of these languages maintains cycles of interaction that are as necessary to us as they are to all animals. Unlike other animals, our bodies also speak symbolic language.

Symbolic language is a social phenomenon. Like all of the major transformations in the history of life, and all of the emergent systems in the history of the universe, the human mind is the product of previous systems sampling the environment they co-created to maintain novel social unities. Group selection is a rare, but exceedingly powerful process by which an emergent symmetry is propagated in certain environments of this universe.

245 Our bodies symbiotically include more prokaryotic cells that reproduce their own DNA through binary fission than they include our own eukaryotic cells that reproduce our own DNA through mitosis and meiosis. Most of these prokaryotic symbionts live in our guts and on our skin.

The socio-symbolic mind persisted first as a social regularity, a way to extend adaptive relationships between groups of humans and their ever fluctuating environments. As such, the symbolic mind existed first in the socio-symbolic network - in the group system of vocal cuing that synchronized cooperative endeavors. Over time this social system of sampling and maintaining adaptive group connectivity to its environment became one of the principle constraints on individual selection. Individual selection differentially preserved individual predispositions to maintain socio-symbolic interactivity through traditional mechanisms like the Baldwin effect, stereotypy and genetic assimilation. The result was that individuals surviving within socio-symbolic networks became capable of maintaining simplified versions of the socio-symbolic network. They preferentially internalized highly adaptive external behavior. The obvious place to host the lion's share of the adaptations necessary to maintain such a symbolic mind was that most plastic of organs - the brain.

Like genes, brains are constrained by their environments. Both genes and brains present ubiquitous amounts of variation from which positive couplings can be selected to persist. Genes present this variation through mutation, ontogeny, recursivity (emphasized in mitotic reproduction) and recombination (emphasized in meiosis).[246] Brains present this variation through all of the above as well as through the massive interactivity of neurons that either learn to support adaptive behaviors or go extinct. Brains are a newer addition to the positive symmetry process here on Earth, but they essentially do the same thing over lifetimes as genes do over generations. Brains and genes both extend the organism's ability to survive current conditions by maintaining what has worked in the past (e.g. memory or DNA).

[246] Genes actually mutate far less than we would expect in a system of that complexity which is why they were earlier referred to as 'the conserved core of what has worked before.' However, genes do mutate. Mitotic genomes also mutate but they are additionally recursive because they have learned a kind of 'cut-and-paste' way to simultaneously access many sequences on many chromosomes and can thereby use the same stretch of code for many different purposes. This complexification is even more evident for meiotic genomes that use the mutation and recursivity substrates of their predecessors to generate variety, but additionally split and mix the genomes from two different organisms. Each reproductive system has added levels of constrained release of energy to expand the variety of phenotypes from which emergent organisms can be selected for persistence.

They also both extend the organism's access to what will work in the future (e.g. prediction or adaptive futures).

At different levels, extending adaptive connectivity to the future is what genes, brains and organisms in general do. Existence is the business of surviving entropy and chaos. Existence is selection for persistence. Over the last four billion years genes have learned some impressive tricks to extend organismic phase space. Case in point, over the last two million years some of those genes have been selected to maintain particularly large brains in one species of ape. This species rediscovered an age old, but nonetheless impressive trick for extending its organismic phase space - cooperation. Each of the previous major transitions occurred in unique ways to unique organisms, but in each case the same process of natural selection led to the same kind of transformation - cooperative unification. In each case social interaction was selected to conserve more and more information until the group could cooperate to the degree that the information of its social organization was reproduced over generations. In each case a different communicative medium was differentially preserved to maintain the unified cooperative group.

The human mind was a social construct first. The symbolic cues the group maintained to sync to ever more distant and cryptic saliencies were socially maintained and success over time ensured that they naturally radiated, differentiated and specialized. The growth of the symbolic halo was fueled by the evolution of a human symbolic sense and the result was the expansion of the socio-symbolic mind. Like all senses, as symbolism evolved it became an ever richer and more nuanced way to predict the future. Strong selection for individual abilities to incite other's to broadcast their own reactions to their own sense halos in order to expand one's own power of prediction continuously selected those of our ancestors who best constrained their here-and-now reactions and could more and more extensively engage in socio-symbolic cooperation. The result was the dissociation of a second 'symbolic' point of view from the classical here-and-now point of view. The symbolic point of view is located in the socio-symbolic ambient, at other's points-of-reaction, not in one's own here-and-now ambient. The sampling and maintenance of this 'other' point-of-reaction that existed within a socially maintained 'virtual' environment of symbolic information was the origin of the symbolic-self.

The symbolic-self is an emergent property of the socio-symbolic network. Only within a network of social symbolic interaction is it adaptive to maintain a symbolic reference to your 'self.' Because individuals can now visit your point of view it is now important to maintain distinctions about whose point of view the system is referring to - as Pinker's and Bloom's seminal article implied, language is good at telling us, "who did what to whom, when, where and why…."

The symbolic-self is a symbolic point of view. It is the virtual location from which the symbols you are using are interacting. It is the symbolic place you inhabit when you travel your concept-map by symbolic transformation. Symbols represent points of symbolic interaction just like your body represents a point of physical interaction. Just like your embodied-self identifies physical saliencies and transforms/travels to sync them from its physical here-and-now, your symbolic-self identifies symbolic saliencies and transforms/travels to sync them from its symbolic point-of-transformation. Thought did for us what echolocation did for bats and whales, it was a new sense that evolved to help us navigate a new environment - in our case, a socio-symbolic environment.

Synchrony is order over time and the more one can predict the longer the synchrony lasts. Prediction is the power to extend order over time. It is survival. It is persistence. Natural order is the extension of synchrony. It is propagating symmetry. The symbolic-self is just part of another rule-set at another level of interaction that extends synchrony, produces adaptive futures, creates meaning and propagates symmetry.

The socio-symbolic network was maintained because it extended order over time for groups of proto-humans. The symbolic-self was maintained because it extended order over time for individuals. Each is a different level of how we propagate human symmetry.

Children are born into the socio-symbolic network. They do not need to create symbolic meaning they only have to learn to constrain behavior enough to maintain it. Part of learning to maintain a personal symbolic network (i.e. a concept-map) is learning to maintain a symbolic-self, a point of view from which symbolic saliencies can be made to transform symmetrically within a symbolic environment - within a concept-map.

As children we all start to maintain symbolic symmetry as a declarative memory system. This system maintains the concept-map that we first inherit from our socio-symbolic ambient and that we continually potentiate throughout our lives. At first we learn to travel our culture's socio-symbolic network and we make decisions according to how that network comes to us pre-potentiated. As adults we start to forge our own connections, learn our own trade, make our own decisions. The ways we change our personal connective potentiation during our lifetimes (e.g. what we learn, what we say, what we do) feedback into our social network. We influence and change what others think and do just as they influence and change what we think and do.

In this model the individual Locative Mind is the personal concept-map - the way people potentiate symbolic, rational, social interaction.[247] The group mind is the socio-symbolic network - the way everyone in a group interacts together to potentiate their cultural socio-symbolic network. The two levels of interaction feed back into each other. Children inherit the socio-symbolic network and as adults they potentiate their own concept-maps that then feedback and potentiate the socio-symbolic network, which in turn propagates new children. It is an inter-level virtuous cycle like nucleosynthesis. The human socio-symbolic inter-level feedback cycle is another kind of recursivity - another kind of extension of order. It is reproduction - symbolic reproduction.

DNA is an older medium of reproduction and it is potentiated in much the same way as the socio-symbolic network is. We could imagine the socio-symbolic network as 'cultural DNA.' Where DNA is the bio-chemical code of what-has-worked-before to maintain cellular 'culture,' the socio-symbolic network is the behavioral code of what-has-worked-before to maintain human culture. Imagine the individuals as codons. No one codon is essential to the function of the system as a whole. However, together the differentiated pattern of how these codons are combined to express different levels of functionality are what constitute the organism. In a similar way individuals

247 For more on the connection between rational and symbolic behavior, see Appendix 3.

are born into a system that differentiates how they combine with others to express the different social functions that constitute their culture. In both cases mutation can happen to either present more variation for selection or it can kill the system. If too many people in a culture do not functionally support its operations, that culture gets 'cancer' and fails. On the other hand, a single Charlemagne or Einstein in the right environment can 'mutate' the pattern of the code and initiate emergent cooperative organizations. In both cases differentiation at the individual level feeds back into the overall organization itself just as the overall organization constrains the expression of the individual. DNA and symbolic language are not the same, but they both express the universal hallmarks of symmetry and selection for persistence of emergent social unities.

The human mind is two tiered. The primary tier is at the social level. Mind started as a social symmetry and remains a social symmetry. The secondary tier of the human mind is personal. The personal mind is derived from, maintained and constrained by the social mind. Traditionally we have thought of this problem the other way around.

If one wanted to study the human mind today, one would study the brain. The scientists who are looking for the answers to the question of how we make meaning are working in medical laboratories dissecting rat brains, culturing neurons or taking images of people's brains as they perform psychology tests in giant magnets. Obviously the brain is a crucial substrate for language and mind. However, if we are to uncover the mysteries of mind we will need to look beyond the brain. Most of what we think of as mind emerged, and is maintained, beyond the brain.

# Conclusion

*"The best way to show how something is possible is to show how it actually exists."*

*John Searle (1984, pg. 24)*

**FIGURE 25: GLOBAL INTERNET CONNECTIVITY**

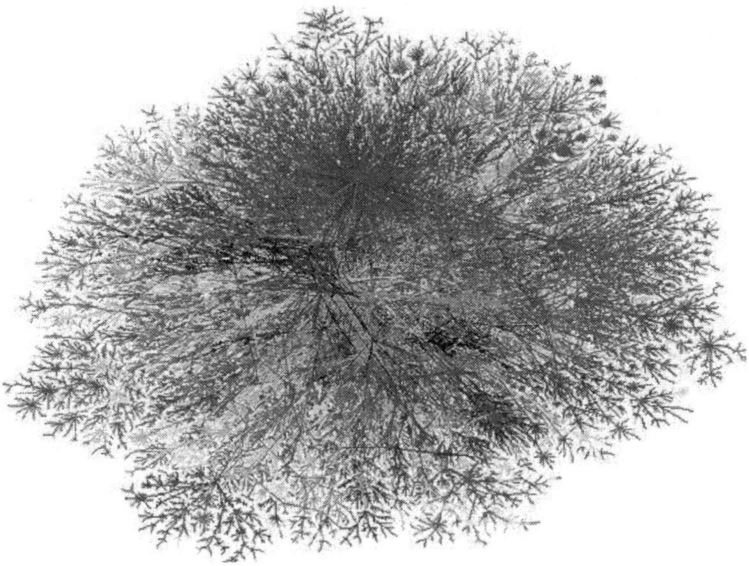

Source: Jurvetson (2004)

For thirteen billion years every case of truly novel emergence has been social. If we consider the human mind to be truly novel we should at least be willing to investigate the possibility that its emergence is a social phenomenon. This investigation proposes that the symbolic language that defines the human experience is an emergent rule-set for the persistence of an emergent product structure, namely socio-symbolic culture. To understand our personal experience (i.e. our own minds) we must first understand the history of social emergence and then understand how we locate and orient ourselves within the specific incarnation of cooperative unity we are part of. Part of this understanding involves not thinking of ourselves as different from all the other levels of persistent interaction from which we were selected.

What makes you and I special is just another level of constrained release of energy, another language that produces and extends our adaptive futures. By reproducing ourselves into this universe, we are expanding the adjacent possible - just as the first cells, the first nucleated cells and the first animals did. The human mind is special and it is unique, but it is also just one of many major transitions to have emerged over the last thirteen billion years.

The language (i.e. rule-set) that reproduces the human mind is socio-symbolic. The language that reproduces the multicellular animal is meiotic. The language that reproduces protists is mitotic and that reproduces simple cells is binary fission. The language that reproduces the molecule is chemical and the language that reproduces the atom is quantum. The universe has a long history of selectively maintaining novel communicative interaction that reproduces novel levels of order. The human mind is part of this long history of selecting for new languages that reproduce emergent levels of cooperative systems.

Ultimately, the transition from animal to humanity shows the hallmarks of a major transition. Human societies are much larger than even the largest of metazoan societies. They have increased their range of action out to Mars and forward to projects that will end in coming centuries. This means that in general, the cycle times of human social projects are now much longer than the daily, hourly cycles of the birds and the bees.[248] Each of these hallmarks of a major evolutionary transition was enabled by human societies' new reproductive force, symbolic language. In fact, the radiation of the

[248] We may process information and energy much faster than our ancestors did, but our projects are now spread out over such enormous socio-economic production groups that the labour, information and energy it takes to make one computer can only be calculated over generations. Most animals start over, basically from scratch, with each generation.

human socio-symbolic system is so extensive that it is now on the verge of connecting the whole world in a real time social media network. What makes us special is what made the other major transitions special, the canalization of an extended platform for cooperation.

"The best way to show how something is possible is to show how it actually exists" (Searle, 1984, p. 24). Following John Searle's suggestion, this investigation has sought to outline how we fit into this universe by tracing the ontology of the universe. It has sought to explain how every class of 'thing' we know of can be different, but also the same. Every 'thing' has a distinct structure, but each thing is also part of the same pattern of development, the same process. The best way to show how this is possible is categorically not by reducing the universe to 'things' that 'act.' The two are really one that our language scissors have split. Things are actions. They are transformation with preservation of structure. Things are relationships. They are processes. To fully appreciate this realization you must begin to explain things as they are, as dynamic symmetries, not static things. Transformation is the rule of the universe. Transformation with preservation of structure (i.e. symmetry) is the rule of the things in this universe.

The human mind is a process, not a thing. In fact, all things are processes. The human mind is a set of persistent relationships selected from a background of ubiquitous and varied interaction on many levels of organization. If we wish to understand the process of mind we have to start examining the conditions and interactive backgrounds that select for positive symmetry relationships. We have to identify the conditions that propagate symmetry. This means that environments (i.e. interactive backgrounds) will need to be brought within the scope of scientific investigation. Despite two thousand years of science's greatest efforts to exclude them, noisy backgrounds are the conditions from which novel structures emerge.

Luckily for us, environments of ubiquitous and varied interaction under some general conditions statistically include some regularities. These regularities are the rules that maintain emergent order in those parts of the environment. Emergent systems both change the environment they emerge in and sample that environment for further regularities - further emergent systems. Interactive environments of ordered systems under some general conditions include even more regularities ... and so on and so on. This is lucky for us because we are out on the very end of the miniscule distribution of complex systems in this universe. We are highly improbable products of those rare, but continually extending, environments that propagate regularities. In fact, each of us individually *is* a particularly energetically dense 'environment that propagates regularities.'

We are composed entirely of systems that came before us. Each of these systems is also composed entirely of systems that came before them. Each emergent system, including us, was selected for social persistence from a background of ubiquitous and varied individual interaction. Systems are composed of preexisting systems cooperating to a degree of persistence that includes at least re-creation, if not reproduction.

For biological systems reproduction is the language of emergent systems. Binary fission is the language of bacterial persistence. Mitosis is the language of protist persistence. Meiosis is the language of metazoan persistence. Socio-symbolic enculturation is the language of human persistence. In each case a new way to store and access what-has-worked-before has been added to the emergent organism's web of maintenance and reproduction. These new reproductive languages are preserved (however and whenever they occur) because they extend the adaptive futures of their organisms - they expand its phase space. They present new reactive faces and propagate emergent symmetries. They build new environments in which new relationships can flourish. Thus, these areas of propagating symmetry, "persistently increase the diversity of what can happen next" (Kauffman, 2000). It's not that the old organisms do not work, au contraire, the world is still full of them. It is that the new organisms can find and exploit new niches that often require the maintenance of extended sensitivity to ever more distant and cryptic regularities - regularities that aren't already spoken for in other relationships.[249] If they are to persist, these emergent communal unities often must extend their syncing mechanisms, their sense halos, the

[249] The comment refers to the point that Maynard Smith and Szathmáry (1995) make about emergent biological systems. They say that it may be very difficult for new systems to emerge when all of the resources are already claimed by extant organisms competitively maintaining certain relationships. It may be that because of this competition, emergent organisms are often forced to keep searching for improbable solutions until they can maintain highly novel, highly extended relationships that previous organisms couldn't maintain

ways they build the environments they can persist in and, ultimately, their behavioral phase spaces.

Once a language emerges to reproduce a social unity, the emergent system typically radiates to the extent that its environment can support. It samples its environment and fills its phase space. In doing so it must find a balance with the extant systems in that environment, be they biotic or abiotic. These can be trying times. It is very likely that millions of emergent organisms have tried and failed to find that balance. Like Hollywood's emerging child stars, too much success can be as disastrous for newly emergent organisms as no success at all.

Margulis & Sagan (1986, p. 109) speak of an "Oxygen Holocaust" that occurred two billion years ago when oxygen exhaling blue-green algae changed the atmosphere from anaerobic (i.e. oxygen less) to aerobic (i.e. oxygen full). This overpopulation event unbalanced the biosphere and caused massive death and destruction to the organisms that existed then. This was fantastic for us because it also created an oxygen rich atmosphere in which aerobic metabolic cycles capable of producing up to fifteen times more energy than fermentation became part of our adjacent possible. From the ashes of this holocaust rose the energetic protist from which we are all descended.

We should consider this story a cautionary one. If we fail to find a balance with the environment that supports us, we too will experience a great dying. From the ashes of this dying may rise a new human. However, the other possibility is all too real. We may not survive. "As biologist A. Meredith suggests, the pattern of sudden appearance, expansion, and then disappearance in the fossil record has much historical precedent and is ominous" (Margulis & Sagan, 1986, p. 228).

As for the mystery of the mind, it will not remain. It is already going the way of the mystery of the humors or the mystery of the heart. Like these other historical mysteries, the idea of mind will be divided up into literary, psychological, sociological and physiological explanations, each to give birth to their own phylogenies of differentiated and specialized terms. There

(Beeby & Kaeser, 1990). In these cases natural selection differentially maintains those emergent organisms that extend environmental connectivity.

will be brain experts (neuroscientists and neurosurgeons) and there will be mind experts (gurus and psychologists). On the street we will continue speak of the mind like we still speak of the heart. In some contexts we will refer to a physical thing that we own, in others we will refer to an ethereal source of our motivations.

Mind can be defined as (i.e. differentiated from all other experience as) the literal 'sense' of experiencing reality through others. Mind was accomplished by differential selection for using others to extend sensory awareness. Mind came from a shared symbolic network that trained personal locative senses and eventually personal Locative Minds. The human mind can extend our symbolic reactive face (i.e. our symbolic-self) beyond our bodies, into the virtual symbolic environment that is socially maintained. The way our mind extends our ability to interact successfully with distant (both in time and space) saliencies is reminiscent of the evolution of other new senses, like echolocation.

The mind is like a sense because it extends our ability to read distant and cryptic saliencies like the other senses do, but unlike other senses, it achieves this extension by locating our reactive face in a virtual world of social representations. This virtual world maps to the physical world with one important difference, interaction is not constrained by physical time and space rules. As humans sampled interaction in this socio-symbolic world they could expand the rate and efficiency of sampling not only because adaptive relationships could be sampled outside of here-and-now time and space constraints, but because the social system itself could maintain and reproduce those relationships that proved adaptive.

Symbolic representation was originally selected for coordinating adaptive social reactivity, but as it became part of the human adjacent possible - as it became increasingly tightly coupled and internalized - it was selected for how well it differentiated and specialized how the symbolic network interfaced with individuals. Part of this differentiation and specialization was the representation of a remote 'self' in the symbolic environment. This symbolic-self is different from the embodied-self because it motivates reactivity by sensing through virtual representations in the symbolic network. It is the self to which others refer. It is the self that others maintain within your social context. It is the point of view, or 'location,' from which all symbolic saliencies transform in your personal concept-map. The symbolic-self can reduce, represent and embody other people, as well as things, that act in our socio-symbolic world much like our embodied-self acts in its physical world.

What the Locative Mind does is begin to displace adaptive reactivity from individual functionality to group functionality. It connects individuals in a way that allows them to access a shared history and present a shared adaptive face to their environment. It begins to move the survival of the individual organization to the survival of the group organization. It distributes decision making over networks of individuals to offload the work of designing global competence onto their neighbors. We use others to gift us with "competence designed elsewhere." Whether those others are in our past or right next to us; they build our cars, protect our borders and train our worldview. Our connectivity via language and its shared socio-symbolic network is a massive "transmission of design." The Locative Mind allows each of us to interface with this network through every concept we hear, use and think of. This connectivity functions to tune the global network by having each of us activate, rehearse and eventually potentiate the network itself by varying its pre-tuned associative pathways to help guide our individual actions. The way we interact symbolically with our neighbors is the background interaction from which coordinated social functionality is selected. Our Locative Minds locate us within a socially distributed system of meaning that we transform by using it to constrain personal actions that statistically produce (i.e. predict) ever more adaptive futures for ourselves - and our groups.

In a way we are re-playing out a thirteen billion year old universal drama. We are canalizing component interaction to maintain a group organization. As a group organization we are not just the sum of the individuals connected. Together we interact with our environment as no other system in thirteen billion years has. Humanity presents a new reactive face to this universe. Thirteen billion years ago, atoms could've said the same thing. But we are downstream. We are high dimensional constrained systems maintained and interacting within a far more high dimensional constraining environment. We interact uniquely. We discover distant regularities in time or space and we use them to create technologies and ideas that sample ever more distant and novel adaptive futures. Like the first atoms and the first solar systems, we are changing what we touch and we are reproducing these adaptive interactions to touch ever more and more distant systems.

Together we are including more of the universe. We are expanding human experience. As a group we are emerging as a new universal system. This new system is propagating ever more symmetry as our collective humanity meets and transforms the universe around us. Beautifully, we are propagating into the universe.

But we are also capable of negative symmetry. We have a choice. Learn to maintain and reproduce that which maintains and reproduces ourselves - propagate ourselves into the universe as cooperators - or allow ourselves to die off and another level of propagating symmetry to emerge. The universe gives us the same choice it gives everything- cooperate or die.

# Epilogue: The Origin Of Religion

*If man first directed his eyes to the heavens, it was not to satisfy a merely intellectual curiosity. What man really sought in the heavens was his own reflection and the order of his human universe. He felt that his world was bound by innumerable visible and invisible ties to the general order of the universe.*

*Ernst Cassirer (1965, p. 48)*

This investigation is a modern origin story. It helps us understand not just that, but also how, we are part of the warp and woof of this universe. In Cassirer's words it seeks, "in the heavens" how our, "world was bound by innumerable visible and invisible ties to the general order of the universe." For tens of thousands of years religions have done the same for people all over the world. Perhaps the best way to test this origin story is to see how it explains our own cultural origin stories, our own religions. The sheer amount of material this investigation has covered has precluded the use of extended examples to illustrate its arguments. This choice has sacrificed some measure of beauty for truth, but perhaps this sacrifice can be ameliorated here at the end. Let us see how religion fits into the Symmetry and Locative Mind framework. Presumably we would like to find some proof in the pudding.

Religion is a pervasive feature of human culture. Anthropologist Ian Tattersall (1998) notes that, "religion in some sense is one of the earliest special proclivities that we are able to detect in the archeological record of modern humans" (p. 201). Like all complex behavioral traits religion should have some kind of adaptive function. On some level it should contribute to the maintenance and reproduction of our selves, our groups or our species. Of course it is also possible that religion is simply a side effect of relaxed selection, viral memes or an accident, an original broken symmetry of human social behavior that has been maintained, infallibly, from our East African ancestors. All of these options seem unlikely simply because similar religions have been found in every culture we have ever known, regardless of its state of duress or its other unique features. If the aforementioned options were causal we should expect to see a more random spread of features across religions. We should expect a more random diversification of different types of religions over the world. Instead, all cultures have religion and all religions have familiar features. Religion seems to have been consistently selected for certain characteristics. For some reason this religion structure is the first, most ubiquitous and probably most distinctive characteristic of human societies. Why should this be so?

Socio-symbolic communication amongst our ancestors probably began over two million years ago. The flourishing of modern human culture that is known as the great leap forward is much more recent, around fifty thousand years ago. At the great leap forward we begin to find widespread evidence of modern symbolic culture - human culture. Evidence for human culture is embodied in the simultaneous abrupt expansion of cave art accompanied by expanding tool technologies in upper Paleolithic Europe.

The cave art indicates that these people were capable of abstract representation. The tools indicate that these people maintained conceptual categories for the kind of tool they wanted to make. These early humans were no longer just making the same cutting tool their father showed them to make. They regularly experimented with tool design to achieve 'imagined' functions (Leakey, 1994). These new tool makers were thinking something like, "a barb will make this kind of point better at holding the fish after it has been struck because I can imagine how the barb will hold it." Abstract ideas of functionality were now leading to radical tool innovations instead of just slightly better adzes. It is in this flowering of modern human culture that Tattersall finds the first evidence of religion.

Something happened around the time of the great leap forward that enabled both abstract conception of tool functionality and abstract conception of a 'self' within a mythical, storytelling, social context. Most likely many things happened, but the Locative Theory of Mind suggests a specific change may have been key to that conjugation. It suggests that the great leap forward was enabled by the first groups of humans to successfully internalize the socio-symbolic network in their 'selves.'

The Linguist Derek Bickerton (2003) suggests that the great leap forward was caused by the emergence of functional syntax. This is consistent with socio-symbolic internalization. Syntax enables symbolic recursivity. It is a rule-set that allows the symbolic system to propagate itself. It allows us to say one thing after another ad infinitum. Part of that propagation could have been coupled and canalized in the form of Bickerton's emergence of syntactic recursivity. The canalization of syntax may have been the Rubicon that transformed a loosely constrained cooperative group to a tightly constrained reproductive group. This transformation would be analogous to loosely constrained groups of cooperative autocatalytic cycles canalizing into proto-RNA/DNA or loosely constrained groups of cooperative cells canalizing into, first, mitochondrial eukaryotes and, second, multicellular organisms. A background of constant yet varied transformations is necessary for the selection of novel persistent systems. Recursivity is essential to symbolic re-creation. It is the 'preservation of structure' in language 'transformation.' Recursivity is reproduction and reproduction is necessary for the persistence of any emergent system.

Syntactic recursivity may have been key in changing a kind of quirkily designed and sloppy socially maintained system of cueing into an internal, self-referential, autopoietic system of reproduction. One can imagine a similar transformation from autocatalytic soups to proto-cells nearly four billion years earlier. Free living autocatalytic molecules would have needed to sample the interactive background enough to find a bounded cooperative arrangement that internalized them within a regular, reproducible system - a proto-cell. Like a recursive molecular code (DNA or RNA), a recursive syntactic code would have enabled the ubiquitous and varied symbolic interactions from which new systems like symbolic selves could be sampled, maintained and reproduced. Perhaps the people of the great leap forward were the first individuals who could maintain symbolic selves capable of travelling within their own personally maintained (and recursively extendable) concept-maps. They were the first humans to internalize the socio-symbolic network.

They were probably not the first talkers, but they may have been the first true 'thinkers.'

Successfully internalizing the socio-symbolic network reverses the previous process of collective unification by now deemphasizing group symbolic functionality and emphasizing individual symbolic functionality.[250] Individuals that can learn to maintain the socio-symbolic network internally can begin to weave futures in which their symbolic selves interact in not-here and not-now contexts. Perhaps more importantly, by displacing their symbolic selves from their embodied selves (e.g. talking or thinking), they have to imagine other things as if having motive powers. Nouns now need verbs.

If I can hold a fish with a hooked finger, then a barb should be able to hold a fish with a 'hooked finger.' If I can run then Fred can run. If Fred can run then the river can run. If the river can run then the river can run away with my son. The power to leave the embodied-self and travel symbolically means that any 'thing' that is symbolically represented can be seen as a locus of transformation much like you are. 'Rivers' and 'tears' can both 'run.' Lakoff's image schemas motivate symbolic transformation in the same way that your body motivates physical transformation. Language animates things.

When individuals began to internally maintain the group symbolic system, they animated the things their symbols represented. Historically, animism is commonly recognized as the most characteristic feature of ancient religions. Animism is the attribution of humanlike powers to nonhuman entities. It is the druidic worship of special groves of trees and the Shinto worship of special water features as if they were cognizant - as if they had hopes, desires and fears. Animism is the personification of the devil in the serpent and of Thor in the lightning. It can be found in the animal headed gods of ancient Egypt and the various human-animal mythological beings common to all ancient cultures. Hunter-gatherer societies uniformly

[250] The proposal here is that socio-symbolic language probably emerged around two million years ago. At that time it was an adaptation that shifted individual decision making to group decision making. Much later, around fifty thousand years ago, this socio-symbolic network became successfully internalized in individuals. This later adaptation had the complex side effect of reversing this process by reemphasizing individual decision making.

have some sort of animistic religion. Likewise, modern religions still maintain vestiges from their animistic origins - like burning bushes that talk.

One of the most striking images from the great leap forward is "The Sorcerer." The image reproduced below was inspired by the same image in Leakey (1994).

**FIGURE 26: THE SORCERER**

Source: Rush 2006

The sorcerer is half man and half animal. This kind of anthropomorphic image was common in upper Paleolithic cave art (Leakey, 1994, p. 116). Obviously, the sorcerer represents some sort of animistic attribution of human qualities to an animal (and/or vice versa). The appearance of animistic art in the great leap forward indicates that individuals were finally able to motivate those things around them symbolically. Tool making became an abstract activity because individual toolmakers could imagine the tool *doing* things. They could shape tools that they imagined would hook, dip, grab, or hold. What happened was that the socio-symbolic process of differentiating and specializing group adaptive reactivity resulted, after almost two million years of selection, in the successful internalization (or canalization, or reproduction) of symbolic concept-maps in individuals.

What does this mean for religion? Everything. Language motivates symbolic representations of 'things' to interact as we do. It anthropomorphizes our ambients. What previously was just a halo of proximate saliencies that triggered the four 'F' responses now became a timeless and spaceless halo of all the saliencies your group ever had cause to imagine and maintain. Each of these saliencies could now symbolically transform the world on its own behalf, just as you do. The trees could complain, the buffalo could be offended, and the Earth could enjoy the blood of virgins. How would you react to a self-motivated ambient? One thing our ancestors almost certainly did was what they did best, talk about it.

Storytelling is probably one of the oldest human social mnemonics.[251] Storytelling is socio-symbolic reproduction, but it also trains the symbolic capacity of individual brains. It is recursivity at the social level, but it is also an elaboration of the pingback. The call-response pingback sought to make sense of an extended environment through others. As this process developed, lexicons grew and individuals could maintain longer focus on symbolic interaction and storytelling evolved. What did we tell stories about? We told stories about what was around us, about the things upon which our lives depended. This process started by symbolizing the saliencies within the group sense halo, what was important around us. However, as these representations interacted, differentiated and specialized they would have found

[251] In fact, storytelling might be considered one of humanity's first external mnemonic technologies. It could be the great grandfather of hieroglyphs, phonetic alphabets, printing presses and social media. In a sense these emergent steps in group information storage mirror the steps our neural nets must have taken to support an emergent declarative memory system.

no space or time constraints. Symbolic interaction would have led from the transformation of a limited group sense halo to an ever growing, ever inclusive group symbolic halo where not-here and not-now saliencies could be maintained indefinitely, externally and socially - simply by storytelling and rehearsal.

The natural features represented in this growing halo of symbolic awareness, the 'things' around us, would have had to be anthropomorphized to be symbolized. Symbolically, a 'volcano' must do something, and that something can only be imagined relative to how we do things. Thus it must be 'upset' if it erupts (e.g. 'eruptar' is the Spanish verb for 'burp'). This investigation proposes that the natural feedback development of storytelling combined with the functional necessity of anthropomorphizing 'things' to symbolize them is at the heart of religion. Nietzsche (2007) said it over a century ago, "'Reason' in language: oh what a deceptive old witch it has been! I fear we shall never be rid of God so long as we still believe in grammar" (p. 20). Language itself motivates religion.

Biologist and Anthropologist David Sloan Wilson (2002) argues convincingly for the adaptive functions of religion. "Around the world and across history, religions have functioned as mighty engines of collective action for the production of benefits that all people want" (p. 187). His argument is that successful religions create groups that cooperate to a degree that ultimately helps them maintain and reproduce as groups. In general, cooperation is characterized by higher levels of synchronous connectivity between individuals - by more inclusion of more components within more cycles of maintenance and reproduction. Historically, religious cooperation has meant exile, conquer or kill the nonbelievers (i.e. non-cooperators) and give the believers (i.e. cooperators) a functional code to live by (for 'code' read the Talmud, the Bible, the Koran, the Buddhist scriptures - all of which can be seen as cultural rule-sets or behavioral boundary conditions). His examples and contexts are mostly drawn from recent western religions, but when you frame shift his argument to the origin of religion you get a glimpse of an interesting process.

Undoubtedly, storytelling trains brains symbolically. It expands, reinforces and maintains declarative memory systems both in terms of personal concept-maps and socio-symbolic networks. This training must have been critical, especially in the period leading up to The great leap forward. Those groups that could train their children to engage in the imaginative process of symbolic representation better were the groups whose powers of symbolic representation grew the quickest. This ability to reproduce adults who could create, manipulate and maintain long term adaptive futures were those groups whose plans for the future bore the most fruit. Those groups multiplied. Any way to canalize storytelling as an iterative training method for their children would have been differentially preserved - classic convergence. The result of this adaptive pressure to emphasize and canalize storytelling practices (e.g. storytelling in the form of ritual, rehearsal, teaching, indoctrination, etc.), combined with symbolic anthropomorphization, would have been a cultural phenomenon very close to what we think of as religion today. Many of us would find it humorous to propose that religious indoctrination exercises the imagination. In modern times scientific and philosophical indoctrination have taken over the vanguard of training the culturally enlightened, but this was not always so. In the beginning, the best and brightest were taught to imagine themselves as sorcerers on cave walls.

An essential corollary to religion in any form is the idea of a soul. Outside of particular religious trappings, this idea that each of us has some kind of transcendent essence that is distinct from the body and, in most religions, continues to exist after death, is part of all major religions. It is the 'Atman' in Buddhism and Hinduism and the 'hun' and 'po' (i.e. yang and yin) of Taoism. On some level, some concept of life after death is one of the very few plausible justifications for the emergence of elaborate ritual burials that all of our immediate ancestors participated in.[252] The question is, why would such

[252] Although the evidence is controversial, Neanderthals may also have ritually buried their dead, implying the possibility of some form of symbolically motivated 'soul' awareness. Perhaps even some form of nascent symbolic capacity. This is definitely consistent with current estimates of Neanderthal, Sapiens ancestry. If the Homo lineage has been developing socio-symbolic networks for two million years and Neanderthals and Sapiens shared ancestors a mere eight hundred thousand years ago then both were limitedly symbolic at the time of divergence. Neanderthals may simply represent the socio-symbolic group capacity without ever having evolved the individual concept-map. They symbolically cooperated, but had limited ability to think on their own. For a good review of Neanderthal burial evidence and possible symbolic capacity see Wynn & Coolidge (2012).

diverse religions with such diverse origins all infallibly include some kind of immortal soul within the logic of their mythological frameworks?

All healthy humans have the feeling that there is something about them that is above and apart from their corporeal experience. This feeling is what religions invariably must justify if they are to be culturally maintained. But what is this feeling? Why do we have it? Where does it come from? As a corollary of religion, the 'soul feeling' comes from language itself.

Where are you when you are reading *All the Pretty Horses*? Why are you crying when John Grady Cole is spurned, knifed and then hunted down? How do pastors transport the faithful to rapturous heights? When we talk or read or watch plays we allow words to transport us to experiences from other places, other times, other points-of-view. We lose ourselves in the virtual interactions that language creates. We travel away from the here-and-now by embodying our motivational point of view in the symbolic references that need our motive force to interact. This symbolic-self dissociates from the embodied-self and travels in a parallel environment of socio-symbolic associations. It travels from concept to concept and is thus connected to all that is conceived. If the concept of 'universe' exists then the symbolic-self can travel the associative pathways that are connected to that concept. 'Distant planets,' 'aliens' and 'black holes' can be imagined.

The symbolic-self can represent and motivate these concepts to transform in logical ways, in ways that our embodied selves have evolved to intuitively understand. The soul feeling that we all have is in this dissociation. It is in the dissociation we all experience between our embodied-self making sense out of our here-and-now environment and our symbolic-self making sense out of our socio-symbolic environment - our imaginative experiences. We all have the ineffable feeling that we can exist in other times and other spaces. We have that feeling because we regularly do exist out there, symbolically.

Once invented, getting good at language is a natural end. It opens up a massive phase space of possible adaptive futures. The means by which we have achieved this end, like all novel universal transformations, is riddled with spandrels, sloppy fits and broken symmetries. The canalization of a cultural adaptation that trained children's imaginations seems inevitable. Novel systems must reproduce to persist. However, the particular features of religions are often the results of quirky historical accidents. The anthropomorphic motivation of symbolic symmetry by our own corporeal symmetry may be somewhat less than inevitable, but this remains to be seen.[253] Considering both of these factors together, it does seem that something like religion was a convergent inevitability when symbolic communication became the medium our groups used to present unified reactive faces to their ambients.

In the beginning, religion in the form of iterated storytelling gave its groups a reproductive learning platform to indoctrinate powers of imagination in their individuals. Over time, these religions would culturally compete, leaving disproportionate numbers of those religions whose socio-symbolic code of cooperation could connectively synchronize their constituents and coordinate unification better. Two side effects of this natural process of development would be;

1) explanations for why our ambients appear animated (i.e. the supernatural) and
2) why we personally feel connected to all times and all spaces (i.e. the soul).

Both of these phenomena are directly related to how language functions. This functionality is biological. It has a four billion year history of comparable biological transformations and a thirteen billion year history of comparable universal transformations.

Like all emergent symmetries, language creates new meaning. It creates a new form of adaptive relationship to its environment - a new way to persist. Language builds new phase space - new adaptive futures for us. This new meaning still embodies selection for persistence, but we experience it through very limited bodies with very limited histories and interactive capacities.

[253] Emergent levels of constraint can be more or less dependent on their substrate interactions. In this case it is not necessary that the super-ordinate system of language necessarily takes the symmetry rules of the subordinate behavioral system. However, language was differentially preserved because it represented things and actions in others. It represented their behavior. In this sense language's super-ordinate symbolic logic is mapped from the subordinate behavioral logic and maybe its particular structure is inevitable. This is a difficult problem and it requires further investigation.

Religion is just one part of the emergent 'meaning' of language. Religion allows us to see this meaning darkly, through supernatural glasses, but it also permeates all of our experience. To be human is to anthropomorphize, to have out of body experiences, to have religious experiences. To be human is to be connected to all concepts, to feel that you can transform beyond the here-and-now. To be human is to have a soul and to be connected to everything out there - because symbolically, we are.

The End

*"I look at it as absolutely certain that very much in the Origin will be proved rubbish; but I expect and hope that the framework will stand."*

*Charles Darwin (as cited in Gould, 2002, p. 2)*

# Appendix 1: Complexity, Information, Entropy, Energy and Order in a Scale Free Universe

## A Scale Free Universe

Coming to terms with scale limits is like coming to terms with prejudice. Both beliefs limit our understanding of the universe. Enlightened people no longer believe in prejudice, but the same cannot be said for scale limits. Judging from the literature, many scientists and science writers believe that there is nothing smaller than elementary particles and nothing bigger than the Big Bang.

In experiment after experiment people of different races have turned out to be equally capable and intelligent. Presumably the statistically equal distribution of intelligence not only applies to people over geographical space, but also over historical time. Presumably, people in the seventh century Persia were no stupider than people in Philadelphia today. They may have known less, but they were no less capable of knowing.

What does this tell us about scale in our universe? Let us take a look at the bottom end. Right now we believe that the bottom end of the scale is Planck's constant and the sixty or so elementary particles. Everything that we know of is made of sixty or so tiny energy structures (Gell-Mann, 1995). We can't *see* any of these structures, but we can measure their effects with high tech tools. By categorizing these effects we have been able to give these structures distinct names like 'electrons' and 'quarks.' In 2010 when this investigation was written, the bottom end of the scale of causality was these sixty or so quanta, but this was not always the case.

At the beginning of the nineteenth century John Dalton re-proposed a classic type of elementary particle.[254] He proposed that each chemical

[254] Democritus originally proposed the concept of atoms.

element was composed of tiny distinct atoms. This was revised at the beginning of the twentieth century when the hydrogen nuclei was thought to be an elementary particle. However, in 1911 Ernest Rutherford discovered that most nuclei contained protons and neutrons. In 1964 Murray Gell-Man and George Zweig discovered that these protons and neutrons were composed of quarks. In just over one hundred and fifty years the bottom end of our scale of causality went from elements, to atoms, to nuclei, to protons, to quarks. The logical question this incites is; where does this all end? What indications do we have that our current quanta are truly at the bottom of causality?

Science is founded on replication. If an experiment can be replicated it can be verified. If it is true that scientists in the twentieth century consistently found smaller scale levels, what about the nineteenth century? What about the eighteenth, seventeenth and sixteenth centuries? Can the twentieth century experiment be replicated? Do the equally clever scientists of each century equally believe that they know what the materia prima of the universe is? Have they all equally been proven wrong by the next generations of scientists? What would we find upon careful inspection?

We would find that the scientists of each century were all more or less equally intelligent (we wouldn't want our experiment to be inherently prejudiced?). We would find that they were all equally convinced that their system explained causality because they are famous for claiming so. We would find that they were all proven more right than their predecessors and more wrong than their antecessors. And finally we would find that they were all equally restricted by the technology of their age because proof depends on exactness of measurement and this has continually improved.

If this investigation were of a different character we could take the time to conduct this experiment. From classical elementalism to medieval alchemy to enlightenment chemistry to modern cosmology, each age has been populated by intelligent men that believed they knew the materia prima at the bottom of causality. All of these men built theories that were not only logically consistent, but were also constrained by their technology. All of these theories were eventually improved when technology improved, and improvement in each case has meant defining smaller, more precise parts that compose the physical world around us.

This historical fact implies that unless we believe ourselves smarter than others, or we have invented perfect measurement technology, we should be very careful in our belief that we know the bottom end of causality. It means that future generations armed with better technology will continue to find smaller, more precise causal patterns. Quarks will be found to contain galaxies of smaller pattern, perhaps the miniscule dimensions that string theories posit. Furthermore, there is no evidence that we know the top end of causality either. Every time we build a stronger telescope we see bigger, more distant patterns. Recently we discovered that the previously known universe is really just a tiny part of a vast universe of unknown dark matter and dark energy. Historically, there seems to be every reason to believe that the Big Bang universe we see now will soon be recognized as just another location within a far vaster 'multiverse.'[255]

The beliefs that everything is within the Big Bang and that nothing is smaller than quanta are probably wrong. We may or may not live in a scale free universe, but there is no reason to believe that the current measurable boundaries are the true boundaries. The history of science has consistently reminded us that there is no special place in space-time from which to measure scale. The discoverer of one of our current scale limits, Max Planck himself, "was warned by a professor of physics that his chosen subject (physics) was more or less finished and that nothing new could be expected to be discovered" (Kragh, 2002, p. 3). That was in the nineteenth century.

Humans have a natural inclination to limit their thought to within the paradigms that map their own location, size and time scales. Revolutionaries like Galileo, Newton, Planck, Einstein and Schrödinger proved these scale-limits wrong at every turn. Enlightened people should not believe in prejudice, whether it be against colour of skin, country of origin, or the fine graining of space-time.

For this reason, this investigation presumes a scale free universe. It takes Nobel Laureate Robert Laughlin's position that the reason quanta are all 'wave like' is because they too are collective phenomena. They too are composed of many parts interacting in concert, like waves. Like everything else in

255 This reference to multiverse is not specifically referring to parallel universe theories, only to an expanded, yet unknown, conception of the universe.

this universe, quanta are *communal.* One of the longest running experiments known to science could be framed as, 'Do we know the materia prima of the universe?' Every century that scientists repeat this experiment the result is the same, a resounding 'No.' If a millennia of historical experiments are any indication, there is every expectation that quarks and electrons will soon be found to be the results of interactions at lower levels of causality. The quark itself will become understood as a "void for the greatest part, only interwoven by centers of energy" just as its antecedents were (Bertalanffy, 1968, pp. 226 - 227).

As our telescopes push the limits of our universe ever outward, our microscopes and particle accelerators push the limits ever inward. At every scale level we find pattern. There is no indication that the current level is special or definitive.

## Complexity and Information

So how does transformation occur in a scale free universe? How do you talk about what caused what and when if everything we currently know of is made of some unknown level of fine grained energy relationships? We cannot. Until we develop better measurement technology we will not know for sure what is down there. What we can say for sure is that the billions of different 'things' that we currently know constitute our universe, are all composed of what looks like the same matter-energy. We all know that a teaspoon is different than a bluebird, but we also know that the two are made of exactly the same matter-energy. What makes the bluebird different from the teaspoon, is not different elementary particles, it is how the very same elementary particles maintain different energy relationships. In fact, different energy relationships are what constitute the different forms of matter.

This investigation pursues how these different relationships were sampled, maintained and reproduced from a completely undifferentiated Big Bang to the vastly differentiated universe of today. However, presuming a scale free universe has important ramifications on how we choose to measure this differentiation.

How can we measure the accrued differentiation of the universe? Can we call this measure complexity, and if so, is the universe getting more or less complex over time? It started with a tiny uniform 'hot soup' of matter-energy and, "from so simple a beginning endless forms most beautiful and most wonderful have been, and are being evolved" (Darwin, 1859, p. 490). Obviously it is getting more complex - isn't it? Well, this would depend on how we define 'complexity.'

Common sense definitions of complexity usually rely on measures of more and different. The more, different elements a system has, the more complex it is. Biologist Steven J. Gould (2002) refers to this measure of complexity when he describes complexity as the, "number and form of components" (p. 1264). By his definition the more different components an organism has, the more complex it is.

On the other hand, physics and computer science definitions of complexity often refer either to information theory or algorithmic complexity. Peter Grünwald and Paul Vitányi (2008) summarize the relationship between the two nicely:

> Both theories aim at providing a means for measuring 'information.' They use the same unit to do this: the bit. In both cases, the amount of information in an object may be interpreted as the length of a description of the object. In the Shannon approach, however, the method of encoding objects is based on the presupposition that the objects to be encoded are outcomes of a known random source—it is only the characteristics of that random source that determine the encoding, not the characteristics of the objects that are its outcomes. In the Kolmogorov complexity approach we consider the individual objects themselves, in isolation so-to-speak, and the encoding of an object is a short computer program (compressed version of the object) that generates it and then halts (p. 2).

Note first that the terms 'information' and 'complexity' appear correlative here. They both measure information even though Kolmogorov calls his measure complexity. Shannon developed his definition first and his solution was to measure information as a relationship between the outcome that does occur and the probability distribution of what could have occurred. He called this statistical relationship 'entropy' because entropy, as previously defined by Boltzmann, was also a measure of a future outcome based on its probability in that environment. What Shannon was essentially able to do was to show how any universal system could be defined as a communication process. This process was the selection of a signal (what does occur) from a limited set of noise (what could occur).

Kolmogorov was unsatisfied with this solution. He was a computer scientist and thus wanted an absolute, computable measurement not a probability. He got it by limiting the definition of the system to what was inside it. For him, any system, or entity, is only as complex as the minimum number of bits needed to describe its components. By eliminating the need to measure a system's relationship to its environment Kolmogorov invented a way to effectively *count* what something *was*. The result was that Kolmogorov complexity is essentially a measure of how random something is. This is because Kolmogorov complexity does not count the regularities, they are compressed, thus his measure is mostly a count of the random irregularities within a system.

Kolmogorov's and Gould's ideas of complexity are both similar in that they say nothing about how a system came to be structured as it is, but once it exists, they can measure its complexity. They both measure a static amount of structure. Shannon does not treat system structure as static, but as a dynamic relationship that unfolds over time. Unfortunately, Shannon's solution just pushes the static assumption a little further down the road by assuming the 'known random source" from which the outcome can be selected. To define Shannon information one needs to have absolute statistical knowledge of the closed system within which the signal is selected, its 'environmental' or 'noise' or 'entropic' potential. Kolmogorov and Gould count static amounts. Shannon counts statistical potential within a static limit.

Now consider how complex complexity can get. The universe contains no closed systems and no system in the universe is static. So how does one use such metrics to count what does actually occur. It turns out that both

Gould and Kolmogorov metrics are great for measuring systems that already exist and can be easily idealized to static structures. For these measurements, one just needs to stop the universe (e.g. make it absolutely static), draw a box around what *you think* the system is, and then count. One can measure dynamic transformations, but only in limited contexts (e.g. by mathematically idealizing them into static bits or statistically static states). In Shannon metrics the box is extended to be arbitrarily drawn around the environment the system is contained within.

Before turning to the limits of static measurements in a dynamic universe, let us briefly consider how to decide what to count. A deep flaw in all of these measurements is that they all require a human to decide what is countable and what is not. Do we need to count gluon states to determine the complexity of a one kilogram rabbit? If we did, then a one kilogram rock could be more complex than our rabbit. It could have more and different quark and gluon relationships. And what about one kilogram of nebulae vapor?

Presumably, when Gould speaks of organismic complexity as a count of more and different parts he is only considering biological parts: organelles, cells, organs, etc. If Gould wants to count detail up to and including organelles, he chooses to ignore the molecular details within each organelle that makes it act uniquely in some situations. Gould is *choosing* to ignore fine grained complexity. Kolmogorov complexity also depends on an arbitrary human decision on how to represent the minimum constitutive part of the system to be computed. What exactly does each bit include, or more interestingly not include, and how could that fine grained and chaotic complexity influence the future.

This brings us back to the inability of *amounts* of static complexity or information to measure a dynamic universe. Both Gould and Kolmogorov complexity metrics are fundamentally limited to measuring structure that has already evolved and are arbitrarily bounded. They contribute little to predicting future structure. Shannon information can be used to make limited predictions because it describes a process of how statistical relationships unfold over time. Numerous attempts have been made to use Shannon information to predict the kind of transformational structures that actually populate our universe, but little consensus has been reached (Grassberger, 1986; Prokopenko, Boschetti, & Ryan, 2009).

Peter Grassberger (1986) does a wonderful job of illustrating the problem of defining information complexity with these three images and the following comment:

**FIGURE 27: GRASSBERGER COMPLEXITY EXAMPLES**

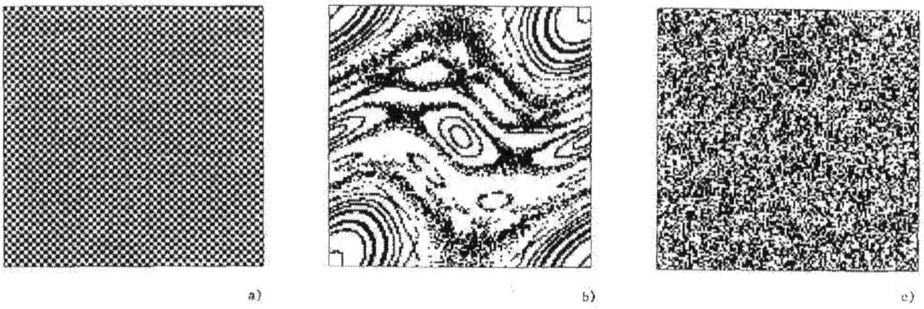


Source: Grassberger (1986, p. 910)

> Compare now the three patterns shown ... [above]. Fig. lc is made by using a random number generator. Kolmogorov complexity and Shannon entropy are biggest for it, and smallest for Fig. la. On the other hand, most people will intuitively call Fig. lb the most complex, since it seems to have more "structure." Thus, complexity in the intuitive sense is not monotonically increasing with entropy or "disorder." Instead, it is small for completely ordered and for completely disordered patterns, and has a maximum in between (Hogg and Huberman 1985). This agrees with the notion that living matter should be more complex than both perfect crystals and random glasses, say.
>
> The solution of this puzzle is the well-known ability of humans to make abstractions, i.e., to distinguish intuitively between "important" and "unimportant" features. For instance, when one is shown pictures of animals, one immediately recognizes the concepts "dog," "cat," etc., although the individual pictures showing dogs might in other respects be very different. So one immediately classifies the

> pictures into sets, with pictures within one set considered as equivalent. Moreover, these sets carry probability measures (since one expects not all kinds of dogs to appear equally often, and to be seen equally likely from all angles). Thus, one actually has ensembles: when calling a random pattern complex or not, one actually means that the ensemble of all "similar" patterns (whatever that means in detail) is complex or not complex. After all, if the pattern in Fig. lc were made with a good random number generator, the chance of producing precisely Fig. lc would be exactly the same as that to produce Figs. la or lb. (namely 2 -n, where N is the total number of pixels). If we call the latter more "complex," it really means that we consider it implicitly to belong to a different ensemble, and it is this ensemble that has different complexity (Grassberger, 1986, p. 909).

Clearly, measuring complexity and information is an unresolved problem. Has the universe gained complexity in the sense of Kolmogorov randomness or in the sense of Gouldian numbers of different parts? And if the universe provides so many natural joints at which to cut it into understandable pieces, how do we choose which joints to use in which situations? Who gets to choose? Perhaps Shannon information allows for a prediction of what might emerge, but it too depends on what we *choose* to look for, and even more importantly, it still limits its predictive power to the *known* limit of the *chosen* system. Clearly, we do not yet know how to measure change in this universe.

This investigation proposes to redefine complexity as a relative, not absolute, measurement. Essentially, complexity is a measure of difference between systems, but systems are scale free so what is environment to one system is component to another. What is a regularity (e.g. signal) in one system can be noise at a smaller scale level. A cell is a regular component within your body which itself is made of many different fluids, boundaries and cells doing many different noisy activities. That same cell is a noisy environment for its own regular organelles which are in turn noisy environments for regular proteins, and on to molecules, atoms, protons and quarks. The only rule to relative complexity is that environments are always more noisy than the systems that are maintained and reproduced within them.[256] Outside is noisier than inside.

This definition is not, in fact, a definition. This kind of proposal is a way to start exploring what a new definition could be. Much of this investigation explores a different way of understanding relative change. It explores how to compare change at different scale levels. However, this exploration is descriptive, not deterministic. This investigation argues for a new way to understand universal evolution by *showing* how it evolved, by showing how atoms, prokaryotes and minds were selected for persistence, not by a deterministic algorithm or formula that *proves* it.

As mentioned, this is just a beginning. It is important to impress upon the reader the faults of the current framework and the need for an improved framework. To lay this foundation of doubt we need to take another short peek at another aspect of the complexity of 'complexity.'

## Entropy, Energy and Order

Another problem with measuring complexity can be approached through a related concept, entropy. Entropy was first defined by Rudolf Clausius in 1867. He was interested in generating work from steam engines and found that, *"heat can never pass from a colder to a warmer body without some other change"* (Clausius, 1867, p. 117). Over time this principle was found to be universal and it became known as the second law of thermodynamics. The second law states that in closed systems, randomness always increases. As a result many theorists think of entropy as the inevitable loss of order (Carroll, 2010, p. 30; Kauffman, 2008, p. 13; Lederman & Hill, 2004, p. 183; McFadden, 2000, p. 132). Everything just naturally falls apart.

If a fundamental law of this universe is that things fall apart then how does this apply to the Big Bang? What fell apart between then and now? What was 'lost' between the Big Bang and now? The second law is often

[256] This definition is based on a per unit volume comparison. This means that per unit volume, internal systems contain a higher density of regular energetic relationships than an average random sample of the same per unit volume from its external environment. This idea is based on Eric Chaisson's (2001) "energy rate density" metric.

interpreted as saying that 'order' was lost. But how does the loss of order lead to humanity? To know this we need to examine what order means in a thermodynamic context. Thermodynamics is the study of heat exchange (thermo = heat, dynamic = change). Thermodynamic order is a measure of heat change. This makes sense because the founders of thermodynamics were interested in producing work through heat exchange. The problem is that the heat differentials necessary to make work can never be permanent. Entropy always reduces the heat differential over time to bring the system to equilibrium and thus the impossibility of a perpetual motion machine. The second law states that order in the form of heat differentials is always lost.

So what is heat? Heat is a macroscopic measurement of microscopic motion. Groups of atoms and molecules that vibrate more, are hotter. When they vibrate they move, but relative to what? They move relative to each other. Since Einstein, there has been no fixed point from which to judge motion. Matter in motion is only 'in motion' relative to other matter. If motion is relative then heat is relative. Relative to what? Relative to everything else in the closed system, in the box. This, of course, begs the question of how fine grained you choose to measure differences inside your box, or between the box and its environment, its 'boundary.' Something can only be hot if something else is cold. There must be a relative differential to make both heat and motion meaningful. More relative motion in the box is 'hotter' only at the level of fine graining the box defines and only if there is a 'cooler' exterior. There is neither heat nor motion if there is no measurable difference created by delineating a boundary that defines this difference. Heat and motion are relative to how you fine grain the measurements that define them. If everything is the same temperature then everything is moving at the same speed relative to everything else. This is called equilibrium. It is the state towards which entropy relentlessly moves. But where do we stop counting what is counted as 'everything'? Where do we draw the fine graining line?

In a nutshell, if both motion and heat are relative measures then the only really important measurement in entropy is difference. Entropy inexorably lowers heat/motion differentials. This puts us in a strange position for using entropy to describe the difference between the Big Bang and now. The initial state was a tiny hot soup of energy. There were no heat differentials in the initial state. Everything was moving equally fast. Almost fourteen billion years later there are heat differentials everywhere. My coffee is moving faster than this sofa, the sun is moving faster than the Earth and your brain is hopefully moving faster than your toes. Differences are everywhere now. The question is, if there were no differences at the beginning and differences in temperature are always lost as per the second law, why do we have more differences now? Why isn't the universe one big cooling gas cloud or crystal?

Scientists like Eric Chaisson (2001) and Fred Spier (2010) talk about the rise of complexity in the universe. Other scientists like Sean Carroll (2010) talk about the loss of order. No one denies the second law, but our understanding of how it applies to a universe that appears to be 'complexifying' is mysterious. This problem is often related the 'arrow of time' and it remains one of the great intractable problems in science. Many theorists try to get around this problem by saying that complexity (gain of order) is built up in some places only by displacing masses of entropy (loss of order) that makes the overall net loss of order fit the second law. This solution has two flaws.

First it presumes that the universe is moving into empty space and that empty space can be used as an entropy 'garbage can.' The problem is that the universe is not like a cupcake with a firecracker in it, where parts of the cupcake move into the air around it. The universe isn't moving into an absolute Newtonian empty space, away from a fixed point source. The universe is expanding, it is growing, everything relative to each other. Everything is accelerating away from everything else in proportion to its distance. The universe is like bread dough that is expanding to become a loaf. There is no empty space that can be filled with disordered matter-energy. Everything around us, including the space, was all here at the beginning. Empty space isn't rushing into our universe to fill the ever widening gaps. Where would it come from?

The second problem with displacing entropy to explain the growth of complex order is even more impossible to explain away. No matter how you cut the cake, the result was that the universe moved from a thermodynamically undifferentiated state to a thermodynamically differentiated state. According to the second law of thermodynamics, entropy, this should never happen. Correction, this should never happen in a 'closed' space.

But what if the universe were scale-free? What if the only box put around the universe was the illusory one we imagine by stopping our measurements at certain scale-levels? What if the only thing that 'closes' the universe system is the limits of our tests?

When Claude Shannon was developing information theory he was looking for a term for the measure of uncertainty in a message. He was going to call this measure the 'uncertainty function' but John Von Neumann recommended otherwise,

> You should call it entropy, for two reasons. In the first place your uncertainty function has been used in statistical mechanics under that name, so it already has a name. In the second place, and more important, no one really knows what entropy really is, so in a debate you will always have the advantage (John Von Neumann as cited in Tribus & McIrvine, 1971, p. 180).

The problem has always been that concepts like entropy, complexity and order have been hard to define. The result has been unclear concepts used to describe theories that are unclear about the big picture. We have no theory of everything. To remedy this problem this investigation works towards a redefining of complexity. As with 'entropy,' we are looking for some way to measure the change from the Big Bang until now. This is extremely difficult and it is, to a deep extent, the impetus behind this investigation.

One thing that does seem likely is that information will play a key role. Shannon's information theory is the only known metric for dynamic change. Canonically, Shannon information is understood as the measure of the reduction of uncertainty by a receiver. Where entropy is the increase in uncertainty, or randomness, or noise within a spectrum of interaction; information is the increase in certainty, or regularity or predictableness within a spectrum of interaction. Information is the regularities within a less regular (i.e. noisy) environment (Deacon, 2007a). This investigation tries to forge a bond between complexity and information so that we can understand the evolution of the universe not just as a loss of order, but also as a gain of order, as defined by complexity and information. This investigation casts entropy and the loss of order as just one side of the coin the universe flips. The other side of entropy is positive symmetry and the gain of order. Both entropy and positive symmetry are universal and both are statistical steps into an unknown future.

So now back to the question. How do we measure the change in the universe between the Big Bang and now? Is the universe more complex or less complex now? The answer depends on scale. If you presume that there are no scale levels smaller than quanta then you can presume the initial state of the Big Bang was a uniform mix of energy, a point source. You just say there cannot be any differences between the different things in this point source because there are no different locations that can maintain different states within it. Everything is the same because it is all in the same place. The amount of information needed to describe that point source is minimal. That is what a point source is, simple, easy to describe, it contains just one bit of information. Unfortunately, information can only exist as a relative measure between existing states and possible states. It only exists as measured against a noisy environment. What is the environment for the universe? The universe was a point source relative to what?

Supposedly, this simple Big Bang 'bit' contained the whole universe. The question is, can we really believe that it was simple, that it had no fine-grained structure, the entire universe? This investigation presumes that if we were the size of a quark component and looked into the initial state we would find a whole universe of pattern. This is what a scale-free universe means, pattern at all levels. The result of this point of view is that there was an unknown level of differentiation packed into that tiny Big Bang. This unknown level of differentiation has changed into the current, known level of differentiation over the last fourteen billion years. Fortunately, this presumption ends entropy's exclusivity.

Physical entropy is a one way process. It goes from heat difference to heat equality, from order to randomness, from information to noise. If the sender were the Big Bang and the current universe were the receiver, and entropy were the *only* rule, then the initial state should have been highly differentiated and rich in information. Over the last fourteen billion years we should have lost that information. We should be noise. We should be a gas cloud. Obviously, we are not.

The problem of defining complexity is the problem of, 'compared to what?' At what scale level do we arbitrarily stop counting differences and how do we choose to box, or limit, what we want to measure. Just because we cannot see fine grained pattern yet does not mean it does not exist. To presume that it does not exist because our current technology has limits is to commit the prejudice of previous centuries.

Was the initial state more complex than the current state? We don't know. It didn't have more matter-energy. It had exactly the same amount of matter-energy. Did this matter-energy have more different relationships than current matter-energy? This depends on the scale-level we measure those differences on. The bottom line is that a scale-free universe makes measures of complexity relative, not absolute. My ninety kilogram body may be considered more complex than a ninety kilogram rock because I contain more movement differentials. However, it may be considered less complex than ninety kilograms of vapourized rock because as this vapour disperses it will enter billions of increasingly randomly differentiated heat states. My body maintains a relatively limited, predictable and repetitive number of heat states. Does this mean that vapour is more complex than I am? When I imagine the pattern that constitutes me, it is not Grassberger's perfect checkerboard image to the left, nor is it the maximum randomness image to the right. It is somewhere in between.

This investigation denies that the simplistic point of view represented by 'the growth of entropy' as represented by cosmologists like Carroll and 'the growth of complexity' as represented by Spier and Chaisson. In their places it takes a new look at how those relative differences that define our current universe came to exist. Entropy is only a *part* of this story. Complexity is only part of this story. This investigation introduces new ways to think about these concepts.

One of the keys to this new approach is reimagining order. Order is not some absolute measure of heat difference or complexity or randomness or information. Order is also relative. In this investigation order is defined as the kind of persistent differences that have appeared since the Big Bang, absent any fatuous comparisons with an unknown initial state. Order here is defined as those regularities in relative motion differentials that have accrued since shortly after the initial state. At some point shortly after the initial state there began to persist those certain regularities of relative motion that constitute matter as we know it. These regularities did not exist previous to this point and they continue to exist up until now. These regularities are what this investigation calls 'order.' The addition of billions of emergent types of regularities to this initial 'order from the unknown' is the story of humanity.

# Appendix 2: Socio-symbolic Efficiency

It should be remembered that the saliencies that potentiate animal memory systems have to be filtered from a massively complex, parallel, analogue and simultaneous background of environmental interaction. Like any signal, saliencies must be separated from ambient noise. This means that learning to recognize adaptive triggers within the environment is probably hard won. A massive amount of information must be sampled by the senses. Once sampled, this information is then analyzed by the brain and selected for behavior. The resulting saliencies are then remembered as those trigger patterns that become associated (i.e. synced) with adaptive reactions.

Even though animals like us inherit a phenotype that is predisposed to sense the right kind of saliencies, much of their brains must be dedicated to filtering the environment for these cryptic triggers. Furthermore, the adaptive reactions to these triggers are constrained to passive associative links of limited extendibility. They are passive because only what is presented by the environment can trigger memory informed behavior. Their extendibility is limited because of the Lyapunov horizon that makes prediction beyond a certain point in the future impossible.

In contrast, the symbolic saliencies that potentiate declarative memory systems are pre-filtered by the social network. They are historically pre-selected for being 'meaningful' by the social network in the form of a pre-potentiated, culturally inherited associative system. These saliencies, or words, are represented digitally and serially, both of which seriously reduce the noise (and chaotic propagation) within the symbolic environment from which adaptive links can be selected. Thus, everything represented within the system has already been selected and maintained for its adaptive functionality and the linking of one concept to the next is highly simplified.

As a result, we should expect the sampling of the symbolic environment to be much more efficient than the sampling of the physical environment.

This efficiency can be linked to the discovery of the phonetic alphabet by the Greeks, or the constant discoveries of newer and more efficient computer programming languages. Each discovery enables the cycle time and the reproducible fidelity of the rules that govern transformation with preservation of structure (i.e. symmetry) to increase. Each allows more of the environment to be sampled faster for emergent positive symmetries and those positive relationships found to be reproduced with more fidelity. The Greek phonetic alphabet enabled easy reading and writing, both of which have vastly improved how we pass on information and resulted in the first truly educated societies.[257] Likewise, each new computer language simplifies and speeds up the work of training computers to enable our overall social, political, economic and scientific connectivity. In the case of the emergence of symbolic symmetry, this efficiency could have translated into the freeing up of extra brain resources available for the creation and maintenance of the associative links between the saliencies (e.g. words or concepts) themselves.

Symbolic memory systems need not filter their environment. Their symbolic environment is pre-filtered to be highly simplified, non-chaotic and meaningful (i.e. adaptive). Instead, when you focus on symbolic communication you can dedicate considerable metabolic resources to extending the connectivity of the saliencies within it (e.g. extending the Lyapunov horizon). Instead of continually deciding what to pay attention to and ho w to capture/couple to it, the socially preprogrammed symmetry of language allows you to sample more relationships, more efficiently and thus travel farther, faster and with less effort. What is being suggested is that structurally, language increases the efficiency by which humans can sample and maintain novel adaptive behavior.[258]

An example of this difference would be a deer browsing. A deer browsing is constantly aware of its surroundings. You can see it constantly sampling

[257] This may be taken as 'western biased' and contentious to some scholars of ancient or far eastern cultures, but I think it is warranted. The Greeks produced the first examples (since the rise of agriculture) of societies where large portions of the population were not only systematically educated, but also politically and philosophically self-determining. It seems to be the first example of modern intellectual life.

[258] It is commonly know that declarative (semantic) memories are far easier for humans to form and keep than other kinds of non-declarative memories. See Parkin (1993).

the air for odors and rotating its ears for sounds of movement. Presumably, it is dedicating most of its attention to dangers and opportunities in its here-and-now. On the other hand, humans eat in socially created and controlled environments (e.g. around a fire, in a dining room). Conversation in this environment is the sampling of the socio-symbolic environment for both shared and individual adaptive opportunities (and dangers) that can range from what should be done tomorrow, to grandpa's tales of youth meant to potentiate future decisions in those eating with him (e.g. children, spouses, etc.). It would seem that humans not only use symbolic networks to sample possible futures, but they also maintain cultural contexts in which the exploration of possible futures and the potentiation of adaptive decision making can be extended. We build and maintain social inner spaces where we are safe to explore each other's adaptive utility and maintain unified social futures (e.g. like families, corporations and chat rooms).

Where the deer is almost fully occupied with making here-and-now decisions that limitedly extend its immediate future, humans are additionally engaging in sampling each other's futures to potentiate not only their own but their partner's and community's distant futures. It may be exactly this positive recursive feedback made possible by the simplified symbolic environment that allows humans to potentiate each other's adaptive utility and extend their power of creating adaptive futures to such a novel and unusual degree. We spend an inordinate amount of time potentiating our own and each other's 'what if' concept-maps. We broadcast our own reactive face and remember others' broadcasted reactive faces in order to potentiate future cooperative utility. We are continually priming each other's decision making abilities by potentiating how our own and our group's concept-maps enable and constrain connectivity so that when we make decisions about our future those decisions have already been adaptively vetted.

Where the deer spends all its time thinking about what to do in its immediate environment, we create and maintain a simplified group environment (i.e. an 'expanded internal' or an 'inner space') that allows us to train each other and ourselves to think about what to do and how to use each other to do it in all the spatial and temporal environments we can imagine. We are constantly potentiating an ever-growing web of personal and social adaptive futures. Like the compounding of ontogenetic potentialities that mitosis adds to binary fission by accessing the genome from multiple simultaneous points, symbolic language compounds the adaptive potentialities that human behavior can have by accessing the socio-symbolic network from multiple individuals.

# Appendix 3: Rational vs. Instinctual Behavior

Non-symbolic behavior can be thought of as 'instinctual' because instinct can be seen as a catch-all term for the many non-symbolic systems that affect behavior.[259] There are various genetic or epigenetic sources, or levels, of control upstream to symbolic behavior, but to include them all in this discussion would be needlessly confusing. Thus, for the purposes of this appendix, 'instinctual behavior' can be easily counter posed to 'symbolic behavior' as an inclusive term for all behavior constraints upstream from symbolic behavior. Also, because endocrine mediated 'emotional' constraints are some of the most important and proximate upstream constraints, instinctual behavior can be considered loosely synonymous with the idea of emotional behavior.

Animal behavior is instinctually mediated. Human behavior is both instinctually and symbolically mediated. Considering that we are basically animals with an extra level of constrained release of energy called language, it is not surprising that both conversation and thought are mixtures of instinctual and symbolic mediation. What this means is that just because language is an extraordinarily powerful system to sample and maintain novel adaptive futures does not mean that it is exclusively used as such. Much to the contrary, most conversation is instinctually mediated.

First of all, the difference between instinctual and symbolically mediated behavior will be treated as the difference between rational and irrational behavior here. Human behavior is a mixture of instinctually primed reactivity and socio-symbolically primed reactivity. 'Rationality' is a judgment that

[259] To some degree language itself is instinctual due to recent strong selection for symbolic predispositions, but here we assume that the vast majority of our genetic and epigenetic instincts were formed previous to those few and recent symbolic instincts. The separation of symbolically mediated behavior from instinctually mediated behavior is proposed here as more of a heuristic tool than a true distinction.

humans make exclusively about human behavior, whether that behavior is physical or symbolic. Thus, an argument can be judged rational or irrational just as an act can be judged rational or irrational as long as the judgment is in a human context. Additionally, rationality need not be absolute. Actions, like arguments, can be more or less rational and this often varies depending on who judges them. This investigation proposes that rationality can be seen as the measure of how dependent the act or argument is on the socio-symbolic system, as opposed to the instinctual systems. What this means is that rational acts are judged as socio-symbolic choices and irrational acts are judged as instinctual choices. When we say that a behavior is rational we mean that it reflects how we all agree symbols interact to make sense of our socio-symbolic environment. When we say a behavior is instinctual, or emotional, or irrational we say that it reflects a different level of transformational logic. Instinctual behaviors evolved to constrain how our embodied selves interacted to navigate our physical environment. Rational behaviors evolved to constrain how our symbolic selves interacted to navigate our socio-symbolic environment.

An argument is often considered irrational if it is emotionally (e.g. instinctually) guided.[260] Emotions prime us individually. They are exclusive to the individual. They mediate behavior to respond adaptively to our here-and-now environment by comparing our responses to our personal and phylogenetic histories. Symbols prime us socially. They mediate our behavior to respond adaptively to a timeless and spaceless social environment where meaning is preserved as social functionality. The amount an act or argument is primed and structured by emotional/instinctual logic is how irrational it *can* be. Conversely, the amount an act or argument is primed and structured by socio-symbolic logic is how rational it is.[261]

That said, most face to face conversation is emotionally guided. To some degree the participants either want, or do not want, to engage each other depending on their moods. This emotional context triggers largely

[260] Emotions can be seen as a certain class of instinctual behavior in the sense that they are heavily associated with the endocrine system (less with neurons) and they are not symbolically determined or controlled.

[261] For the moment please refrain from reading the typical good and bad value judgments into 'rational' and 'irrational.' In this context rational has its own separate value scale just as irrational does.

rehearsed verbal exchanges that only need to be minimally tweaked on the way out of the mouth. If it's a cocktail party amongst friends the exchange is usually some version of:

"Hey Harley, how's it going?"

"I'm doing well Francine, you still seeing Dominic?"

If one or the other is in a bad mood or they are in a different social context (e.g. strangers in a bar), the exchange can be very different, but equally composed of the recognizable phrases we all use to engage in the four 'F's.[262]

Robin Dunbar was largely right to identify language as a kind of social grooming. Most conversation is exactly that, an instinctually mediated way to maintain cooperative networks. In much the same way that isolated neurons or Asian fireflies light up to their own rhythms when alone, but sync when in groups; humans symbolize (i.e. think) to their own rhythms, but sync those symbolizations when in groups (Strogatz, 2003). However, this is not exclusive to humans. Ants, sardines and some yeasts all physiologically mediate their individual behavior to maintain cooperative networks. What is exclusively human is *how* we maintain synchronicity, *how* we cooperate. What is exclusively human is socio-symbolic behavior - rational behavior.

Rational conversation or thought is not typically instinctually guided. It is guided by the concept-map itself. It is a process of unpacking conceptual buoys and allowing their metaphorical relationships to both guide and build the structure of thought. The rules for transformation in this process are the weightings of the associations between concepts and the cultural and personal history of how those weightings were adjusted over time. What decides the logic of rational acts or arguments is the logic of the socio-symbolic system and its pre-potentiated connectivity structure. [263] This is complicated when there are multiple participants who bring differently adjusted cultural socio-symbolic networks to bear on the interaction. No one should envy U.N. mediators and translators.

[262] The four 'F's can be seen as subcategories of cooperation. They are each versions of co-operation with different saliencies in our environments (i.e. food, predators or conspecifics).

[263] This implies that some of us act irrationally because we have built concept-maps that don't correlate to the social norm, the socio-symbolic map. In this case there is something fundamentally wrong with the structure of our concept-map. Others act irrationally because they react on instinctual, non-symbolic levels or they just make a mistake using an otherwise highly correlated concept-map.

The thread of rational thoughts or conversations is maintained by how concepts relate to each other while the thread of irrational thoughts and conversations is maintained by how organisms relate to each other and their environments.

Fundamentally, there are two kinds of symbolic expression. Instinctual thought or conversation is motivated by how the organism has developed to react to its physical surroundings. This kind of thought or conversation can be referred to as irrational when in conflict with rational behavior. Rational thought or conversation is based on how the symbolic-self interacts within its socio-symbolic surroundings. This kind of thought is what sets humans apart from other animals.

# Appendix 4: Creativity

*Man is double. There are two beings in him: an individual being which has its foundation in the organism and the circle of whose activities is therefore strictly limited, and a social being which represents the highest reality on the intellectual and moral order that we can know by observation - I mean society. This duality of our nature has as its consequence in the practical order, the irreducibility of a moral ideal to a utilitarian motive, and in the order of thought, the irreducibility of reason to individual experience. In so far as he belongs to society, the individual transcends himself, both when he thinks and when he acts.*

*Emile Durkheim (1915, p. 16)*

One thing that is important to recognize is that both instinctual and rational (i.e. socio-symbolic) behavior provide adaptive functionality to us (see Appendix 3 for more on instinctual vs. rational behavior). Even though the symbolic system is superimposed over the instinctual system as an added level of constrained release of energy, it is still dependent on the instinctual substrate that gave birth to it. Historically it is just another differentiated and specialized function of that substrate. Like the other instinctually primed brain-senses, language is maintained to extend adaptive reactivity. What this means is that important human functions like creativity are founded in both, and often in the interaction of the two.

We often divide creativity along the lines of rational and irrational, symbolic and emotional. Artists are often thought to use a kind of embodied emotional intelligence to create meaningful novelty while philosophers and scientists use more symbolic systems of logic to create novelty. This is both true and false. How artists access novelty may be greatly dependent on embodied intelligence of the type associated with emotional, instinctual and irrationally guided behavior. However, good art expresses a truth about us that we as a social group value. Equally, scientists and philosophers are in the business of expressing rational truths that we, as a group, also value (excluding perhaps religious fundamentalists).

What is different in these two cases is not the expression of truth, nor its importance to how we contextualize ourselves to create our futures, both do that. What is different is how each endeavor accesses novel and important adaptive truths about us. One uses the instinctual substrate more and the other uses the symbolic superstrate more. However, the elaboration of these truths requires both levels of interaction. In the end, each of us learns to access our futures using differently weighted mixtures, feeding back differentially, at different times. Creativity is often the feedback between the two levels that produces science and art from which the society as a whole selects which novel products will be preserved and taught to the next generations.

C.P. Snow divided western thought into two cultures, the sciences and the humanities. This old division can be seen through new eyes if one considers its origin in the origins of language and mind themselves. Value and meaning in both the sciences and the humanities has been accumulating for millennia but in different ways. Where the humanities add new information individually the sciences add it socially. Value in the humanities is a matter of taste; it is a personal point of view that results from individual selection. What is considered good can change quite drastically between different individuals. Value in the sciences is a matter of consensus, it is a social point of view that results from group selection. What is considered good is a shared consensus on what the truth is. Science aims to unite all its participants into one community that shares one rational point of view. The humanities have no such aim. There is no shared language like mathematics that binds the humanities into a logical whole. Creativity in the humanities is a matter of expressing personal interpretation. Creativity in the sciences is a matter of adding to a social body of knowledge.

What is rational in the humanities can be argued from many points of view because its practitioners are not constrained to speaking the same language. Rationality in the sciences is constrained to the language of mathematics and experiment. The goal of science is one unified rational point of view, the one supported by physical evidence and mathematics. The result is two cultures. One limited to a unified rational consensus and the other endlessly divisive and creative. One champions the group and the other the individual. One sits lightly in us internally, but creates a very compelling social bond. The other sits deeply in us internally and reinforces our individuality. Both were born in the original split of the rational group from the emotional individual. Both were born to serve us in different environments.

Innovators like Einstein and Picasso blazed new trails. They made new connections - new meaning. By expanding the socio-symbolic network they expanded our capacities to build adaptive futures for our 'selves.' These new connections can become preserved in the brain as neural-endocrinal reactivity to physical experience or as neural-endocrinal reactivity to socio-symbolic experience. They can be preserved in the emotional associations we remember by maintaining and reproducing Guernica or in the symbolic associations we remember by maintaining and reproducing $E=MC^2$. Both of these systems feedback and depend on the other.

If the science of chemistry and optics had not led to the technical perfection of camera portraiture, would Guernica exist? If art like Guernica didn't exist, would neuropsychologists be trying to map 'emotional intelligence'? Like so many other systems, the ways that humans sample and maintain both socio-symbolic rational and personal emotional adaptive futures feedback into each other. Social rationality and personal emotions characterize two levels of selection that both exhilarate and plague us as we simultaneously step into our adjacent possibles, both as individuals, and as a group.

# Works Cited


Anderson, P. (1972, August 4). More is different: Broken symmetry and the nature of the hierarchical structure of science. *Science, 177*(4047), 393-396.

Ash, J., & Gallup, G. G. (2007, June 1). Paleoclimatic variation and brain expansion during human evolution. *Human Nature, 18*(2), 109-124.

Bak, P. (1996). *How nature works: The science of self-organized criticality.* New York, NY: Copernicus.

Balsters, J., Cussans, E., Diedrichsen, J., Phillips, K., Preuss, T., Rilling, J., & Ramnani, N. (2009, October 14). *Evolution of the cerebellar cortex: The selective expansion of prefrontal-projecting.* Retrieved November 4, 2012, from http://research.yerkes.emory.edu/Preuss/2009-Balsters_Evolution_ of_cerebellar_cortex-19857577.pdf

Barbour, J. (1999). *The end of time: The next revolution in physics.* New York, NY: Oxford University Press.

Barrow, J. (1994). *The origin of the universe.* New York, NY: Basic Books.

Beeby, R., & Kaeser, H. (1990). Metabolic constraints in evolution. In J. M. Smith, & G. Vida (Eds.), *Organizational constraints on the dynamics of evolution* (pp. 57-77). Manchester: Manchester University Press.

Ben Underwood Foundation. (2009). *Echolocation: What is it anyway?* Retrieved July 17, 2011, from www.benunderwood.com/echolocation.html

Bergen, B. K. (2013, February 25). How the brain makes meaning (BSP94). *Brain Science Podcast.* (V. Campbell, Interviewer)

Bertalanffy, L. V. (1959, August). An outline of general systems theory. *The British Journal for the Philosophy of Science, 1*(2), 134-165.

Bertalanffy, L. V. (1968). *General system theory.* New York, NY: George Braziller.

Bickerton, D. (2003). Symbol and structure: A comprehensive framework for language evolution. In M. H. Christiansen & S. Kirby (Eds.), *Language evolution* (pp. 77-93). Oxford: Oxford University Press.

Bickerton, D. (2009). *Adam's tongue: How humans made language, how language made humans.* New York, NY: Hill and Wang.

Blakeslee, S. (2007, October 18). The body has a mind of its own (BSP23). *Brain Science Podcast.* (V. Campbell, Interviewer)

Bobe, R., & Behrensmeyer, A. K. (2004, May 20). The expansion of grassland ecosystems in Africa in relation to mammalian evolution and the origin of the genus *Homo. Palaeogeography, Palaeoclimatology, Palaeoecology, 207*(3-4), 399-420.

Brown, R. (1965). *Social psychology.* New York, NY: Free Press.

Brown, W. (2009, October 9). Did my neurons make me do it? (BSP62). *Brain Science Podcast.* (V. Campbell, Interviewer)

Buss, D. M., Haselton, M. G., Shackelford, T. K., Bleske, A. L., & Wakefield, J. C. (1998). *American psychologist.* Retrieved December 12, 2010, from http://www.sscnet.ucla.edu/comm/ haselton/webdocs/ spandrels.html

Buzaki, G. (2006). *Rhythms of the brain.* New York, NY: Oxford University Press.

Calcott, B., & Sterelny, K. (Eds.). (2011). *The major transitions in evolution revisited.* Cambridge, MA: MIT Press.

Campbell, V. (Producer). (2007a, May 17). *"Memory: From mind to molecules" (BSP 12).* [Audio podcast]. Retrieved from http://brainsciencepodcast.com/bsp/memory-from-mind-to-molecules-bsp-12.html

Campbell, V. (Producer). (2007b, September 22). *A review of "The body has a mind of its own" (BSP 21).* [Audio podcast]. Retrieved from www.brainsciencepodcast.com/bsp/2007/9/22/body-maps-a-review-of-the-body-has-a-mind-of-its-own-bsp-21.html

Campbell, V. (Producer). (2008, October 4). *Review:"Principles of brain evolution" by George F. Striedter (BSP 47).* [Audio podcast]. Retrieved from http://www.brainsciencepodcast.com/bsp/review-principles - of-brain-evolution-bsp-47.html

Carroll, S. (2010). *From eternity to here: The quest for the ultimate theory of time.* New York, NY: Dutton.

Cassirer, E. (1965). *An essay on man.* New Haven, CT: Yale University Press.

Chaisson, E. J. (2001). *Cosmic evolution: The rise of complexity in nature.* Cambridge, MA: Harvard University Press.

Chaisson, E. J. (2010, May 17). Energy rate density as a complexity metric. *Complexity, 16*(3), pp. 27-40.

Cheney, D. L., & Seyfarth, R. M. (1990). *How monkeys see the world.* Chicago, IL: The University of Chicago Press.

Chomsky, N. (1976). *Reflections on language.* London: Temple Smith.

Chomsky, N. (2006). *Language and mind.* Cambridge: Cambridge University Press.

Christian, D. (2011). *Maps of time: An introduction to big history.* Berkeley, CA: University of California Press.

Christiansen, M. H., & Kirby, S. (2003). From language learning to language evolution. In Christiansen, M. & Kirby, S. (Eds.), *Language evolution* (pp. 272-294). Oxford: Oxford University Press.

Christiansen, M. H., & Kirby, S. (Eds.). (2003). *Language evolution.* Oxford: Oxford University Press.

Clark, K. (1969). *Civilization: A personal view.* New York, NY: Harper & Row.

Clausius, R. (1867). *The mechanical theory of heat: With its applications to the steam engine and the physical properties of bodies.* London: John Van Voorst.

Cohen, J., & Stewart, I. (1994). *The collapse of chaos: Discovering simplicity in a complex world.* New York, NY: Penguin Books.

Corballis, M. C. (2011). *The recursive mind: The origins of human language, thought and civilization.* Princeton, NJ: Princeton University Press.

Cox, B., & Forshaw, J. (2011). *The quantum universe: And why anything that can happen, does.* Cambridge, MA: Da Capo Press.

Damasio, A. (1989). Time-locked multiregional retroactivation: A systems level proposal for the neural substrates of recall and recognition. *Cognition, 33*(1-2), 22-62.

Damasio, A. (1999). *The feeling of what happens: Body and emotions in the making of consciousness.* Orlando, FL: Harcourt.

Damasio, A. (2010). *Self comes to mind: Constructing the conscious brain.* London: Random House.

Darwin, C. (1859). *On the origin of species by means of natural selection, or the preservation of favoured races in the struggle for life.* London: Murray.

Dawkins, R. (1980). *The selfish gene.* New York, NY: Oxford University Press.

Dawkins, R. (1999). *The extended phenotype: The long reach of the gene.* New York, NY: Oxford University Press.

Dawkins, R. (2008). *The god delusion.* New York: First Mariner Books.

Deacon, T. (1997). Evolution and intelligence: Beyond the argument from design. In A. Scheibel, & W. J. Schopf (Eds.), *The origin and evolution of intelligence* (pp. 103-133). London: Jones and Bartlett Publishers.

Deacon, T. (1998). *The symbolic species: The co-evolution of language and the brain.* New York, NY: W.W. Norton & Company.

Deacon, T. (2007a). *Shannon-Boltzmann-Darwin: Redefining information part 1.* Retrieved November 23, 2013, from http://teleodynamics.com/wp-content/uploads/sites/16/2013/10/ ShannonBoltzmannDarwinPt11.pdf

Deacon, T. (2007b). *Shannon-Boltzmann-Darwin: Redefining information part 2.* Retrieved November 23, 2013, from http://teleodynamics.com/wp-content/uploads/sites/16/2013/10/ ShannonBoltzmannDarwinPt21.pdf

Deacon, T. (2010a, February 14). *On the human: Rethinking the natural selection of human language.* Retrieved January 21, 2011, from http://nationalhumanitiescenter.org/on-the-human/2010/02/on-the-human-rethinking-the-natural-selection-of-human-language/

Deacon, T. (2010b, August 25). *Language and complexity: Evolution inside out.* Retrieved November 2, 2010, from https://www.youtube.com/watch?v=OT-zZ0PMqgI

Deacon, T. (2013). *Incomplete nature: How mind emerged from matter.* New York, NY: W.W. Norton & Company.

Dennett, D. (1991). *Consciousness explained.* New York, NY: Back Bay Books.

Dennett, D. (1996a). *Darwin's dangerous idea: Evolution and the meanings of life.* New York, NY: Simon & Schuster.

Dennett, D. (1996b). *Kinds of minds: Towards an understanding of consciousness.* New York, NY: BasicBooks.

Dennett, D. (2003). *Freedom evolves.* New York, NY: Viking.

Dennett, D. (2007). *Breaking the spell.* New York, NY: Penguin Books.

Deutscher, G. (2005). *The unfolding of language: The evolution of mankind's greatest invention.* London: Random House.

Diamond, J. (1992). *The third chimpanzee: The evolution and future of the human animal.* New York, NY: Harper Perennial.

Diamond, J. (2006). *Collapse: How societies choose to fail or succeed.* New York, NY: Penguin Books.

Donald, M. (2001). *A mind so rare: The evolution of human consciousness.* New York, NY: W.W. Norton & Company.

Dobzhansky, T. (1973, March). Nothing in biology makes sense except in the light of evolution. *The American Biology Teacher, 35*(3), 125-129.

Dunbar, R. (1996). *Grooming, gossip, and the evolution of language.* London: Faber and Faber Limited.

Dunbar, R. (2003). The origin and subsequent evolution of language. In Christiansen, M. & Kirby, S. (Eds.), *Language evolution* (pp. 219-234). Oxford: Oxford University Press.

Durkheim, E. (1915). *The elementary forms of the religious life.* London: George Allen & Unwin Ltd.

Ehret, W. F. (1960). *College chemistry.* New York, NY: Appleton-Century-Crofts.

Eisler, R. (1987). *The chalice and the blade: Our history, our future.* New York, NY: Harper One.

Emes, R. D. & Grant, S. G. (2012 July). Evolution of synapse complexity and diversity. *Annual Review of Neuroscience, 35,* 111-131.

Field, M., & Golubitsky, M. (1992). *Symmetry in chaos.* New York, NY: Oxford University Press.

Fitzgerald, J. (2011, March 19). *Things that might have been otherwise.* Retrieved November 04, 2012, from http://johnfitzgerald.me.uk/ 2011/03/19/ daily-photo-140-sand-fractals/

Francis, B. (1988). Where is my mind? [Recorded by Pixies]. On *Surfer Rosa* [LP]. Boston, MA: Rough Trade.

Frith, C. (2009, May 10). How our brain creates our world (BSP57). *Brain Science Podcast.* (V. Campbell, Interviewer)

Gell-Mann, M. (1995). *The quark and the jaguar: Adventures in the simple and the complex.* London: Abacus.

Gell-Mann, M., & Lloyd, S. (2003, June 26). *Effective complexity.* Retrieved June 2011, from http://citeseerx.ist.psu.edu/ viewdoc/ summary?doi=10.1.1.113.5604

Gerhart, J., & Kirschner, M. (1997). *Cells, embryos, and evolution: Toward a cellular and developmental understanding of phenotypic variation and evolutionary adaptability.* Malden, MA: Blackwell Science Inc.

Goldschmidt, R. (1960). *The material basis of evolution.* Paterson, NJ: Pageant Books.

Goldstein, R. E. (2006). *Volvox carteri (darkfield).* Department of Applied Mathematics and Theoretical Physics, University of Cambridge. Retrieved July 20, 2013, from http://www.damtp.cam.ac.uk/user/ gold/pix/volvox_darkfield.jpg

Goodwin, B. (1994). *How the leopard changed its spots: The evolution of complexity.* New York, NY: Charles Scribner's Sons.

Google Images. (n.d.). *Molecules and atoms.* Retrieved July 20, 2013, from http://goo.gl/ZD9cV3

Google Images. (n.d.). *Colorful sea sponges.* Retrieved July 20, 2013, from http://goo.gl/jo1D8Z

Gould, S. J. (1977). *Ontogeny and phylogeny.* Cambridge, MA: Harvard University Press.

Gould, S. J. (1980). *The panda´s thumb: More reflections on natural history.* New York, NY: W.W. Norton & Company.

Gould, S. J. (1989). *Wonderful life: The burgess shale and the nature of history.* New York, NY: W.W. Norton & Company.

Gould, S. J. (1993). *Eight little piggies: Reflections in natural history.* New York, NY: W.W. Norton & Company.

Gould, S. J. (1996). *Full house: The spread of excellence from Plato to Darwin.* New York, NY: Three Rivers Press.

Gould, S. J. (1998). *Leonardo's mountain of clams and the diet of worms: Essays on natural history.* New York, NY: Three Rivers Press.

Gould, S. J. (2002). *The structure of evolutionary theory.* Cambridge, MA: Harvard University Press.

Gould, S. J., & Vrba, E. S. (1982). Exaptation: A missing term in the science of form. *Paleobiology, 8(1),* 4-15.

Grant, S. (2008, December 5). Surprising discoveries about synapse evolution (BSP51). *Brain Science Podcast.* (V. Campbell, Interviewer)

Grassberger, P. (1986, September). Toward a quantitative theory of self-generated complexity. *International Journal of Theoretical Physics, 25*(9), 907-938.

Grünwald, P., & Vitányi, P. (2008, February 1). *Shannon information and Kolmogorov complexity.* Retrieved February 2011, from http://arxiv.org/pdf/cs/0410002v1.pdf

Gutenberg. (2011). *Mackenzie's voyages.* Retrieved May 19, 2011, from http://www.gutenberg.org/files/35658/35658-h/35658-h.htm

Harper, D. (2013). *System.* Retrieved December 31, 2013, from http://www.etymonline.com/index.php?allowed_in_frame=0&search=system&searchmode=none

Hauser, M. D., Chomsky, N., & Fitch, T. T. (2002, November 22). The faculty of language: What is it, who has it, and how did it evolve? *Science, 298,* 1569-1579.

Hawkins, J. (2008, May 30). On Intelligence (BSP38). *Brain Science Podcast.* (V. Campbell, Interviewer)

Held, R. & Hein, A. (1963). Movement-produced stimulation in the development of visually guided behavior. *Journal of Comparative and Physiological Psychology, 56*(5), 872-876.

Hurford, J. (2011). *Linguistics from an evolutionary point of view.* Retrieved November 18, 2011, from http://www.lel.ed.ac.uk/~jim/handbook.phil.ling.pdf

Irvine, W. (1955). *Apes, angels, and victorians.* Alexandria, VA: Time Life Books.

Johnson, M. (1987). *The body in the mind: The bodily basis of meaning, imagination and reason.* Chicago, IL: The University of Chicago Press.

Jurvetson, S. (2004, October 17). *Internet splat map.* Retrieved March 17, 2014, from http://www.flickr.com/photos/ jurvetson/916142/

Kauffman, S. (2000). *Investigations.* New York, NY: Oxford University Press.

Kauffman, S. (2008). *Reinventing the sacred: A new view of science, reason and religion.* New York, NY: Perseus.

Kirby, S. (2007). *The evolution of language*. Retrieved June 18, 2009, from http://www.lel.ed.ac.uk/~simon/Papers/Kirby/The%20 Evolution%20of%20Language.pdf

Kragh, H. (2002). *Quantum generations: A history of physics in the twentieth century*. Princeton, NJ: Princeton University Press.

Kuhn, T. S. (1970). *The structure of scientific revolutions* (2nd ed.). Chicago, IL: The University of Chicago Press.

Lakoff, G. (1987). *Women, fire, and dangerous things*. Chicago, IL: The University of Chicago Press.

Langer, S. K. (1972). *Mind: An essay on human feeling, Vol. 2*. Baltimore, MD: The Johns Hopkins University Press.

Laughlin, R. B. (2005). *A different universe: Reinventing physics from the bottom down*. New York, NY: Basic Books.

Laughlin, R. B., & Pines, D. (2000, January 4). The theory of everything. *Proceedings of the National Academy of Sciences, 97(1)*, 28-31.

Layzer, D. (1975, December). The arrow of time. *Scientific American, 233(6)*, 56-68.

Leakey, R. (1994). *The origin of humankind*. New York, NY: Basic Books.

Leakey, R., & Lewin, R. (1992). *Origins reconsidered: In search of what makes us human*. London: Abacus.

Lederman, L. M., & Hill, C. T. (2004). *Symmetry and the beautiful universe*. New York, NY: Prometheus Books.

Leibniz, G. W. (1962). *Leibniz: Discourse on metaphysics. Correspondence with Arnauld, and monadology*. Chicago, IL: The Open Court Publishing Company.

Lewin, R. (1992). *Complexity: Life at the edge of chaos*. New York, NY: Macmillan Publishing Company.

Li, G., Wang, J., Rossiter, S. J., Jones, G., & Zhang, S. (2007, September 19). *Accelerated FoxP2 evolution in echolocating bats*. Retrieved April 28, 2011, from www.plosone.org/ article/info:doi%2F10.1371%2Fjournal.pone.0000900

Lieberman, P. (2001). *On the subcortical bases of the evolution of language*. Retrieved August 8, 2012, from http://edoc.bbaw.de/ voltexte/2006/188/pdf/25nzvtXxXJA_188.pdf

Luhmann, N. (1995). *Social systems.* (J. Bednarz, with D. Baecker, Trans.) Stanford, CA: Stanford University Press.

Lundmark, C. (2001, November). Evolution of mammalian brains. *BioScience 51(11)*, 992.

Lynch, G. (2008, October 17). Brain evolution (BSP48). *Brain Science Podcast.* (V. Campbell, Interviewer)

Lynch, G. & Granger, R. (2008). *Big brain: The origins and future of human intelligence.* New York, NY: Palgrave Macmillan.

Margulis, L. (1998). *Symbiotic planet: A new look at evolution.* New York, NY: Basic Books.

Margulis, L., & Sagan, D. (1986). *Microcosmos.* New York, NY: Touchstone.

Mashankar, A. (n.d.). *The universe.* Retrieved July 20, 2013, from http://www.dpssecunderabad.in/Students_Projects/Project3/past.html

Maturana, H. R., & Varela, F. J. (1980). *Autopoiesis and cognition: The realization of the living.* Dordrecht: D. Reidel Publishing Company.

McFadden, J. (2000). *Quantum evolution: The new science of life.* New York, NY: W.W. Norton & Company.

McMenamin, M., & McMenamin, D. (1994). *Hypersea: Life on land.* New York, NY: Columbia University Press.

Milliken, R. G. (1984). *Language, thought, and other biological categories: New foundations for realism.* Cambridge, MA: MIT Press.

Montagu, A. (Ed.). (1980). *Sociobiology examined.* New York, NY: Oxford University Press.

Morowitz, H. (2002). *The emergence of everything: How the world became complex.* New York, NY: Oxford University Press.

Nagel, T. (1974, October 04). What is it like to be a bat? *The Philosophical Review, LXXXIII*, 435-450. Retrieved November 25, 2010, from instruct.westvalley.edu/lafave/nagel_nice.html

NASA, ESA and the Hubble Heritage Team (STScI/AURA) (2011). *A "rose" made of galaxies highlights Hubble's 21st anniversary.* Retrieved July 20, 2013, from http://hubblesite.org/ newscenter/archive/releases/2011/11/image/a/

National Geographic. (2010). *The genographic project.* Retrieved December 14, 2010, from https://genographic.national geographic.com/genographic/lan/en/atlas.html

Neurotic Physiology. (2011, May 04). *Scientopia*. Retrieved September 13, 2012, from http://scientopia.org/blogs/scicurious/2011/05/04/ science-101-the-neuron/

Newton, S. I. (1848). *Newton's principia: The mathematical principles of natural philosophy*. New York, NY: Daniel Adee.

Nietzsche, F.W. (2007). *Twilight of the idols with the antichrist and ecce homo*. Hertfordshire: Wordsworth Editions.

Noe, A. (2009). *Out of our heads: Why you are not your brain and other lessons from the biology of consciousness*. New York, NY: Hill and Wang.

O'Callaghan, J. (2012, November 18). *Spaceanswers*. Retrieved July 20, 2013, from http://www.spaceanswers.com/solar-system/ 1554/ why-do-planets-have-magnetic-fields/

Panksepp, J. (2010, January 13). Affective neuroscience (BSP65). *Brain Science Podcast*. (V. Campbell, Interviewer)

Parker, A. (2003). *In the blink of an eye*. Cambridge, MA: Perseus Publishing.

Parkin, A. J. (1993). *Memory: Phenomena, theory and experiment*. Oxford: Blackwell Publishers.

Parsons, T. (1951). *The social system*. London: Routledge & Kegan Paul.

Parsons, T. (2012). An outline of the social system. In C. Calhoun et al. (Eds), *Classical sociological theory* (pp. 502–521). Chichester: John Wiley & Sons Ltd.

Pfeifer, R., & Bongard, J. (2007). *How the body shapes the way we think: A new view of intelligence*. Cambridge, MA: MIT Press.

Pinker, S. (1995). *The language instinct: How the mind creates language*. New York, NY: Harper Perrenial.

Pinker, S., & Bloom, P. (1990). *Natural language and natural selection*. Retrieved December 13 2011, from http://pinker.wjh.harvard. edu/articles/papers/Pinker%20Bloom%201990.pdf

Plato. (n.d.). *Plato: Timaeus*. Retrieved May 2, 2012, from http://www.perseus.tufts.edu/hopper/text?doc=Perseus%3Atext%3A1999.01.0180%3Atext%3DTim.%3Apage%3D31

Polanyi, M. (1968, June 21). Life's irreducible structure. *Science, 160(3834)*, 1308-1312.

Potts, R. (1998a). Environmental hypotheses of hominin evolution. *Yearbook of Physical Anthropology, 41*, 93-136.

Potts, R. (1998b). Variability selection in hominid evolution. *Evolutionary Anthropology, 7(3)*, 81-96.

Prokopenko, M., Boschetti, F., & Ryan, A. J. (2009, September). An information-theoretic primer on complexity, self-organization and emergence. *Complexity, 15(1)*, 11-28.

Ritmos Sud Americanos (2013). *Danza ayllas de los mineros de Nuantajayal.* Retrieved March 10, 2014, from http://pacoweb.net/Danzas/ImageDanzas/Danzas1.jpg

Rumbaugh, D. M. (Ed.). (1977). *Language learning by a chimpanzee: The lana project.* New York, NY: Academic Press.

Rush, J. A. (2006, May 06). *Clinical anthropology.* Retrieved February 07, 2012, from http://goo.gl/C9zztx

Savage-Rumbaugh, S. (1997). Why are we afraid of apes with language? In A. B. Scheibel, & J. W. Schopf (Eds.), *The origin and evolution of intelligence* (pp. 43-69). London: Jones and Bartlett Publishers.

Schrödinger, E. (1944). *What is life?* Retrieved November 28, 2011, from http://whatislife.stanford.edu/LoCo_files/What-is-Life.pdf

Science Photo Library. (n.d.). *Yukon River delta: Satellite image.* Retrieved January 28, 2012, from http://www.sciencephoto.com/media/136726/enlarge

Searle, J. R. (1984). *Minds, brains and science.* Harvard University Press.

Seyfarth, R. M., & Cheney, D. L. (1997). Communication and the minds of monkeys. In A. B. Scheibel, & J. W. Schopf (Eds.), *The origin and evolution of intelligence* (pp. 27-40). London: Jones and Bartlett Publishers.

Shapiro, L. (2011). *Embodied cognition.* New York, NY: Routledge.

Sheldrake, R. (1981). *A new science of life: The hypothesis of morphic resonance.* South Paris, ME: Park Street Press.

Simmons, N. B., Seymour, K. L., Habersetzer, J., & Gunnell, G. F. (2011, March 14). *What the fossil record tells us about the evolution of flight and echolocation in bats.* Retrieved July 28, 2011, from http://www.youtube.com/watch?v=Nh9Xt2B9FHs

Smith, J. M., & Szathmáry, E. (1995a). The major evolutionary transitions. *Nature, 374*, 227-232.

Smith, J. M., & Szathmáry, E. (1995b). *The major transitions in evolution.* New York, NY: Oxford University Press.

Smolin, L. (2001). *Three roads to quantum gravity.* New York, NY: Basic Books.

Speakman, J. (2001). The evolution of flight and echolocation in bats: Another leap into the dark. *Mammal Review, 31(2),* 111-130.

Spier, F. (2010). *Big history and the future of humanity.* Oxford: Wiley-Blackwell.

Stanley, S. M. (1996). *Children of the Ice Age: How a global catastrophe allowed humans to evolve.* New York, NY: Harmony Books.

Stewart, I. (1995). *Nature's numbers: The unreal reality of mathematical imagination.* New York, NY: Basic Books.

Stewart, I. (2007). *Why beauty is truth.* New York, NY: Basic Books.

Strogatz, S. (2003). *Sync.* New York, NY: Hyperion.

Studdert-Kennedy, M., & Goldstein, L. (2003). *Launching language: The gestural origin of discrete infinity.* Retrieved November 5, 2011, from http://goo.gl/Lm86Tk

Swain, K. (2009). *Norwalk virus.* Retrieved July 20, 2013, from http://udel.edu/~onyx/

Tattersall, I. (1998). *Becoming human: Evolution and human uniqueness.* Orlando, FL: Harcourt Brace & Co.

Tattersall, I. (2002). *The monkey in the mirror: Essays on the science of what makes us human.* Orlando, FL: Harcourt Brace & Co.

Tomasello, M. (2008). *Origins of human communication.* Cambridge, MA: MIT Press.

Tomasello, M. (2009). *Why we cooperate.* Cambridge, MA: MIT Press.

Tribus, M., & McIrvine, E. C. (1971, September). Energy and information. *Scientific American, 225(3),* 179-188.

Triggs, D. (2012). *Orbiserihtru.* Retrieved July 20, 2013, from http://orbiserihtru.wordpress.com/2012/03/16/hello-world/

Uexkull, J. V. (1934, 2010). *A foray into the worlds of animals and humans: With a theory of meaning.* (J. D. O'Neil, Trans.) Minneapolis, MN: University of Minnesota Press.

University of California, Berkeley. (2007, September 6). *Migrating squid drove evolution of sonar in whales and dolphins: Researchers argue.*

Retrieved November 28, 2011, from http://www. sciencedaily.com/releases/2007/09/070905143411.htm

Vygotsky, L. (1986). *Thought and language. Cambridge, MA: MIT Press.*

Waldrop, M. M. (1992). *Complexity.* New York, NY: Touchstone.

Weaver, A. H. (2004, November 10). *Reciprocal evolution of the cerebellum and neocortex in fossil humans.* Retrieved November 4, 2012, from http://www.ncbi.nlm.nih.gov/pmc/articles/ PMC553338/

Weiner, J. (1994). *The beak of the finch.* New York, NY: Vintage Books.

Wikimedia Commons. (2006, August). *Major rivers in West Canada.* Retrieved December 19, 2012, from http://commons.wikimedia.org/wiki/File:Major_Rivers_in_West_Canada.png

Williams, G. C. (1997). *The pony fish's glow.* New York, NY: Basic Books.

Wills, C. (1994). *The runaway brain: The evolution of human uniqueness.* London: Flamingo.

Wills, C., & Bada, J. (2000). *The spark of life: Darwin and the primeval soup.* Cambridge, MA: Perseus Publishing.

Wilson, D. S. (2002). *Darwin's cathedral: Evolution, religion, and the nature of society.* Chicago, IL: The University of Chicago Press.

Wilson, E. O. (1975, 2000). *Sociobiology: The new synthesis.* Cambridge, MA: Harvard University Press.

Wilson, E. O. (2012). *The social conquest of earth.* New York, NY: Liveright Publishing Corporation.

Wynn, T., & Coolidge, F. L. (2012). *How to think like a Neanderthal.* New York, NY: Oxford University Press.

Yong, E. (2009, November 11). *Revisiting FOXP2 and the origins of language.* Retrieved July 15, 2011, from http://goo.gl/mgjdQb

Zweiger, G. (2001). *Transducing the genome: Information, anarchy, and revolution in the biological sciences.* New York, NY: McGraw-Hill.

Printed in Dunstable, United Kingdom

91220040R00315